FOUR
ELEMENTS

FOUR ELEMENTS

WATER
AIR
FIRE
EARTH

Rebecca Rupp

P

PROFILE BOOKS

First published in Great Britain in 2005 by
Profile Books Ltd
3A Exmouth House
Pine Street
London EC1R OJH
www.profilebooks.com

10 9 8 7 6 5 4 3 2 1

Typeset in Palatino by MacGuru Ltd
info@macguru.org.uk

Printed and bound in Great Britain by
Clays, Bungay, Suffolk

A CIP catalogue record for this book is available from the British Library.

ISBN 1 86197 234 2

I am a little world made cunningly
Of elements, and an angelic sprite.
John Donne

For my four: Randy, Josh, Ethan and Caleb

Contents

Part I

How Many Elements?

The improver of natural knowledge absolutely refuses to acknowledge authority, as such. For him, skepticism is the highest of duties, blind faith the one unpardonable sin.

Thomas Henry Huxley
On the Advisableness of Improving Natural Knowledge

Everything in the world, we now know, is made from one or more of a mere ninety-two naturally occurring elements. These distinctive individuals – ranging in size from tiny hydrogen to such behemoths as uranium – are the alphabet of matter: that is, in the same way that arrays of twenty-six letters can produce everything from knock-knock jokes to the tragedies of Shakespeare, the elements, creatively reshuffled, can make a moon rock, a Barbie doll, a monkey, or a B-52. The history of mankind is inextricably interwoven with the history of the elements. We've wondered, worried, and argued about them; we've discovered, detected, and analyzed them; we've torn them apart and synthesized some of our own. Our association with them is simultaneously intimate and remote: the elements are both the esoteric stuff of cyclotrons and the bedrock of daily life. Our civilizations have been built on the elements; and all our technology derives from them. We watch the process of their formation every time we squint into a sunrise or gaze overhead at a night sky full of stars.

The Periodic Table today lists some 118 elements, nature's aforementioned ninety-two and the remaining two dozen or so that we've produced in laboratories ourselves. Such elemental richness, though taken for granted today, is a relatively recent phenomenon. For over two thousand years of human history, as far as anybody knew, the elements numbered just four. These were the classical elements of the ancient Greeks: water, fire, earth, and air.

The idea that there are a limited number of substances that make up all forms of matter originated in the sixth century BCE in Ionia, Greek territory on what is now the southwest coast of Turkey. Its perpetrators were the founders of the school of thought known as natural philosophy, and this – the root of all Western science – was in turn the direct outgrowth of the Greeks' insatiable curiosity about the workings of the world. No question was too small or too large for budding natural philosophers to tackle. What caused the moon and stars to shine? Why was the sea salty? Which was sweeter: honey or figs? (Honey, according to Xenophanes, who put some thought into this.) Was there such a thing as space? Why were stones shaped like seashells sometimes found on mountain tops? And, perhaps most importantly, what was the nature of matter?

It was here on the ancient Turkish coast that the great divide between science and religion first began to yawn. Prior to the sixth century BCE, the events of the natural world were routinely attributed to (or blamed upon) the will of the gods. The Greeks' conviction that there were logical explanations for natural phenomena not only overturned previous modes of thinking, but had dazzling implications for the future. If events had causes other than the fickle fingerings of the supernatural, infinite possibilities opened up for predicting, circumventing, re-directing, controlling, or exploiting them. The natural philosophers were intellectual revolutionaries, star players in a splendid and startling era of cognitive epiphany. Today, in their honor, a Ph.D. – Doctor of Philosophy – is the highest of academic degrees.

The Ionian philosopher Thales, though undoubtedly not the first to ponder the question, generally gets the credit for being the first to speculate about the nature of the elements. Arguably the first of the natural philosophers, Thales was born around 624 BCE in the seaport city of Miletus – a town which, according to legend, had been founded by a band of roving Athenians who, upon arrival, killed all the local males and forcibly married their widows. In revenge, the first generation of Milesian women refused to sit at the table with their husbands or to call them by their given names.

By Thales's time, however, Miletus was peaceful and prosperous; in general, the late seventh and sixth centuries BCE were good times to be Greek. Elsewhere the world was in turmoil. To the east, the Assyrian Empire, beset by angry neighbors, was coming to a deservedly bloody end. In the south, the pharaoh of Egypt was about to invade Judah, where he would kill King Josiah at Megiddo – a site so noted for awful battles that its name has come down to us as Armageddon. By the time Thales was a teenager, Babylonia had risen to power under the warlike Nebuchadnezzar II, immortalized in the Old Testament for his conquest of Jerusalem and enslavement of the resident Jews. In contrast, the Greece of the seventh and sixth centuries BCE was – for the times – a haven of security. The Greek mainland was dominated by hundreds of city states, each a self-contained entity under a ruling tyrant. The term *tyrant*, in the ancient Greek world, had not yet acquired the odious connotation of today, but instead was a general term denoting a non-hereditary ruler, some of whom enjoyed considerable popular acclaim. Cypselus, tyrant of Corinth, for example, was so adored that he was able to dispense with the protection of a personal bodyguard. He ruled for thirty-two years and died peacefully in bed.

The Greeks of Thales's time – with the exception of the insular and ill-tempered Spartans – were for the most part a commercial people, primarily interested in trade. The world, despite some ominous competition from the Persians, was becoming increasingly Greek. Greek colonies surrounded the shores of the Mediterranean, from Asia Minor and the Black Sea to north Africa, Italy, France, and Spain; and Greek triremes plowed the waves, conducting a brisk exchange in grain, olive oil, dried fish, wine, metals, timber, and slaves. In Athens – a thriving urban center with a population of 300,000 – the liberal politician Solon instituted reforms that set the city on the path toward primitive democracy; and the first buildings began to go up on the Acropolis. On the island of Lesbos, the poet Sappho composed love songs; and in Lydia, King Croesus – a monarch so fabulously wealthy that the envious analogy "rich as Croesus" survives to this day – busily minted the world's first coins.

Thales, raised and educated in the cosmopolitan atmosphere of Ionia's largest city, was a mathematician, astronomer, and indefatigable traveler, variously touring mainland Greece, Babylon, and Egypt, where he distinguished himself by calculating the height of the Great Pyramid of Khufu at Giza. He was the first to determine that a circle is divided in half by its diameter and to establish the usefulness of the mathematical proof; and his astronomical expertise – probably picked up in Babylon – carried over at least once into the diplomatic arena: one of the few surviving tales about Thales, recounted 150 years after the fact by Herodotus, tells of his remarkable prediction of a solar eclipse. The timely occurrence of this event during a battle between the Medes and the Lydians so petrified the combatants that they hastily sheathed their swords, settled their differences, and went home. Modern calculations place Thales's eclipse on May 28, 585 BCE, making this the first event in history to which we can definitively assign an exact day. Thales was also a canny politician, who urged the alliance of the Greek settlements along the Turkish coast in defense against the encroaching Lydians; and he had enough business sense to make himself a comfortable fortune in olive oil.

The remains of Miletus today are a landlocked 6 miles (10 km) from the sea. The city died in mud; silt from the winding Meander River gradually clogged the harbor, eventually rendering it unusable. In Thales's day, however, the blue waters of the Mediterranean lapped at Milesian doorsteps. To the casual walker on the shore, water stretched to the distant horizon, as far as the peering eye could see. The wind carried the sound of waves and the very air smelled of water. It may have been this seemingly endless sea that inspired Thales's theory of the elements – or rather element, for in Thales's philosophy there was only one. The fundamental principle of the universe, said Thales, is water. All known natural substances are modifications or permutations of it. Water evaporates, thereby turning itself into air, and falls as rain, striking the ground, where it eventually condenses and hardens into rock. Everything that exists came originally from water and to water all, sooner or later, will return. The Earth itself – you have only to look at it – is a flat slab, floating like an immense log upon an infinite mass of water.

Exactly what Thales thought or said is a matter of best guess; his vision of a water-based universe is preserved only in fragments, in the writings of later philosophers. Not one scrap by Thales himself survives, and it's possible that he, like Socrates, never actually wrote anything down. Greece in the sixth century BCE was largely an oral culture; its literature was intended to be recited, rather than read; and two centuries later finds Socrates bemoaning the invention of the alphabet which, he insisted, only served to weaken men's memories. Much of what we know about Thales comes from Aristotle, who wrote copiously, but long after the fact.

Though he sounds for the most part a practical man, Thales seems to have had his lapses. Two millennia have not sufficed to eradicate the story of his pratfall down a well, taken while strolling through the streets of Miletus, stargazing. He is said to have become hopelessly stuck, only to be rescued from his embarrassing predicament by a slave girl, who – after hauling him out – teased: "Here is a man who studies the stars, but

cannot see what lies beneath his feet!" One hopes he had a sense
of humor about it. In any case, his reputation didn't suffer. On
the famous list of the "Seven Sages of Greece," Thales invariably appears first (beating out Pittacus, Bias, Solon, Cleobulus,
Periander, and Chilo); and when a boatload of Coan fishermen
netted a tripod identified as belonging to Helen of Troy and
were instructed to give it to the world's wisest man, Thales got
it. He lived to be over ninety and died of heat prostration while
attending the Olympic Games.

Thales's conclusion about the elemental nature of the world
attracted considerable opposition; and his own prize pupil,
Anaximander (ca. 610–546 BCE), proved his most formidable
and articulate adversary. Like Thales, Anaximander was a man
of broad interests with a finger in many proto-scientific pies.
He recognized and described the apparent immobility of the
North Star (at that period in history, Thuban, in the constellation Draco, a pole star even more nondescript than the present
day's Polaris); he prefigured Darwin's theory of evolution,
proposing that man may just possibly be descended from fish;
and he is said to have drawn the first known map of the world,
for which accomplishment he is sometimes called the "Father
of Maps." He also rejected the idea that water was the fundamental building block of matter, and scoffed at the hypothesis
that the Earth was a floating slab in a boundless sea.

The Earth, Anaximander pointed out – possibly sweeping a
hand toward the distant horizon – is clearly curved. (Observe
the approach of a distant ship: the tall sails appear first,
followed by the prow, as if the ship were climbing up the side
of a massive hill.) The Earth, he suggested, was cylindrical;
and, rather than bobbing on the waves, it dangled unsupported in space. It was an imaginative idea that no one would
pick up on for centuries. As for water, Anaximander argued,
physical matter was simply too varied to be based on anything
so prosaic. The fundamental element must be less graspable

and more versatile. Anaximander hypothesized that it took the form of an invisible all-purpose plasma which he called *apeiron*, from the Greek for "infinite" or "indefinite," because it was capable of morphing into all earthly materials. The very nature of this undetectable but pervasive mystery substance precluded active investigation. Nobody could prove Anaximander wrong, but then he couldn't prove himself right either.

Just as Anaximander cast off the water hypothesis of Thales, his own follower, Anaximenes (ca. 585–525 BCE), spurned the *apeiron*. Anaximander's indefinite infinite was so nebulous, in Anaximenes's view, that it amounted – like the Emperor's new clothes – to nothing at all. Everything clearly had to be composed of *something*, rather than nothing, and Anaximenes's candidate for that something was air. Like water, there was a lot of it; however – since it covered the entire Earth – it was far better situated than water to be the source of everything else. Furthermore, it was more adaptable. Rarefied – that is, thinned out or expanded – it could turn into smoke, steam, and fire. Condensed, it could become mist and water; condensed even further, mud, dirt, and solid rock. For the shape of the Earth itself, Anaximenes returned to the teachings of Thales. It was flat, he said, like a table.

From Miletus, the debate over the nature of the fundamental element spread to Ephesus, the second most important city in Greek Asia Minor, located some 30 miles (50 km) to the north, and famed for its Temple of Artemis, a structure so impressive that it was eventually referred to as one of the Seven Wonders of the Ancient World. By 500 BCE, Ephesus was also home to Heraclitus (ca. 540–480 BCE), a sage known to history as the Weeping Philosopher from his dismal view of life, the Obscure Philosopher from the near-impenetrable opacity of his writing, and the Dark One, presumably from both. (Socrates, who never managed to finish any of Heraclitus's works, said resentfully that it would take a deep-sea diver to get at the meaning of them.) The prevailing view seems to be that Heraclitus made his texts obscure on purpose, to ensure that only the brightest and best of the upper classes could possibly understand them. He was by all accounts a thorough curmudgeon and adamant elitist, contemptuous of the common man, and vitriolic about his philosophical peers. "Learning of many things does not teach one sense; else it would have taught Hesiod, Pythagoras, Xenophanes, and Hecateus," sniped Heraclitus unpleasantly. Across millennia, one can picture the sneer.

The philosophy of Heraclitus centers around the permanence of impermanence – that is, the inevitability of change. He coined such (surprisingly straightforward) statements as "Nothing endures but change" and "No man can step in the same river twice," as well as the somewhat contradictory "One day is like any other," which is the sour sort of sentiment one might expect from Heraclitus. No man can step in the same river twice in Heraclitian terms because, with the passage of time, neither the man nor the river remains the same. This inspired a contemporary joke about a debtor who tried to weasel out of his obligations, on the grounds that he was no longer the same man who had contracted the debt.

Heraclitus's insistence on the preeminence of change led him to conclude that fire – that most changeable of elements – was the fundamental material of the universe. The world, according to Heraclitus, arose from fire, which turned into water, and then into earth. Earth, in its turn, liquefied to form water, which then became fire once more, and so on, in an endlessly repeating cycle. Human consciousness itself was a form of fire, and the more you had of it, the better. Sleep caused the fire to burn low; and alcohol, being wet, threatened to put it out altogether. Many of the surviving Heraclitian pronouncements are ominous warnings about the evils of wine.

While Thales postulated a world based on water, Anaximenes on air, and Heraclitus on fire, Xenophanes favored earth. Xenophanes was born around 570 BCE, just down the road from Ephesus in Colophon, a city which, along with Smyrna, Rhodes, Salamis, Chios, Argos, and Athens, claimed to have been the birthplace of Homer. He seems to have spent much of his adult life wandering from place to place, though some suggest he eventually settled in Italy, in the Greek colony of Elea, where a lot of Ionian philosophers ended up after the Greeks were trounced by the Persians in the mid-sixth century BCE. Many of Xenophanes's pronouncements have a pointedly reproachful tone: if only his fellow Greeks had been less frivolous, more abstemious, and less obsessed with sporting events, the Persians wouldn't be here now. He managed to imply that his countrymen had picked up a lot of their bad habits from the neighboring Lydians, a feckless people given to purple clothing, perfume, and fancy hair-dos.

Xenophanes is primarily described as a poet; however, like all the natural philosophers, he had a multiplicity of interests, among them geology. Pondering the question of why fossilized sea shells can sometimes be found on mountain tops, Xenophanes concluded that the Earth is not immutable, but changes slowly over long periods of time. Thus, where today we see mountains, there once was ocean; and the stony shells must be the remains of ancient marine creatures, long ago buried in mud. Though the shape of the Earth may shift, change, flood, flatten, upheave, and rearrange, however, its essence remains the same, in the same manner that clay can be shaped and baked into a pot but remains, at rock-bottom, clay. For such reasons, Xenophanes postulated that the fundamental element of all matter must be earth.

Empedocles, the next prominent philosopher to deal with the troublesome question of fundamental substances, was born to a wealthy family of aristocrats in Acragas on the then-Greek island of Sicily around 494 BCE, the year of the Battle of Marathon. Greater Greece, throughout Empedocles's childhood, was embroiled in the Persian Wars. War had begun in Ionia, with the revolt of the Persian-dominated Greek city states, among them the early philosophical centers of Miletus and Ephesus. The resident Greeks objected to the reigning Persian tyrants, who – in the modern sense – were tyrannical, and to the Persian taxes, which were extortionate. The revolt was initially unsuccessful – the Persians thoroughly crushed the recalcitrant colonists and sacked Thales's home city of Miletus. The distraught Ionians, however, managed to persuade the Athenians to come to their aid, with some help from an Athenian poet named Phrynichos whose sensational operatic tragedy, *The Fall of Miletus,* was filled with heart-wrenching descriptions of Persian atrocities.

The Persian Wars continued for another 16 years – Empedocles, as a young adult, must not have been able to remember a time when the Persians weren't attacking someone somewhere – and Acragas participated in a number of fierce battles, primarily with the pro-Persian Carthaginians. The Persians were finally defeated at the famous Battle of Salamis, a victory so definitive that the overconfident King Xerxes – who had ordered a marble throne to be set up on a cliff top so that he could view the fun – picked up the shattered remains of his navy and went home. The removal of the Persian threat made way for what became known as the Golden Age of Greece.

Empedocles seems to have made the most of it. In his heyday, he was given to parading through the streets of Acragas sporting a gold girdle, bronze sandals, and a laurel wreath, trailed by a sycophantic train of youthful attendants. He insisted that he was a divine being, temporarily doing a stint on Earth as penance for eating meat in a previous life; and his celestial origin, he claimed, gave him the power to control the weather and raise the dead. One account of his death holds that, after having announced that he was about to be taken up to heaven by the gods, he threw himself into the crater of Mount Etna in the hope that his admiring public would believe that his story was true. The trick was revealed, unfortunately, when a spurt of gas tossed up one of his distinctive bronze sandals. For all that, he was a brilliant man; and his concept of the fundamental nature of matter was by far the closest to date to a modern theory of elements.

Empedocles synthesized the confrontational one-element-takes-all theories of his predecessors into a coherent whole, evolving a diplomatic four-fold model in which water, air, fire, and earth all played equal and interactive parts. These essential four, Empedocles argued in *Tetrasomia*, or *Doctrine of the Four Elements*, either singly or in combination, account for all matter on Earth. Things take on different forms when their component elements separate and rearrange, variously directed by the force of Love, which brings elements together, and Strife, which tears them apart. The proportions in which the elements combine are fixed and definite for any particular substance: bone, for example, according to Empedocles, is composed of fire, water, and earth, in the unvarying ratio of 4:2:2. The theory showed a surprising grasp of the basics: that is, all matter is assembled from a finite number of basic and irreducible elements; and these, combined in specific proportions, make up all the substances that exist. There's some debate over how much credit Empedocles should get for this; skeptical modern historians intimate that Empedocles, an advocate of Pythagorean mathematical mysticism, simply made a lucky shot in the dark. At the very least, however, it was the right idea, even if for the wrong reasons, and later philosophers ran with it.

So what did the Greeks mean by "element"? Nothing, according to some. "The terms translated as 'matter,' 'substance,' 'attribute,' and 'element,'" writes G. E. R. Lloyd in *Early Greek Science*, "were all introduced into philosophy in the fourth century, and it is inconceivable that any of the Milesians used them." On the other hand, the gist of the Greeks' argument is clear: the natural philosophers were obviously questioning the origin and fundamental composition of matter. Empedocles refers to water, air, fire, and earth as *rhizomata* – roots – which he defines as both eternal and indestructible, the sole sources of "all the mortal things that appear in their countless numbers." Later Greeks adopted the term *stoicheion*, which survives today in the science of stoichiometry, the study of quantitative relationships in chemistry. Our modern word element comes from the Roman *elementum*, which even the authoritative *Oxford English Dictionary* (*OED*) admits is "a word of which the etymology and primary meaning are uncertain." Still, in some sense *elementum* and its kin must have denoted a basic building block; and, in modern definition, element variously indicates a unit in a series, a component of a group, or a rudimentary constituent, as in Euclid's *Elements of Geometry*.

The Greek concept of a small number of basic elements that comprise all matter is found in cultures around the globe. The ancient Chinese postulated a cosmos based on five elements – fire, water, earth, wood, and metal. The Indians originally hypothesized three – fire, water, and earth – later expanded to include air and the nebulous ether, or void. Frequently such elemental principles were associated with a host of other basic triplets, quartets, or quintets, such as sets of colors, symbols, seasons, planets, compass directions, jewels, smells, organs of the body, moral virtues, and even poisons. The Chinese element fire, for example, was traditionally paired with the color red, the self-immolating phoenix, and the planet Mars; water with black, the tortoise, and Mercury.

Pythagoras and his followers, whose passion for numbers amounted to quasi-religious mania, were entranced by numerical sets, particularly sets of four, since the first four numbers – one, two, three, and four – added together equaled the "sacred" number ten. The world, to the observant Pythagorean, seemed to fall naturally into sets of four: not only were there four elements, but four essential figures (pyramid, cube, octahedron, and icosahedron), four prime faculties (reason, knowledge, opinion, and sensation), four societies, four seasons, four ages of man, and four parts of living things. The Pythagoreans also devised the *quadrivium*, later the basis of the medieval university degree, which divided all mathematics into four parts: arithmetic, geometry, astronomy, and music.

The cult of four extended to medicine under the aegis of Hippocrates and his followers, who associated the four elements with the body's four essential fluids, or humors. Black bile was associated with earth; yellow bile with fire; blood with air; and phlegm with water. While today to be in a good or bad humor indicates merely a passing pleasant or peevish mood, to the ancients the relative balance among the humors was a life or death affair. Individuals in blooming health had equal portions of the four; when the balance tipped in favor of one humor or another, however, the result was debility and disease. The presumptive excess (bad) humor then had to be eliminated from the body to restore the sufferer to equilibrium. Doctors, attempting to do so, prescribed such humor-reducing methods as blood-letting and the vigorous administration of emetics and purgatives, often to fatal effect. Such practices persisted well into the nineteenth century. One victim was George Washington, whose death in 1799 was hastened, if not assured, by his physicians, who bled him repeatedly and dosed him with calomel.

Element today, in a chemical sense, means a substance made of only one kind of atom. Pure gold, which consists of nothing but gold atoms, is an element, as are the atomically singular hydrogen, iron, and mercury; butter, polypropylene, and paint, however, all shameless atomic conglomerates, are not. According to the *OED*, our word atom comes from the ancient Greek *atomos*, meaning indivisible or uncuttable. It refers, in a philosophical and scientific sense, to the ultimate particle of matter: "a hypothetical body," says the *OED* cautiously, "so infinitely small as to be incapable of further division." An alternative definition – the *OED* inveigles you into these – comes from the Latin and means "in a twinkling of an eye;" thus the atom, in medieval terms, was the smallest formal measure of time, equal to 15/94 of a second – or 375 atoms to the minute, 22,560 atoms to the hour. There was also a period in the seventeenth century when the word atom functioned as a verb – you could, for example, in a fit of rage, atom things into dust, which concept still, depending on the things, retains considerable appeal.

The idea that all matter is made of indivisible atoms is also Greek, and also from Miletus, devised by the philosopher

Leucippus and his better-known pupil Democritus in the mid-fifth century BCE. Democritus, the atom's prime original proponent, was born around 470 BCE in Abdera in Thrace, a rough hilly region of the Greek mainland, north of Macedonia, bordering the Black Sea. The Abderans seem to have been an uncouth lot – to Greek urbanites, *Abderite* was synonymous with hick, redneck, or yokel – which may be why Democritus, a bright boy with a passion for education, grew up with a chip on his shoulder. In adulthood, he was known as the Laughing Philosopher or the Mocker, from his habit of cackling contemptuously at the stupidities of his fellow man. Socrates – a formidable adversary – doesn't seem to have liked him, though that may have been the inevitable result of their widely differing philosophies. Democritus's view of the world was purely mechanical – even the mind, he argued, was made of atoms; while Socrates had little interest in the workings of natural world, preferring to concentrate on personal ethics via the incessant questions that invariably obliterated his opponents and earned him the nickname "the Gadfly." Socrates's intellectual pursuits were ultimately fatal; when one of his students was convicted of treason, Socrates was accused of corrupting the Athenian youth and condemned to death by means of the infamous cup of hemlock. Democritus, grousing and preaching atoms, lived unmolested to a crochety ripe old age.

Democritus wrote over seventy books, of which almost nothing survives, though his theory – occasionally accompanied by unfriendly personal references – was preserved in later works by others. From these we know that the universe, according to Democritus, was comprised of minuscule atoms which filled empty space, colliding, clumping, and separating to make all forms of matter. Furthermore, atoms of different elements were physically different: for example, the atoms of each of the classic four elements had characteristic textures and shapes. Water atoms, in Democritus's scheme, were smooth and cool like tiny pearls; earth atoms, rough and lumpy like miniature pebbles; fire atoms, barbed, thorny, and painful.

As conceived by Democritus, Greek atomism, despite the

silly-sounding fire atoms, was surprisingly clever and sophisti-cated, based on astute observations of everyday change. Grape-vines, for example, sprouted, produced shoots, leaves, flowers, and fruit, and then, the harvest over, reverted to leafless sticks. Infants outgrew their cradles, became children, adolescents, and adults, then aged, died, and decayed, reduced to dust and ash. Things continually came into being and just as continu-ally disappeared; and their constituent substances clearly had to come from somewhere and go somewhere. Democritus hypothesized that such flow could be explained by atoms perpetually being gained or lost, assembling themselves into one form only to disperse again and eventually form another. It was a breathtaking mental leap and a landmark in the annals of science. Like so many such, at the time it was largely ignored.

Plato, a disciple of the Democritus-loathing Socrates, none the less adopted a brand of atomism at his famous academy in Athens. Born in 427 BCE to an aristocratic Athenian family, Plato, as a teenager, fought in the Second Peloponnesian War, returned home to study with Socrates, and – after his mentor's execution – toured the Mediterranean, visiting Egypt and Italy. One story claims that during this period he was captured by pirates and held for ransom. Presumably he was well-equipped to cope: Plato is a nickname, referring to his broad shoulders – rather than an etiolated intellectual, Plato seems to have been built like a football player. His real name was Aristocles. Plato's school – arguably the world's first univer-sity – was founded in 387 BCE in a suburb of Athens, a lovely spot, walled and planted with sacred olive trees. The area was named after Academus, a legendary Athenian hero – hence the term academy, which has since come to mean an institution of learning. The school offered students a comprehensive curric-ulum in astronomy, biology, mathematics, political theory, and philosophy – with an emphasis on mathematics, which Plato considered the most elegant manifestation of pure thought. Inscribed above the school's doors was the intimidating motto "Let no one ignorant of geometry enter here."

Accordingly the Platonic theory of the elements was

dominated by geometry. It is best described in the *Timaeus*, written around 355 BCE – a late work, penned when Plato was in his seventies. Like many of Plato's best-known writings, the *Timaeus* is presented as a conversation, in this case a four-way debate among Socrates, Critias, Hermocrates, and Timaeus, this last a Pythagorean philosopher, mathematician, and scientist, and the inventor of the pulley. As the title suggests, Timaeus holds the floor. The four elements, Timaeus hypothesized, are aggregates of atoms whose shapes, rather than pebbles, prickles, or pearls, are those of regular polyhedra. A polyhedron is a solid composed of identical regular polygons, of which just five are possible: the cube, composed of 6 identical squares; the tetrahedron – a 4-sided pyramid of equilateral triangles; the octahedron (8 equilateral triangles); the icosahedron (20 equilateral triangles); and the dodecahedron (12 pentagons). The crucial five were discovered by the secretive Pythagoreans, but are generally known as the Platonic solids, since Plato took the trouble to describe them in print. Of these, earth, "being the most immobile of the four bodies and the most retentive of shape," was assigned the stolid cube; fire, the sharpest and most mobile of the bunch, the prominently pointed tetrahedron; air, the intermediate octahedron; and water, in recognition of its size, the bulbous icosahedron. The leftover dodecahedron was used, Timaeus explains somewhat confusingly, "for embroidering the constellations on the whole heaven," which, according to some interpretations, has something to do with the twelve-part zodiac. Since each polyhedron could be shown to be composed of triangles, Plato suggested that one element could be transformed to another by a mathematical process of triangular rearrangements, but the crucial details of just how this took place in nature remained obscure.

Plato's polyhedral scheme of the universe was seized upon some 2,000 years later by the German astronomer Johannes Kepler, who had a mathematical revelation concerning it while giving a classroom lecture at the University of Graz in 1595. Kepler realized – literally between one chalk stroke and the next – that the five Platonic solids and the six known planets were intimately interrelated. "It is amazing!" Kepler wrote ecstatically. "Although I had as yet no clear idea of the order in which the perfect solids had to be arranged, I nevertheless succeeded in arranging them so happily that later on, when I checked the matter over, I had nothing to alter." The planets, in Kepler's happy scheme, revolved in circular orbits circumscribing the shapes of Plato's polyhedra, each nested within the other like complex cosmic Russian dolls. Kepler began with the Earth's orbit ("the measure of all things"), around which he circumscribed a dodecahedron. The circle that precisely encompassed the dodecahedron's vertices, he explained, corresponded to the orbit of Mars. This in turn was circumscribed with a tetrahedron, whose vertices defined the orbit of Jupiter; and around this was positioned an immense cube, which established the orbit of Saturn. Within the Earth's dodecahedron, he inscribed an icosahedron, which delineated the orbit of Venus; and inside that an octahedron, mapping the orbit of Mercury.

With all the enthusiasm of a religious convert, he embarked upon an ambitious project to build a model of his Platonic solar system in the form of an enormous mechanical drinking cup, constructed in the form of seven interlocking basins circumscribed by geometric shapes, each basin representing a planetary orbit. The cup was intended to deliver, via concealed pipes, an array of astronomically appropriate beverages: aqua vitae from the Sun, water from the Moon, brandy from Mercury, mead from Venus, vermouth from Mars, and white wine from Jupiter. Saturn, a planet with evil astrological connotations, was to dispense bad beer. Kepler wrote a letter proposing that he build a version in silver for Frederick, Duke of Wurttemberg – "The Almighty granted me last summer a major inventum in astronomy, after lengthy toil and diligence; which whole work and demonstration thereof can be fittingly and gracefully represented by a drinking cup of an ell [45 inches or 1.1 meters] in diameter which then would be a true and genuine likeness of the world and model of the creation insofar as human reason may fathom." The Duke, inured to monetary requests masquerading as expensive presents, wrote back, "Let him first make a model of copper;" and the cup – for lack of funds – was never finished. A detailed drawing and description of it appeared in *Mysterium cosmographicum* (*The Cosmic Mystery*), Kepler's explanation of his divinely geometric universe, published in 1596 in the teeth of united opposition from the entire faculty of the University of Tübingen, Kepler's alma mater.

Kepler ultimately abandoned his elaborate polyhedral cosmography in favor of elliptical planetary orbits, which had the advantage of corresponding to the observational data – thus establishing his famous first Law of Planetary Motion. "Ah, what a foolish bird I have been," he wrote in 1609. Plato never admitted as much, and presumably his theory of poly-hedral elements was still being taught in 529 BCE, nearly 900 years after his death, when his Academy's doors were finally closed by order of the Christian Emperor Justinian. It is this date that traditionally marks the beginning of the dismal European period known as the Dark Ages.

The four-element model as promulgated by Plato was passed on to Plato's most famous student, Aristotle (384–322 BCE), who arrived at the Academy from Macedonia at the age of seventeen – an effete young man, said to have spoken with a pronounced lisp, whose intellectual brilliance soon earned him the title "the intelligence of the school." Aristotle even-tually added a fifth element to the classic Greek quartet, an elusive and insubstantial material hypothesized to make up the heavens. This he initially called the *ether* – hence our modern word ethereal, which refers to anything other-worldly. Later philosophers called it the *quintessence* – "fifth element" – which word today is still used to mean the most essential feature of a substance or the purest embodiment of a given quality. ("You have escaped the quintessence of bores," wrote Sir Walter Scott, who hadn't, to a friend in 1823.) The heavenly fifth element was markedly different from the worldly four; and was said by Aristotle to contain embedded within it a multi-plicity of concentric crystalline spheres variously carrying the Moon, Sun, planets, and stars. This pretty concept – a leftover from Pythagoras – would stymie the progress of astronomy for centuries, though in recompense it did leave us the evocative expression "music of the spheres," which refers to the sublime melodies supposedly generated as these spheres rotated about the Earth in harmonious synchrony.

The remaining four Aristotelian elements were confined to the planet, upon which they were hypothesized to arrange themselves in concentric layers, with elemental earth positioned solidly at the center, water wrapped around it, air encircling water, and fire – the most volatile of the four – in the outermost tier, the last before the heavenly ether. As an extra wrinkle, each element was imbued with associated qualities of hot, cold, wet, or dry; and it was postulated that interchanges among these qualities allowed one element to transmogrify into another. Air, for example, being hot and wet, could be cooled, thereby converting it to water, which was cold and wet; water in turn could be dried, converting it to cold dry earth. The suggestion that one element could be readily changed into another was to cause endless trouble in centuries to come, as medieval alchemists, extrapolating from Aristotle, struggled to transmute lead into gold.

Aristotle's concept of the elements dominated Western science for the next 2,000 years – largely because, sometime in the second century BCE, some fifty volumes of his books were thoughtfully buried. The texts were unearthed 200 years later by trench-digging Roman soldiers under the command of General Lucius Cornelius Sulla; and the general, after chopping down the Academy's sacred olive trees to make siege engines, sacking Athens, and completing the subjugation of Greece, took the books back to Rome. There, while he otherwise occupied himself by seizing control of the Roman Republic and executing thousands of his political opponents in a bloodbath of fearsome proportions, he had the books preserved, copied, and distributed. They were lost again to Western Europe following the fall of Rome to the illiterate Visigoths, but survived in Arabic translations to resurface unimpaired in the eleventh and twelfth centuries.

The sheer magnitude of Aristotle's output – and the surviving works are believed to represent only about a third of his total – left medieval scholars in awe. Aristotle, just by being there, in quantity, acquired a voice of near-omnipotent authority heretofore unprecedented in Western scientific thought. Throughout the Middle Ages, *Aristotle dixit* – "Because Aristotle said so" – was the ultimate trump card in argument; similarly, because Aristotle said not was a clincher for any refutation. Unfortunately Aristotle rejected any number of things, among them the intellectual capacity of women ("the deliberative faculty is absent in the female"), the immorality of slavery ("a natural institution"), and the existence of the atom.

An early sixteenth-century painting by Raphael titled *The School of Athens* pictures the famous Academy as a vast and crowded marble arcade where Plato and Aristotle hold center stage. Plato – whose features are those of Leonardo da Vinci – is draped in red and clutches a copy of the *Timaeus*; Aristotle, in blue, brandishes the *Nichomachean Ethics*. A dozen other famous philosophers cluster round, among them Pythagoras, Empedocles, Socrates, Ptolemy – shown from the back, twirling a globe – and Heraclitus, in purple, looking sulky. There's some debate over the characters' identities – a bearded sage in a cap with earflaps, for example, is said to be either the Greek geographer Strabo or the Persian Zoroaster – but most critics agree that Democritus, the ancient world's prime proponent of the atom, simply isn't there. As a legacy of the School of Athens, the anti-atom Aristotle ruled; and it may be due to his prolonged and overwhelming influence that the first mention of atom in English only appeared in 1678, and then was accompanied by the adjective "spurious."

The intellectual stultification of the European Dark Ages did not extend to the Middle East, where in that period science steadily flourished. It was among the Arab alchemists that the ancient four-element concept slowly began to evolve. The best known of these early Arab researchers was Abu Musa Jabir ibn Hayyan, known in the West as Geber, from his Latin translator's best guess at the spelling and pronunciation of his name. Geber was born around 760 CE and lived the bulk of his life in Baghdad during the magnificent reign of the near-legendary caliph Harun al-Rashid, the rich and disillusioned monarch to whom the beautiful Scheherazade told her 1001 tales in *Arabian Nights*. Baghdad in Geber's time was new. The city was built, like Brasilia and Washington, DC, from scratch, but upon completion was a far more exotic proposition: a vast triple-walled metropolis filled with rose gardens, spice-scented bazaars, towering mosques with glistening minarets, and immense marble palaces furnished in gold, silver, silk, and brocade. An awed visitor of the next century described a pond of silver mercury on which floated golden boats, a tree of gold and silver in which perched jeweled mechanical birds, and a menagerie of a hundred lions, each with its own personal attendant.

By the ninth century, Baghdad was the acknowledged richest city in the world, as well as a renowned center of learning and the arts; when it finally fell to the Mongols in 1258, so many books were hurled into the Tigris from ransacked libraries that the river ran black with ink. Among these drowned manuscripts were doubtless Geber's alchemical treatises, of which he wrote many. His masterpiece, compiled at the culmination of his career and modestly titled *The Sum of Perfection*, included descriptions of all his major chemical experiments, which showed that Geber, a skilful technician, had come within a hair's-breadth of understanding the processes behind simple chemical reactions. It also contained an explanation of his theory of elements. All metals, wrote Geber, were composed of varying proportions of two elements – sulfur and mercury. Thus any metal, in theory – for example, lead – could be separated into its component sulfur and mercury, which could then be recombined in a different ratio to generate another metal – say, gold. Effective recombination required a catalyst, which Geber referred to as *al-iksir*, or elixir, a mysteriously

elusive substance the search for which was to frustrate generations of future alchemists, who never found it.

The sulfur–mercury theory was expanded upon in the
next century by the Persian physician Al-Razi – Latinized
as Rhazes – in his masterwork, *The Secret of Secrets*. Al-Razi
was an obsessive list-maker, and his book is essentially an
immense collection of elaborately annotated lists of laboratory
apparatus (including beakers, flasks, phials, basins, braziers,
tongs, spatulas, hammers, sand and water baths, alembics,
cucurbits, crystallization dishes, and mortars and pestles),
laboratory techniques, recipes, and chemicals. Chemicals
were variously classified as metals, vitriols, boraxes, salts, and
stones, depending on such properties as taste and solubility – a
first tentative groping toward modern theories of compounds
and elements. Al-Razi added salt to Geber's elemental duo:
all solids, he hypothesized, were made from various combinations of sulfur, mercury, and salt, which respectively contributed combustibility, volatility, and substance.

The Greek four elements and the Arabic three were creatively melded in the early sixteenth century by the Swiss
alchemist Paracelsus, much in the same accommodating
manner that Empedocles had amalgamated the one-element
theories of his philosophical predecessors into the four-part
doctrine of earth, air, fire, and water. Paracelsus was born in
Einsiedeln, Switzerland, in 1493, and christened with what
may be the most flamboyant name in the history of science:
Philippus Aureolus Theophrastus Bombast von Hohenheim.
From his middle name comes the common noun *bombast*,
meaning inflated braggadocio, which was by all accounts his
defining feature. This was exemplified by his brazen adoption
in the 1520s of the name Paracelsus, or "better than Celsus," a
nominative demonstration of one-upmanship over the Roman
physician Celsus whose theories were enjoying a contemporary academic popularity. He could as easily have chosen
Paragalen, after the famous second-century Greek physician,
whose works he publicly and dramatically burned (in a brass
pan with an explosive handful of sulfur and saltpeter) to show

his contempt for the outmoded beliefs and practices of the past.

Paracelsus seems to have been educated piecemeal, while wandering from university to university across Europe, though he claimed to have received a medical degree from the University of Ferrara in Italy, an event of which the university has no record. He was convincing enough about it, however, to acquire the post of town physician in Basel – the site of the aforementioned book-burning – where he so impressed the public that he was admiringly designated "the King of All Knowledge." His pupils adored him; his detractors sneered that he consorted with rabble and lived like a pig. His long-suffering secretary Oporinus, who was in a position to know him best, tended toward the latter view, writing, "Often he would come home tipsy, after midnight, and throw himself on his bed in his clothes wearing his sword, which he said he had obtained from a hangman."

When not brawling, drinking, or insulting the establishment, Paracelsus applied himself to chemical research. Oporinus, who served as unhappy guinea pig, wrote, "His kitchen blazed with constant fire; his alkali, oleum sublimate, rex praecipitae, arsenic oil, crocus martis, or his miraculous opoldeltoch or God knows what concoction. Once he nearly killed me. He told me to look at the spirit in his alembic and pushed my nose close to it so that the smoke came into my mouth and nose. I fainted from the virulent vapor." Based on such experiments, Paracelsus determined that, though the Greek four were indeed the fundamental components of all matter, earth, air, fire, and water in turn were composed of the three Arabic "principles," mercury, sulfur, and salt. A common contemporary explanation of the four-element theory cited a "green stick" which, upon burning, separated into its four basic components, variously released as smoke (air), flames (fire), oozing sap bubbling from the cut ends of the stick (water), and ash (earth). In Paracelsian terms, the volatile smoke was mercury; the flames, sulfur; and the ash, salt. Mercury was also considered a prime constituent of water, and was believed to function as a cohesive principle – that is, water, which was made of mercury, claimed Paracelsus, was the reason that things stuck together.

Paracelsus died in 1541 at the age of forty-eight, in a tavern – possibly of a fall, though some accounts claim an attack by assassins. Some of his innovative medical theories were ultimately vindicated: he argued, for example, that mental illnesses were physiological in nature, rather than the result of demonic possession; and he planted the seed for modern pharmacology with the idea that diseases had specific causes and should thus be treated with specifically targeted drugs. His take on the elements, on the other hand, turned out to be bombast.

"As for the greene sticke," wrote Robert Boyle crisply in 1658, "the fire does not separate it into elements; but into mixed bodies, disguised into other shapes . . ." Boyle was an Irish aristocrat, born in Lismore Castle in 1627, the fourteenth child and seventh son of the impressively wealthy Earl of Cork. He was by all accounts a brilliant child, fluent in Greek and Latin by the age of eight and in French by the age of twelve. At fourteen, he was living with a tutor in Florence, learning Italian and studying the works of Galileo; and by the time he was in his early twenties, he had settled in England, established a laboratory, and embarked upon a life of prolific scientific research. His best-known work is *The Sceptical Chymist*, published in 1661.

In personality, Boyle was the diametrical opposite of Paracelsus, being pious, reserved, and something of a wet blanket – he had "nothing of Frolick or Levity in him," one contemporary remarked. As a young man in Italy, he confessed to having "in his Governor's Company" visited the "famousest Bordellos;" however, he emerged still staunchly in possession of "an unblemish't Chastity." This last may have had something to do with a monumental thunderstorm he encountered as a schoolboy in Geneva, a storm of such terrifying proportions that he experienced in the midst of it a religious epiphany. Thereafter, Boyle was a zealous, though somewhat nervous,

Christian. He was also arguably the world's first chemist; and is often credited with coining the first recognizably modern definition of an element.

The crucial definition appears in Part VI of *The Sceptical Chymist*, a dense and enormous critique of traditional alchemy, in which scientific arguments are couched in the form of a four-sided argument among Carneades (a skeptical chemist), Themistius (an Aristotelian), Philoponus (a Paracelsian), and Eleutherius (undecided). Carneades, Boyle's skeptical alter-ego, carries the day, defeating both Aristotelian four-element and Paracelsian three-principle theories, and eventually winning over the waffling Eleutherius. Elements, explains Carneades/Boyle, are "certain primitive and simple, or perfectly unmingled bodies; which not being made of any other bodies, or of one another, are the ingredients of which all those called perfectly mixt bodies are immediately compounded, and into which they are ultimately resolved." Elements, in other words, are substances which cannot be broken down into simpler substances. The output of the much-debated "greene sticke" certainly did not consist of elements, wrote Boyle: the smoke, for example, far from being singular air, could be distilled to yield an oil, which left an "earthe behind it," from which in turn could be extracted a salt. Furthermore, all substances could not, as traditionally hypothesized, be neatly reduced to an elemental four. Gold, for example, seemed to consist of nothing but gold; while "bloud," analyzed, contained not four components, but dozens. In chemical circles, the age of the Greeks' four elements was over; for the next 300 years, their numbers would steadily multiply.

In 1789, Antoine Lavoisier, in his *Elementary Treatise on Chemistry*, published a list of the known chemical elements, variously classified as gases, metals, non-metals, and earths. There were thirty-three of these in all, each ostensibly fulfilling a definition of element that would have made sense to Carneades and

Boyle, though Lavoisier was cautious: "...We express by the terms elements or principles of bodies the idea of the last point reached by analysis, all substances that we have not yet been able to decompose by any means are elements to us, not that we can assert that these bodies that we consider simple are not themselves composed of two or even a greater number of particles, but, since these principles are not separated, or rather since we have no means of separating them, they are to us as simple substances, and we must not suppose them compounded until experiment and observation have proved them to be so."

Of the thirty-three, eight indeed later proved to be decomposable compounds, consisting of two or more elements – among them lime, silica, and magnesia; and two – heat and light – proved not to be elements at all. (Lavoisier, in an uncharacteristic moment of vagueness, had listed both as "imponderable fluids.") Of the remaining twenty-three, ten had been known since ancient times: eight metals (gold, silver, copper, iron, mercury, tin, lead, and antimony), sulfur, and carbon ("charcoal" on Lavoisier's list). Zinc and arsenic had been known since the thirteenth century; bismuth and phosphorus since the sixteenth. Platinum, artifacts of which have been found in South American burial sites dating back 2000 years, is thought to have been formally identified as an element in the early 1700s, reportedly by a Don Antonio Ulloa, who collected some on a beach in Panama. It appears on Lavoisier's list by its Spanish designation *platina*, little silver. Brilliantly blue cobalt had been isolated from copper ore in 1735 by Swedish chemist Georg Brandt; its name comes from the German *kobold*, meaning goblin or evil spirit, since miners often found it associated with toxic arsenic. The name nickel has similarly devilish connotations, derived from *kupfernickel* – devil's or Old Nick's copper – a name bestowed upon its red-brown ore by frustrated copper miners who found that, despite its suggestive copper-like appearance, they were unable to get any copper out of it. In 1754, chemist Alex Fredrik Cronstedt of Stockholm isolated nickel from it instead. Manganese and molybdenum were both first isolated and identified as elements in Sweden, respectively in 1771 and 1781, though both had been previously known in impure form. Manganese compounds – in the form of black powders used by glass-makers to make transparent crystal – had been known since ancient Rome; and a molybdenum ore – molybdenum sulfide, or molybdenite – which looks a lot like graphite, was being used in Lavoisier's time for making pencils.

Hydrogen, one of Lavoisier's three designated gases, was discovered in 1766, by the bizarre but brilliant Henry Cavendish, of whom more later. Stunningly wealthy and pathologically shy, Cavendish had such a fear of women that he would only communicate with his housekeeper by way of notes left on the hall table. He seems to have done little better with men, and when he did manage to utter a few words in his squeaky voice, he insisted that listeners not look him in the face. He refused to sit for a portrait; our best guess at what he looked like comes from a sketch reputedly done without his knowledge at a Royal Society dinner, in which he appears in rumpled violet suit, turned away from the artist. None the less, his scientific accomplishments were impressive, and Cavendish considered the discovery of hydrogen so notable that it inspired one of his rare forays into the public arena. He published a paper about it, titled "On Factitious Airs."

Nitrogen, though studied by both Cavendish and Joseph Priestley, was formally identified by a 22-year-old Scottish graduate student named Daniel Rutherford, who wrote his doctoral dissertation about it in 1772. It appears on Lavoisier's list as "azote", which means lifeless. Oxygen was discovered at least three times between 1772 and 1775, the credit variously claimed by Carl Wilhelm Scheele of Sweden, Priestley, and Lavoisier himself. Lavoisier, in any case, named it, dubbing it *oxygène* or "acid-forming," since he believed it to be an essential component of all acids. The name initially found little favor with the Priestley-supporting English (who patriotically stuck by Priestley's term "dephlogisticated air"), but was eventually adopted after Erasmus Darwin, grandfather of the evolutionary Charles, published an effusive ode to oxygen in *The Botanic Garden* (1791). Tungsten, the most recent of Lavoisier's elements, was discovered in 1783 by the brothers Juan José and Fausto Elhuyar of Spain. Tungsten – from the Swedish for "heavy stone" – was Juan José's choice of name, favored in England and Lavoisier's France. Fausto preferred wolfram – or "wolf dirt" – a pejorative of German tin miners, who found that the stuff interfered with the efficient smelting of tin. In Germany, Spain, and Italy, the Elhuyars' element is still known today as wolfram; and the International Union of Pure and Applied Chemistry (IUPAC) even-handedly recognizes both names.

For the most part, Lavoisier's list was surprisingly accurate and up-to-date. Of the elements known at the time, he missed only chlorine (1774), chromium (1780), and tellurium (1783); and he could hardly have been expected to include zirconium or uranium, both discovered in 1789, his year of publication. By 1808, when John Dalton spelled out the relationship between atoms and elements in his *New System of Chemical Philosophy*, chemists had discovered eighteen more.

Dalton was the son of a weaver, born into a Quaker family in 1766 in the little village of Eaglesfield in Cumberland, where he grew up at the edge of England's Lake District – the lovely landscape of mountains, lakes, and waterfalls that so entranced William Wordsworth and Beatrix Potter. There, by the age of twelve, he was teaching at the local school, a profession he was to pursue in one capacity or another for the rest of his life. There also he first became enthralled with science, first meteorology – ultimately he amassed a 57-year record of daily weather data, totaling some 200,000 separate observations – and then chemistry. In both, he was handicapped by his color-blindness, as evidenced by his bland and clinical description of the aurora borealis, and his inability to effectively utilize colorimetric assays, linchpins of eighteenth-century experimental chemistry. However, he did get a scientific paper out of his condition – his first, titled "Extraordinary Facts Relating to the Vision of Colours," published in 1794. Subsequently color-blindness became generally known as "daltonism."

Dalton's contribution to the theory of elements was a creative mix of Greek philosophy, quantitative chemistry, and simple mathematics. Elements, Dalton explained, were composed of tiny indestructible particles; and each separate element was made of a different and characteristic kind of particle. He called these particles atoms, after Democritus's proposed particles of two millennia past. Chemical compounds, Dalton continued, were created when these atoms combined in specific proportions – an idea that dated to Empedocles. These proportions – as would be expected if compounds were made of tiny particles – were always simple whole-number ratios: 4 to 1, for example, or 3 to 4, but never 4.2 to 1, which would imply a fraction of an atom, which was impossible. Dalton then proceeded to calculate the relative atomic weights of the elements, starting with the lightest, hydrogen – a "calculus of chemical measurement," writes William Brock in *The Norton History of Chemistry*, "that for the first time in history married the theory of atoms with tangible reality." By 1810, Dalton had established atomic weights for twenty different elements, among them oxygen, nitrogen, carbon, phosphorus, sulfur, copper, and iron. The invisible atom at last had size, substance, and character.

To represent his atoms on paper, Dalton developed the first meaningful system of chemical notation: each type of atom was symbolized by a circle carrying "some distinctive mark." Hydrogen, for example, was represented by a circle with a dot in the center; oxygen by an empty circle; carbon by a solid black circle; sulfur by a circle divided into quarters by a cross; and copper by a circle with a tiny c in the center, like a modern copyright sign. Ultimately, however, the circle system proved more artistic than useful, and the chemical community opted instead for the less attractive but far more efficient method-ology of Swedish chemist Jons Jacob Berzelius, still in use today.

Berzelius's was a system of shorthand abbreviations in which each element was represented by the first letter of its Latin name – or first two letters in the case of such first-letter duplicates as calcium (Ca), cobalt (Co), copper (Cu), and chlorine (Cl). The Latin stipulation explains the seeming discrepancies in modern chemical symbols: gold's Au, for example, comes sensibly from the Latin *aurum*, meaning gold; iron's Fe from the Latin *ferrum*; and silver's Ag from the Latin *argentum*. Abbreviated letter symbols strung together quickly revealed the structure of chemical compounds – as in NaCl, sodium chloride or table salt – and subscripts could be added to denote numbers of interacting atoms, as in H_2O (two hydrogens, one oxygen) or H_2SO_4 (two hydrogens, one sulfur, four oxygens). The helpful letter cues acted as memory aids, and the letters, unlike Dalton's circles, were cheap and easy to print, which was a plus from a publication standpoint. Still, the enthusiastic and universal adoption of Berzelius's letters roused the lamb-like Dalton to protest. "Berzelius's symbols are horrifying," he wrote. "A young student in chemistry might as soon learn Hebrew as make himself acquainted with them . . . [they] equally perplex the adepts of science, discourage the learner, as well as . . . cloud the beauty and simplicity of the Atomic Theory."

This last, of course, no abbreviation, no matter how horrifying, could do. Dalton's elegant theory occupies just five pages in his 916-page *New System of Chemical Philosophy*, but those few revolutionized the entire field of chemistry. A dedicated Quaker, Dalton steadfastly avoided his much-deserved public acclaim; when he became a Fellow of the Royal Society in 1822, he was elected *in absentia* at the instigation of his friends. Retiring or not, after the publication of the *New System* universities as far afield as Paris, Berlin, Munich, and Moscow showered him with honors; and in 1832, he received an honorary degree from the University of Oxford, which gave him the right to wear a scarlet doctoral gown. His Quaker sensibilities ordinarily would have prevented him from wearing it; however, to the color-deficient Dalton, the gaudy robes appeared to be an acceptably subdued gray. Two years later, when he was presented at the court of King William IV, he refused to deck himself out in court apparel, but consented to wear the gown, remarking placidly "I see no scarlet here" to the ditherings of worried friends. He maintained his creed of modest simplicity to the end, requesting at the last a quiet Quaker funeral, but his reputation was such that this was not to be. In 1844 when the man the London *Times* now called a "venerable Chymist" died at the age of seventy-eight, 40,000 mourners gathered to bid him good-bye.

Order is heaven's first law.

Alexander Pope
An Essay on Man

Though many of us move through life in an atmosphere of misplaced objects, misfiled documents, and a sort of generalized casual messiness, we are at heart an orderly species. In fact, we're addicted to classification. According to the Book of Genesis, one of Adam's first acts upon hitting the turf of Eden was to name the animals, which implies a rudimentary form of categorization – at the very least, he had to distinguish animals from vegetables and minerals. Entire books have been dedicated to the classification process, among them Carl Linnaeus's *Systema Naturae* (1735), which expanded mightily on Adam's first tentative try, plugging all Earth's animals and plants into their proper kingdoms, phyla, classes, orders, families, genera, and species; and more recently Barbara Ann Kipfer's *The World of Order and Organization* (1997), a massive and eclectic collection of categorized lists, among them the five stages of boiling, the six social orders of ancient Rome, the piece rankings in chess, and the hierarchies of devils, angels, and the Mafia.

The foremost chemical puzzle of the 1800s was the classification of the elements, and throughout the century chemists struggled to package them into logically related groups. In 1811, Berzelius – of chemical symbol fame – proposed that all matter be divided into two parts: imponderables, which included such mysterious entities as electricity, magnetism, light, and heat; and ponderables, which included elements and compounds. Elements (*Simplicia*, in Berzelian terms) in turn could be divided into three groups – oxygen, metalloids, and metals; while compounds were divvied up between minerals and "organized bodies," by which last he meant animals and plants. Elements, within their category, were arranged in order based on their electrolytic behavior, with the highly electronegative oxygen at the top of the list and strongly electropositive potassium at the bottom. Hydrogen, planted solidly in the middle, marked the cut-off point between negatives and positives.

In 1817, German chemist Johann Dobereiner suggested an alternative arrangement, based on the apparent tendency of the elements to fall into chemically similar groups of three. Dobereiner, born in Bavaria in 1780, was the son of a coachman, and his family was too poor to provide much in the way of formal education. However, the bright youngster was apprenticed to an apothecary as a teenager, under whose tutelage he worked hard, read extensively, and attended every public lecture available. Eventually he attracted the attention of a wealthy patron, who finagled a position for him as professor of chemistry at the University of Jena, where, between classes, he concentrated on the problem of the elements. Dobereiner's idea of elemental triplets originated when he noticed that strontium – which shares chemical properties with calcium and barium – fell precisely between the two in atomic weight. Once the first triplet surfaced, Dobereiner soon identified more: chlorine, bromine, and iodine displayed the same pattern of atomic weight sequence and chemical similarities; as did sulfur, selenium, and tellurium; and lithium, sodium, and potassium. Based on these observations, he proposed that all elements naturally occurred in subgroups called triads – a vaguely Pythagorean hypothesis which, though ultimately proved wrong, was beneficial in that it inspired considerable interest in discovering patterns among elements. Almost immediately an exception to the triad rule was found in the form of an elemental quintet. Nitrogen, phosphorus, arsenic, antimony, and bismuth formed a progression in which the atomic weight of phosphorus was midway between that of arsenic and nitrogen; the weight of arsenic fell into place between phosphorus and antimony; and that of antimony lodged midway between arsenic and bismuth.

The first workable table of elements, which merged Dobereiner's scattered triads and the odd quintet into an ordered list of twenty-four, was devised in 1862 by Alexandre-Emile Beguyer de Chancourtois, a geologist of international repute from the Paris School of Mines. In 1862, de Chancourtois was in his forties, and his foray into chemistry was something of an aberration, since his work to date had centered around topography. None the less, he was the first scientist to identify periodicity among the elements, noting that if all were arranged in a row in order of atomic weight, similar properties popped up at regular repeating intervals. To demonstrate this chemical repetition, he invented an elaborate visual aid on which the elements were positioned according to increasing atomic weights in an ascending spiral inscribed on the surface of a cylinder. Elements that lined up vertically on the cylindrical surface had similar chemical properties: one complete turn around the cylinder from oxygen, for example, landed seven elements above it on behaviorally similar sulfur. De Chancourtois referred to his convoluted device as a telluric screw (*vis tellurique*), since the element tellurium fell at the midpoint of the spiral. Though it was an elegant construct with an attention-grabbing name, this first try at a Table of Elements plunged promptly into obscurity. First, de Chancourtois had chosen a geology-based terminology to describe his invention, thus confusing chemists, who didn't know geology, and baffling geologists, who had yet to grasp the importance of chemistry. Still, he might have risen above his impenetrable text if the journal that published his landmark paper had seen fit to publish the accompanying illustration. Unfortunately it did not; and de Chancourtois was forced to watch other, more straightforward, chemists rake in the credit that should have been his due.

Among these was English chemist John Newlands. The son of a London minister, Newlands had spent two years in his twenties quixotically fighting in Italy with Garibaldi's Red Shirts, before returning home to settle into a job as an analytical chemist in a sugar refinery. There he transferred his passion for politics to chemistry – specifically to the question of the relationships among the chemical elements. The elements, if arranged in order of atomic weight, Newlands was soon pointing out in a series of lengthy letters begun in 1863 to the editors of *Chemical News*, fell into a repeating pattern in which each element showed properties similar to that of the elements eight places ahead of it and eight places behind it in the sequence. Newlands likened this eight-fold periodicity to the relationships seen among musical notes in repeating octaves, and accordingly dubbed it the "Law of Octaves." He arranged the fifty-six known elements in an eight-column table which he presented at a meeting of the London Chemical Society in 1866. The Chemical Society had a good laugh at his expense – the octaves comparison seems to have struck a particularly sour note with the audience – and refused to publish his paper. Three years later, Dmitri Mendeleev published his version of the Periodic Table of Elements, vindicating Newlands, who large-spiritedly forwent the opportunity to have a prolonged last laugh. In 1887, the Royal Society very belatedly awarded him the Davy medal for innovative work in chemistry.

Credit for the ultimate solution to the element classification puzzle generally goes to Russian chemist Dmitri Mendeleev. The youngest of fourteen (some sources say seventeen) children, Mendeleev was born in February 1834 in frigid Tobolsk, Siberia, a town of wooden houses with blue-painted shutters, set among birch and fir trees, on the edge of the arctic wilderness. When Mendeleev was a boy, Dostoevsky passed through the town – unnoticed, en route to prison camp – and later Tsar Nicholas II and family would spend an isolated year here, prior to their execution. The Mendeleevs were dogged by bad luck: Mendeleev's father, Ivan, headmaster of the local high school, went blind when Dmitri was a child, leaving the support of the immense family to his wife Maria, who took a job as manager of a glass factory. In 1848, a year of disaster for the family, Ivan died and the glass factory burned to the ground, leaving Maria with no source of income and no money other than the nest egg she had painstakingly set aside for the education of her promising youngest son. The following year, with her two youngest children, Dmitri and Elizabeth, in tow, she left Siberia and set out on the 1300-mile journey to St Petersburg, determined to enter her 15-year-old son in the university. In 1850, the year the young Mendeleev matriculated, both his mother and sister died of tuberculosis, leaving him an orphan and a long way from home. It sounds a lonely existence for a teenaged boy, but – perhaps bolstered by his mother's deathbed admonition to "patiently search divine and scientific truth" – Mendeleev persevered, to graduate first in his class. He spent some years studying abroad, but ultimately returned to St Petersburg, where he settled down to a career in teaching and research.

Mendeleev devised his landmark Periodic Table of Elements while writing *The Principles of Chemistry*, a textbook for his university students, noted both for the clarity of its prose and the extreme length of its footnotes. In the course of its composition, he arranged the known elements, which now totaled sixty-three, in order by atomic weight, and was startled to notice a rhythmically repeating pattern of valences. Valence is an indication of the power of an atom to form chemical bonds; the word comes from the Latin admonition "*Valete!*" meaning "Be strong!" Lithium, for example, will combine with only one other atom – thus it has a valence of 1. Its next-heaviest neighbor, beryllium – the chemical basis of emeralds – can bind to two other atoms and therefore has a valence of 2; next in line are boron (3) and carbon (4). Working his way through the list, Mendeleev found that valences of the ordered atoms rose and fell in periodic fashion – 1,2,3,4,3,2,1, over and over again like the undulating waves of the sea. The moment of revelation, he later explained, had come to him in a dream. Exhausted, he had fallen asleep with his head on his study table and there had a dream "where all the elements fell into place as required. Awakening, I immediately wrote it down on a piece of paper."

The final arrangement of the elements in table form he attributed to solitaire. Mendeleev was a card-player; and he found that the names of the elements, written on cards and dealt out in rows, solitaire-style, could be ordered in horizontal lines by weight and vertical columns by valence, such that elements with similar chemical properties lined up. Furthermore, the chart had predictive value – there were holes in it, for example, where as-yet-unknown elements of specific size and valence should fit. Among these were Mendeleev's "eka-aluminum," hypothesized to occupy the gap between aluminum and uranium, and "eka-silicon," plugging the empty space between silicon and tin.

Mendeleev published his results in 1869 in a paper titled "A New System of the Elements" which, though rapidly translated and made available in Western Europe, was received at best with cautious suspicion. Chemists, blasé survivors of triads, telluric screws, and octaves, had seen such proposals before. The climate changed abruptly, however, on August 27, 1875, when Paul Lecoq de Boisbaudran of Cognac – son of a family of distillers – wrote to the French Academy of Sciences: "During the night before last . . . between three and four in the morning, I discovered a new element in a sample of zinc sulfide from the Pierrefitte mine in the Pyrenees." It was a silvery-white metal characterized by a melting point so low that it turned into a puddle when held in a warm human hand, with an atomic weight of 69 and a valence of 3. It was, in fact, Mendeleev's postulated eka-aluminum; and with its discovery, Mendeleev became the most famous chemist in the world. Lecoq de Boisbaudran named the new element gallium, ostensibly in honor of his native land – Gallia, the Latin name for France. An alternative story holds that he shamelessly named it for himself: Lecoq means "the cock" or "the rooster," and the Latin for rooster is *gallus*; gallium may thus be the only element that perpetuates a self-serving pun. Eka-silicon duly appeared in 1880, isolated by German chemist Clemens Winkler and patriotically named germanium.

In photographs, Mendeleev – who allowed his hair and beard to be cut only once a year, and then with a pair of sheep shears – has the shaggy and wild-eyed look of Rasputin; his rough exterior, however, hid a romantic side. In 1876, he fell passionately in love, divorcing Feozva, his wife of thirteen years, to marry Anna Popova, a beautiful 17-year-old art student. According to the law of the Russian Orthodox Church, re-marriage was forbidden for seven years following a divorce; Mendeleev, who remarried immediately, was thus arraigned for bigamy. Tsar Alexander II summarily dismissed the charges on the grounds that "Mendeleev may have two wives, yes, but I have only one Mendeleev."

The Periodic Table of Elements was to prove an aid and inspiration to all present and future chemists and physicists; and its elucidation was a key event in the history of science. None the less, the Tsar's "one Mendeleev" never won a Nobel Prize for his achievement – though he came close, losing out by only one vote in 1906 to French chemist Ferdinand Moissan, the discoverer of fluorine. The shortsightedness of the Nobel committee, however, has been more than compensated for. Two chemistry journals, a chemical society, a university, and a meteorological institute have all been named after Mendeleev, as well as two towns in Russia, streets in Moscow and St Petersburg, and an underground station on the Moscow Metro. Mendeleev's apartment in St Petersburg, where he completed his work on the Periodic Table, has been preserved as a museum, complete with the study table upon which he had his fateful dream, his books, inkstand, and laboratory equipment, and even his scribbled notes, one page marked with a discolored ring where he set down a glass of tea. Both a moon crater and an asteroid bear his name; and element 101 – first created in a cyclotron in Berkeley, California, in 1955 – was named mendelevium in his honor.

Mendeleev's famous first Periodic Table consisted of eight vertical columns called groups, and twelve horizontal rows called periods. Although there are currently some 600 different Table models in circulation, among them circular, spiral, and pyramidal tables, and even one with extended paddles that looks a bit like a high-tech weathervane, today's most frequently used Table is basically an enhanced and updated version of Mendeleev's original. Roughly rectangular, the Table now features eighteen vertical groups, numbered with Roman numerals, and seven horizontal periods, numbered in Arabic from one to seven. Elements within vertical groups share a valence and are thus behaviorally similar – these, being recognizable relatives, are often referred to as chemical families.

P. W. Atkins, in *The Periodic Kingdom* (1995), describes the Table allegorically in terms of geography and topography – a compelling but somewhat dizzying image that conveys the feel of flying over a vast chemical fantasyland in a small airplane. He begins on the far western "shore" – the left-hand edge – of the Table with Group I, the alkali metals (hydrogen, lithium, sodium, potassium, rubidium, cesium, and francium), a region that proves to be increasingly reactive as one moves from north to south. Rain, hitting the metaphorical ground here, Atkins hypothesizes, would fizz like soda pop in the far northern zone of lithium and sodium, ignite and burn when it contacts mid-region potassium, and literally explode when it strikes southerly rubidium and cesium. (The mechanism is a cleavage of water molecules to release hydrogen gas, generating so much heat in the process that the gas bursts into flame.) None of the alkali metals are found free in nature; all are so volcanically reactive that they are inevitably bound up with something else. In laboratories, the alkali metals – fearsome chemical velociraptors – are cautiously stored in kerosene.

A step in from the seething western coast brings us to Group II, the alkaline earth metals, a still reactive but less violent collection, running from beryllium and magnesium in the north, through calcium, strontium, and barium to the blue glow of radium in the far south. Atkins then visualizes a vast central "Western Desert" of metals, comprising some ten groups in the middle of the Table, a long lustrous region of whites and silvers, punctuated with the reddish sheen of copper, the yellow gleam of gold, and a thick silver lake of mercury. The "Eastern Rectangle" on the right-hand side of the table is a patchwork mélange of nonmetals – Atkins imagines, for example, a soot-black expanse of carbon, a yellow splash of sulfur, invisible gaseous shimmers of oxygen and nitrogen. A bright vertical stripe running down the eastern side of the Table is the land of the halogens, extending from northerly fluorine through the poisonous yellow-green splotch of chlorine, the red lake of bromine, and the deep purple-black of iodine to radioactive astatine in the far south. Finally the far eastern "coast," or far right-hand border of the Table – on my copy it's color-coded in aloof aqua blue – is the region of the so-called noble gases: helium, neon, argon, krypton, xenon, and radon. And poking southward, in a sort of cartoon-balloon extension of the Table – Atkins refers to it as the "Southern Island" – is the antipodal region of the lanthanides and actinides, which last – sometimes referred to as the transuranium or transfermium elements – are offshoots of America's World War II Manhattan Project.

Rapid discoveries of new elements put a strain on Mendeleev's Periodic Table. The lanthanides were particularly problematic since, according to their chemical characteristics, all seemed to belong in exactly the same position on the table as lanthanum, element number 57. Originally known as the rare earth elements, only a handful had been identified when

Mendeleev was creating his elemental matrix: lanthanum itself, plus cerium, erbium, and terbium. These were annoying from the beginning and appear on Mendeleev's original schematic as question marks. During the 1870s and 1880s, seven more rare earths (holmium, ytterbium, gadolinium, praseodymium, neodymium, and dysprosium) came to light – each a triumph of chemical perseverance, since these elements, by virtue of their near-clonal similarity, were extraordinarily difficult to separate. Europium and lutetium were isolated in 1901 and 1907, respectively; and the last of the series, promethium, was discovered in 1945. It was clear that the original Table had to be restructured to accommodate these. Furthermore, there was the question of what to do with the ostensibly unreactive noble gases, the first of which, argon – from the Greek *argos*, meaning lazy – was identified in 1894 by Scottish chemist William Ramsay and English physicist John Strutt (a.k.a. Lord Rayleigh). The noble gases, which appeared to combine with no other atoms whatsoever, were assigned a valence of zero; nobody knew where to put them, and some chemists insisted that they had no business appearing on the Periodic Table at all.

Over a hundred different versions of the Periodic Table were proposed within Mendeleev's lifetime. None quite filled the bill, since the true order of the elements could only be understood in terms of the structure of the atom – an entity which, for millennia, had been steadfastedly envisioned as a tiny but solid particle somewhat akin to a pea. It was not until the twentieth century that scientists proved the putatively solid atom to be both multipartite and practically empty – a discovery with major implications for making sense of the elements. The modern model of the atom pictures a nucleus of subatomic particles – positively charged protons and uncharged neutrons – surrounded by a buzzing haze of much smaller negatively charged electrons, each moving at a blinding velocity near that of the speed of light. Over 99 per cent of the mass of an atom is in the nucleus, but that mass occupies less than one-trillionth of total atomic volume. Thus an atom, dissected, is mostly nothing: envision that primal pea set down in the middle of a football stadium while somewhere in the region of the goal posts a few infinitessimal electrons flash about, each the size of a grain of salt.

For elements, the nucleus is where it's at. Ultimately what makes an element an element is not its atomic weight, as Mendeleev had hypothesized, but the nature of its nucleus – specifically, its resident number of protons, represented by the figure now known as the atomic number. Smallest and simplest of the elements is thus hydrogen, whose barebones nucleus consists of just one single proton. Accordingly, it has an atomic number of 1. Helium, with an atomic number of 2, has two protons in its nucleus; lithium (3) has three; and so on in order, through the as-yet-unnamed element 118, the largest element isolated to date. (Three atoms of it were produced in 1999 in a cyclotron at Lawrence Berkeley National Laboratory in California.)

Steadily increasing proton number explains the ordered horizontal periods of the Periodic Table; the vertical groups, however, are the work of the electrons. In 1913, Danish physicist Niels Bohr correlated the chemical reactivity of elements to the electron configurations surrounding their nuclei. An electrically neutral atom has the same number of electrons as protons, such that negative and positive charges are evenly balanced. Single-proton hydrogen thus boasts a single electron; larger elements may have dozens. Electrons *en masse* bear a superficial resemblance to a maddened swarm of hornets, but Bohr's calculations showed that they actually circle the central nucleus in layered arrays of orbits and sub-orbits, positioned one beyond the other rather like the concentric layers of an onion. Bohr referred to these as shells.

Each shell corresponds to a defined energy level. That closest to the positively charged grip of the nucleus is lowest in energy; electrons found there are of relatively sluggish character compared to their increasingly frenzied cousins moving in increasingly distant orbits. Furthermore, each orbit supports only a limited electron population. The innermost shell, for example, can contain only two electrons; the next, eight; the third, eighteen. The chemical behavior of a given element depends on the electron population of its outermost orbit. Elements with partially filled outermost shells are interestingly reactive; those with fully stuffed outer shells are – in the manner of replete diners – lumpishly inert. This explains the recalcitrant behavior of the noble gases: each has a filled outermost electron shell.

The elucidation of atomic structure revealed the physical underpinnings of Mendeleev's Table – that is, horizontal periods reflected increasing proton numbers; vertical groups, similarities in outermost electron shell configuration. This new conception led to the rearrangement of the Table into an approximation of modern format. The next upheaval occurred in the late 1930s and 1940s when physicists, rather than simply discovering new elements, began manufacturing elements of their own.

There are ninety-two naturally occurring elements, plus another two dozen or so that – on Earth, at least – are solely manmade. The first of these, technetium – the name, appropriately, comes from the Greek *tekhnetos*, meaning artificial – was isolated in 1937 by Italian chemists Emilio Segrè and Carlo Perrier at the University of Palermo from a sample of molybdenum that had been intensively bombarded with deuterons in a cyclotron at the University of California in Berkeley. Prior to this, technetium was essentially unknown on Earth (a trio of German chemists may have glimpsed a nanogram-sized sample in 1925); today it is produced by the ton as a by-product in nuclear reactors, which is a problem since, as radioactive waste, it has to be disposed of safely. Less troublesomely for Earth-dwellers, it's also manufactured by red giant stars.

The scientists of Berkeley – a city famed for its exemplary climate – were the first to deliberately exploit the possibilities of element synthesis by nuclear fusion, bombarding heavy nuclei with particles that, if they stuck, created new elements of greater mass. Neptunium was discovered in this manner in 1940 by Edward McMillan and Philip Abelson, the result of bashing uranium with neutrons. The name most popularly associated with the heavy element manufacture, however, is that of Glenn Seaborg, who shared the Nobel Prize with McMillan in 1951. Seaborg – who began his college career as a literature major – produced the first of his transuranium elements eleven days before Christmas in 1940: plutonium, the stuff of the "Fat Man" atomic bomb dropped with such awful effect on the Japanese city of Nagasaki in August 1945. Neptunium and plutonium, respectively elements 93 and 94, were initially called extremium and ultimium, under the misconception that these were as heavy as elements could possibly get. By 1944, however, Seaborg and colleagues had definitively proved this wrong with the synthesis of americium (95) and curium (96). Due to the secrecy surrounding atomic research during World War II, their results were not revealed until after the war. In November 1945, Seaborg – in what is surely the most unexpected pre-publication announcement in the history of chemistry – spilled the beans on *Quiz Kids*, a children's radio show, when a 12-year-old boy asked, "Mr Seaborg, have you made any more elements lately?" Berkelium (97) and californium (98) soon followed. The new elements – along with thorium, protactinium, and uranium – required a whole new row on the Periodic Table. Collectively they were referred to as the actinides, in reference to their chemical similarity to element 89, actinium.

The names of the elements are something to savor. Spoken, they have a magical cadence all their own: the sonorous bray of strontium, the sci-fi slickness of krypton, the polished elegance of palladium and selenium, the earthy practicality of carbon and tin. Twenty-one of the elements are named for places, including copper, from Cuprum, the Latin name for Cyprus, prime exporter of the metal in ancient times, and magnesium, from Magnesia, a district of Thessaly in northern Greece, source of magnesium-rich mineral ores. Three are named in honor of France – francium, gallium, and lutetium, which last comes from the ancient Latin name for Paris; and three in honor of Germany: germanium, rhenium (from the Rhine), and hassium (from the state of Hesse). Eight are named for territories, countries, or towns in Scandinavia – scandium, thulium, holmium, yttrium, ytterbium, terbium, erbium, and hafnium; and Marie Curie named polonium after her native country, Poland. Europium and americium commemorate, respectively, the continents of Europe and America. Ruthenium, first isolated from a sample of platinum from the Ural Mountains, comes from Ruthenia, the Latin name for Russia; dubnium from the Russian town of Dubna, where scientists first synthesized it in 1967 by peppering americium nuclei with neon ions. Strontium is named for the homely town of Strontian in Scotland, since the original sample from which it was isolated came from an obscure Strontian lead mine.

Ten elements are named for colors. Chlorine comes from the Greek *chloros*, greenish-yellow, for the poisonous shade of its gas; and thallium from the Greek *thallos*, meaning green twig or shoot. Thallium itself is a silver-white metal; the name refers to its spectral line, which is pea-green. Indium, iodine, and cesium respectively refer to shades of indigo, violet-purple, and clear sky-blue – though to the eye of the beholder, only iodine vapor is genuinely purple. Indium, a silvery metal, is named for the distinctive bright purple line of its atomic spectrum; and cesium, which is gold-colored, forms compounds that burn blue. Zirconium comes from the Arabic *zargun*, meaning gold-colored, though purified it's silver; the name refers to the ancient gems called zircons, hyacinths, or jacinths, the most prized of which were pale gold. Rubidium – from the Latin for ruby-red – is a soft white metal so reactive that it must be stored under oil to prevent it from spontaneously bursting into flame. The name refers to its brilliant crimson spectral lines. Rhodium, from the Greek *rhodon*, rose-colored, is a silvery metal, but its salts are deep rose-red. Chromium and iridium are both colorful catch-alls, chromium from the Greek *chroma*, meaning color, and iridium from Iris, the multicolored goddess of the rainbow. Both names refer to the colorful range of the elements' salts.

Two elements – the previously mentioned cobalt and nickel – are named for evil spirits. Nine are named for gods, goddesses, or other mythological beings. Thorium and vanadium, for example, are respectively named for the Scandinavian Thor, god of war, and Vanadis, goddess of beauty. Tantalum is named for the legendary King Tantalus, condemned by the Greek gods to spend eternity suffering from hunger and thirst, with food and water forever just out of his reach. It was named by Anders Ekeberg, its discoverer, who had a particularly difficult time isolating it. The closely related niobium is named for Niobe, Tantalus's daughter. Titanium is named for the collective Titans, the giant offspring of Gaia, the Earth goddess; and promethium for Prometheus, the Titan who stole fire from the gods. Tellurium is named for Tellus, the Roman Earth goddess; silvery selenium is named for Selene, the Greek goddess of the Moon; and helium for Helios, the Sun. Palladium and cerium, which sound as if named for Pallas, goddess of wisdom, and Ceres, goddess of the harvest, actually commemorate asteroids of those names; and mercury, uranium, neptunium, and plutonium are named for planets.

The first element to be named after a human being – unless Lecoq de Boisbaudran really did name gallium after himself – was gadolinium, in 1880, for Finnish chemist Johan Gadolin, who first investigated the mineral gadolinite, subsequently found to contain all fifteen of the so-called rare earth elements. All other eponymous elements are both heavy and manmade: curium (96), einsteinium (99), fermium (100), mendelevium (101), nobelium (102), lawrencium (103) – for E. O. Lawrence, inventor of the cyclotron, rutherfordium (104), seaborgium (106), and bohrium (107). Seaborgium – the only element to have been named after a living scientist – was officially, though somewhat grudgingly, recognized in 1997 by the IUPAC, whose responsibility it is to approve chemical names.

The bulk of the remaining elements are named for some telling property or characteristic. Osmium, for example, which continually oozes evil-smelling osmium tetroxide, comes from the Greek *osme*, smell; bromine, worse, from the Greek *bromos*, stench. At least three elements reflect the travails of the discovering chemist: krypton and lanthanum are both derived from Greek roots meaning hidden or concealed; and the name of the rare earth element dysprosium, whose isolation involved an excruciatingly tedious 58-step process of sequential precipitations, comes despairingly from the Greek *dysprositos*, meaning hard to get. Radium and actinium, which glow in the dark, are named respectively for the Latin *radius* and Greek *aktinos*, both meaning ray. Phosphorus, which both glows and bursts spontaneously into flame, comes from the Greek meaning "bringer of light."

Some elemental names are old. Gold, silver, iron, tin, and lead, for example, are all ancient Anglo-Saxon words, familiar in the time of Beowulf; and sulfur – from the Latin *sulfurium*, meaning just that – was common parlance on the streets of Rome. So was carbon, which comes from the Latin *carbo*, charcoal. Sodium, on the other hand, harks back to medieval-era soda, a multipurpose word referring to a mixed bag of alkaline salts; though one story holds that Sir Humphry Davy, who first isolated it, named it for sodanum, a popular early nineteenth-century headache remedy.

The names of the elements encompass the entire chemical history of mankind, from its tentative and often puzzled beginnings to its sophisticated present. In this sense, a recent proposal that we adopt a standardized scheme, basing elemental names solely on atomic number, is an archaeological shame. The new system was proposed by IUPAC in the 1980s to carry new heavy elements through the difficult and often disputatious period between discovery and formal name approval. Names, in this system, are based on a straightforward atomic number code. The number 0, decreed IUPAC, would henceforth be called nil; 1 would be known as un; 2, bi; 3, tri; 4, quad; 5, pent; 6, hex; 7, sept; 8, oct; and 9, enn. A stringing together of these in proper order plus a final suffix – "-ium" – thus allows the logical naming of any element. Element 110, under IUPAC rules, thus becomes 1-1-0 + ium or ununnilium; element 111 is unununium; element 112, ununbiium. By analogy, it was next suggested that this mellifluous methodology be expanded to *all* elements. Thus hydrogen, with an atomic number of 1, would be a simple unium; helium (2), biium; and so on. Though sensibly reproducible, it's also linguistically tedious; and one can't help imagining a whole roomful of chemists using IUPAC-speak, all ium-ing away at each other like so many out-sized bumblebees. In my opinion, the only element that might benefit from this nomenclature is carbon, element number 6, since there's a certain witchy appeal to the name hexium.

*I do not know what I may appear to the world, but to myself
I seem to have been only like a boy playing on the sea-shore,
and diverting myself in now and then finding a smoother
pebble or a prettier shell than ordinary, whilst the great
ocean of truth lay all undiscovered before me.*

Isaac Newton

Since the Greeks' first inspired groping toward a fundamental theory of matter, our concept of the elements has undergone radical change. One by one the original four were proven to be something else altogether: water, a compound; air, a mixture; earth, a conglomeration; and fire, a chemical reaction. Fundamental substances exist, but they are far more numerous and stunningly more complex than even the wisest of the natural philosophers ever imagined. Our view of elements today is clinically precise: modern technology allows us to peer at the structure of the atom and to take the chemical fingerprints of stars. The elements, in the twenty-first-century world, are the province of empirical science. In the face of their streamlined replacements, the primal four, like old soldiers, should long ago have faded away. Still the Greek elements stubbornly remain, throwbacks to an earlier age, a pentimento behind the logic of the Periodic Table, the intuition that still lurks behind the atoms. The weight of memory is behind them.

The Greek four are the elements of tradition and time. Water, air, earth, and fire dominated human thought for over two millennia. The four have been around long enough to insinuate themselves into our lives, language, art, and literature. Albrecht Dürer, Brueghel the Elder, and M. C. Escher immortalized the four in art; manifestations of the Greek four appear in everything from twelfth-century Persian poetry to the plays of Shakespeare to T. S. Eliot's *Four Quartets*. When beset by wind, rain, and weather, we still speak of battling, braving, or being ignominiously defeated by the elements. King Lear, mad, furiously contended with them.

The four elements abound in proverb and metaphor. We speak resignedly of water under the bridge; we point out that it's an ill wind that blows nobody any good; complain that you can't get blood from a stone; and advise fighting fire with fire. Christopher Marlowe writes of "Nature that frames us of four elements/Warring within our breasts for regiment," explaining that our characters are forged by whichever of the four dominates the rest. This strikes an enduring chord. Earth, air, fire, and water are still said to shape our psyches. The airy are the frivolous protagonists of P. G. Wodehouse novels. The earthy lack polish, but they're dependable and solid: Kipling's Saxon, say, stubborn as an ox in his field. We have fiery tempers, stony faces, or light hearts; we burn with passion, but drown in sorrow.

Most of all, however, the four elements continue to exert their power because it's through them that we first and best experience the world. In a way, as each of us strives to make sense of the complex reality that surrounds us, we retrace the path that the Greeks took. All of us begin at that beginning, and build upon what we discover there. Helen Keller – rendered deaf and blind by a bout of scarlet fever as a toddler – described in her autobiography the miraculous realization she experienced as her teacher, Annie Sullivan, pumped water over her hands from the dooryard pump and then used her fingers to spell the word "water" into Helen's wet palm. "I knew then that 'w-a-t-e-r' meant that wonderful cool something flowing over my hand," Keller wrote in 1902. "That living word awakened my soul, gave it light, hope, joy, set it free!" We all have such moments. Just as Keller's revelatory link between word and water opened before her a rich world of language, so our initial experience of the elements opens before us the rich diversity of the physical world.

We now live in the Information Age. We know that we owe our psychologies, physiologies, the straightness of our teeth, and the color of our eyes to an array of 3 billion nucleotide base pairs, elaborately packaged in the DNA in the nuclei of each of our bodies' 100 trillion cells. We have discovered that the observable universe contains billions of galaxies, each home to multiple trillions of stars; and we know that even the shabbiest pebble and dullest puddle are arrays of spinning particles, each so tiny that 25 trillion of them could dance on the head of a pin. We live in an expanding world of marvels, some so large, so small, or so esoteric that they seem impossible to comprehend. This, of course, is science, but it's not all of science, and it's certainly not where science begins. Science begins where the Greeks began: with the world as we perceive it, with what we see, hear, smell, taste, and touch. Our concept of the world is pieced together from boulders, waterfalls, and meteors, from the ground beneath our feet, the air we breathe, the fires with which we warm our toes, the salt spray we splash through at the edge of the sea. From the elements we've known since the beginning – the Greeks' elements: water, air, earth, and fire.

Part II

Water

Praised be Thou, O Lord, for sister water, who is very useful, humble, precious, and pure.

St Francis of Assisi
Canticle of the Sun

"All earth's powers are due to the gift of water," wrote Pliny the Elder in his *Natural History* in the first century CE, and two thousand years later we've yet to prove him wrong. Water, of all the Greek four, is the signature element of our planet, the element of life. Life began in it, flourished in it, and went forth out of it and multiplied. And wherever it wandered, it took water along with it: nearly 90 per cent of cell protoplasm is made of water, and 70 per cent of each human being. Civilizations grew up around water, and withered away because of its loss. Carried upon water, we've reached the ends of the Earth; contemplating it, we've been moved in equal parts to art, invention, and war.

Our lives and languages are saturated with water. Water, in its many aspects, is the most compelling of human metaphors. It represents change and transformation, redemption and purification, salvation and security, danger and death. We speak of time as a river, and benevolence as rain. Power is wave and waterfall; peace is a forest pool. Anything vast is the sea: we have seas of love and faith, knowledge and ignorance, blood and sorrow. The sky is a sea of stars; the plains, seas of grass; and the deserts, seas of sand – but never the other way around; water is our frame of reference, not earth or air. Water quenches a multitude of thirsts. Spiritual, emotional, and intellectual emptiness are deserts; enlightenment, love, and learning are water.

We celebrate water in paint, poetry, prose, and music. Of all the four elements, water has the most versatile voice. It murmurs, whispers, babbles, chatters, laughs, and plinks; it roars, thunders, crashes, and pounds. Chopin, Debussy, Ravel, Vivaldi, and Ralph Vaughan Williams all wrote compositions in imitation of water; and Handel's *Water Music*, a 25-part orchestral suite composed in 1717 for a boating party on the Thames hosted by King George I, is a piece so popular that it would surely – if there had been such a thing – been elected to the Baroque Hall of Fame. (The king, according to the *London Daily Courant*, was so impressed upon hearing it from his seat on the royal barge that he "caused it to be played over three times in going and returning.") In 1952, John Cage composed a version of *Water Music* that featured, in lieu of Handel's horns and violins, gurgling water poured back and forth between pots. The controversial Cage was a proponent of music as it occurs around us in the sounds of daily life. His most notorious piece, 4'33", consists of four minutes and thirty-three seconds of silence, during which the audience is expected to appreciate the background "music" of the concert hall – a music dominated, at least at the first performance, by the angry fidgets of a roomful of annoyed ticket-buyers trying to sit still. He made his point better with *Water Music*. Water is a music everyone intuitively hears. Water is the song of the planet, as anyone knows who has ever fallen asleep to the sound of rain on the roof.

Water is also about as peculiar as a substance gets, and as such has provided centuries of entertainment for chemists, physicists, and biologists. Chemically speaking, water is not an element at all, but a compound, composed of one atom of oxygen and two of hydrogen, a discovery that dates to the late eighteenth century. The scientist responsible was the afore-mentioned British chemist Henry Cavendish, who, being both immensely rich and intransigently pathological, was able to devote sixty solitary years to scientific research. Among his accomplishments were the discoveries of hydrogen and argon, and the calculation of the gravitational constant, either quite sufficient to confer lasting fame. He is perhaps best known, however, for his extensive work on gases, which led somewhat inadvertently to the elucidation of the true nature of water. In the winter of 1784, while applying a lighted paper to the mouth of a bottle containing hydrogen ("inflammable air") and oxygen ("common air"), Cavendish created an explosion. The bang produced water – Cavendish, who carefully recorded the event in his laboratory notebook, called it "dew" – and subsequent studies confirmed that the water resulted from a combination of the two gases, in a reproducible ratio of 2:1. The data were so startling that the taciturn Cavendish decided to communicate. He confided his results to Joseph Priestley. Priestley promptly told James Watt, who had been working on experiments along the same lines; and Cavendish's secretary, on a flying visit to Paris in 1783, passed the information on to Antoine Lavoisier. Everybody published. Only Cavendish and Lavoisier emerged from the resultant credit scramble, the outcome of which remains somewhat murky to this day. The British tend to favor Cavendish's claim since he performed the crucial experiment; the French side with Lavoisier since he gave the best explanation of the results.

If you remember only one chemical formula from your high-school science laboratory days, it's likely to be the one for water: H_2O. Or then again, maybe not. In the spring of

1997, Nathan Zohner, a 14-year-old junior-high-school student from Idaho Falls, Idaho, distributed a report on the dangers of dihydrogen monoxide ("Dihydrogen Monoxide: The Unrecognized Killer") to his classmates. Dihydrogen monoxide, the report announced chillingly, is responsible for thousands of deaths annually ("through accidental inhalation"), is a prime component of acid rain, and has been shown "to accelerate the corrosion and rusting of many metals." In gaseous form, it causes severe – even fatal – burns. And the stuff is addictive. Dependency on it is irreversible; complete withdrawal results in certain death. The majority of Zohner's fellow students – 43 out of 50, or 86 per cent – promptly signed a petition to ban this hideous stuff, and Zohner used the results to win the Great Idaho Falls Science Fair, with an exhibit titled "How Gullible Are We?" Very, apparently, since hardly any of the students polled realized that dihydrogen monoxide was plain water.

Plain water may be plain, but it isn't simple. Modern research into theories of liquids – a field of truly outrageous complexity – tends to center around so-called "simple" liquids, wholly theoretical substances that consist of innumerable spherical particles reminiscent of infinitesimal identical marbles. Based on the theoretical sizes of the component particles and the theoretical distances between them, scientists can make reasonable guesses about the physical characteristics of the putative liquid: its freezing point, its boiling point, its ability to form attractively shaped droplets. Water, by such criteria, is stunningly odd: its freezing point is lower than predicted; its boiling point, higher. Dripped on to a flat and impermeable surface, such as the toe of a patent-leather shoe, it forms little beaded droplets rather than flat splats; and it travels up the insides of narrow little straws without outside help, a phenomenon known to science as capillary action.

Most of water's oddities are the result of the way it's put together. Modern molecular models of Cavendish's compound are often compared to Mickey Mouse: a fat oxygen head topped with a pair of bulbous hydrogen ears. The hydrogen atoms both share electrons with the single oxygen, thus forging stable and equitable relationships known to chemists as covalent bonds, which are what holds the molecular Mickey head together. Oxygen, however, being much bigger than hydrogen, doesn't play fair, but behaves like an atomic bully, dragging the ostensibly shared electrons much closer to oxygen territory than by all rights they ought to be. This possessive pull gives the water molecule an electrically asymmetric character. The oxygen atom, with its grabby overdose of negatively charged electrons, acquires a slight negative charge; the deprived hydrogens reciprocally acquire a slight positive charge. Water has differentially charged corners, which makes it an electric dipole.

Because of this electrical polarity, water molecules *en masse* are cooperatively attracted to each other. Negatively charged oxygens form tentative connections to the positively charged hydrogens of their next-door neighbors; while their own hydrogens in turn link flirtatiously to oxygens on the opposite side. The result – found in every raindrop, water glass, and water balloon – is a promiscuous network of molecules linking oxygen to hydrogen and hydrogen to oxygen. Water is not a conglomeration of independent individuals; it's a social construct. These intermolecular links – about ten times feebler than the covalent bonds that hold individual water molecules together internally – are called hydrogen bonds. Their elucidation won Linus Pauling a Nobel Prize in 1954.

Water's more blatant peculiarities are largely attributable to hydrogen bonds, whose intrinsic intermolecular clinginess gives liquid water a certain amount of internal structure, a slurpy but detectable skeleton. Water, in effect, is sticky. On a watery surface, untold millions of interconnecting hydrogen bonds create a cohesive skin, such that the tops of puddles, ponds, lakes, and rippling streams act like thin rubbery membranes. This skin, though barely detectable to the human pedestrian sploshing along in galoshes, is sturdy enough to support a host of insects: hydrogen bonds allow water striders, skaters, and beetles to walk on water. Hydrogen bonds hold raindrops together, allow water to climb from underground roots to the topmost branches of very tall trees, and ensure that ice floats. Water, because of hydrogen bonds, is a near-universal solvent: practically everything dissolves in it, including salts, sugars, and amino acids. Lipids, on the other hand, don't, which has considerable implications for the integrity of cell membranes.

Sweating cools us off so efficiently because of hydrogen bonds. Water has a high heat of vaporization: that is, it takes a tremendous amount of energy to convert liquid water into gas. This is due to hydrogen bonds, which are hard to pull apart. For water to evaporate, it has to suck energy in the form of heat from its surroundings. Sweating, panting, or dashing through a lawn sprinkler all cool us off because vaporizing water uses our body heat to tear apart its hydrogen bonds. And not only is water hard to vaporize; it's hard to warm up. Due to its gluey population of hydrogen bonds, water has a high heat capacity: it takes a lot of effort to heat it; and, conversely, once it's warm, it's notably resistant to cooling down. Water in lakes and oceans absorbs heat from the sun and then refuses to let it go. The Earth's water thus becomes a massive heat reservoir, a vast circulating electric blanket, a heat buffer for the globe. It's the same mechanism that, in microcosm, helps maintain human body temperature: our warm blood, which is mostly water, has a high heat capacity. It resists warming up or chilling down; despite the vagaries of the air outside our skins, it maintains the status quo.

We're all here by the grace of water. More accurately, we're here by grace of the hydrogen bond – a feeble connection between water molecules, an evanescent attraction that operates over a space a mere fraction of a nanometer long.

"Man is not an aquatic animal," writes naturalist Hal Borland, "but from the time we stand in youthful wonder beside a spring brook till we sit in old age and watch the endless roll of the sea, we feel a strong kinship with the waters of the world." Of course we do. Thales's choice of water as our primal element was not so much a prescient piece of chemistry as an exercise in common sense. Water is the distinguishing feature of our planet. Our world is soggy; Earth, viewed from space, is wet and blue. Over two-thirds of the Earth's surface is covered by water, a total of 139,781,000 square miles (326,127,000 sq km) – most of it in four interconnected oceans, the Arctic, the Indian, the Atlantic, and the shamefully misnamed Pacific, christened by Magellan, who at first glance thought it peaceful. Given this aqueous immensity, it's easy to see why the ancients believed their world to be an island surrounded by an infinite sea; and why Samuel Taylor Coleridge's Ancient Mariner, becalmed somewhere in the middle of it, wailed of "Water, water, everywhere."

Our vast stores of water may well be unique. Water has been found elsewhere in the Universe, but so far only here on our planet has it been found on the surface in liquid form, suitable for wading in, bathing in, floating upon, and mixing with single-malt Scotch. Its very abundance perversely breeds contempt: traditionally, water, paired with a dismal crust or two, is prison fare.

Though Henry David Thoreau, leading a Spartan life in his hand-built cabin at Walden Pond, insisted that water was the only drink for a wise man, many cultures traditionally have felt otherwise. The ancient Greeks thought water was bad for the disposition. It made people surly, curmudgeonly, and over-earnest. Demosthenes, a teetotaler, according to liberal contemporaries, spent all his time hunched over a desk and never laughed at anybody's jokes. Wine-drinkers, on the other hand, were pleasant people: convivial, creative, passionate, and fond of intellectual discourse. The original Greek *symposium* – now a sober scholarly affair – was a wine-drinking party. The Greek poet Cratinus, in a play tellingly titled "The Wine-Flask," stated that "A water-drinker never gives birth to anything ingenious," and the Roman poet Horace pointed out ominously that no poem composed by a water-drinker could possibly expect to achieve posterity.

Early American colonists, faced with the dreadful prospect of drinking water and nothing but water, sent doleful letters home. Victorian travelers, reduced to imbibing it, felt ill-used; and Emily Post, writing for the harried hostess at the turn of the twentieth century, warned that "A water glass standing alone at each place makes for a meager and untrimmed looking table." She recommended at least two wine glasses per diner, filled with sherry, claret, or at the very least, ginger ale. It was advice to be spurned at one's social peril. "It was a brilliant affair; water flowed like champagne," a nineteenth-century senator wrote bitterly after attending a White House party hosted by temperance advocate Rutherford B. Hayes.

Despite all this shortsighted grousing, the fact remains that water is marvelous stuff and our possession of it is an astounding piece of luck. Current biological theory holds that liquid water is one of the three essential ingredients for life, along with heat and complex carbohydrates. The last two aren't all that difficult to come by in our neck of the Universe, but the first is clearly a rarity. The planets, for the most part, are fried or frozen deserts. Mars, it now appears, was once dripping wet: a host of data, most recently from NASA's twin rovers, the rousingly named Spirit and Opportunity, shows a planet marked by water. Surface channels and gullies suggest bygone rivers; sedimentary-style layers and wave patterns preserved in surface rock indicate past lakes or shallow seas. Opportunity, parachuted on to the Martian Meridiani Planum in early 2004, found itself planted in a vast flat field littered with tiny spherules – NASA observers promptly dubbed them "blueberries" – that proved to have been fashioned like pearls from layer upon layer of deposited sediments. Their mineral structure suggests a Martian land of salt lakes, filling, drying, and filling again, in a repetitive cycle that may have gone on for tens – even hundreds – of thousands of years. Still, though water may remain somewhere beneath the cold red surface – scientists now speculate about buried Martian aquifers – not so much as a liquid puddle has been seen still lying about on the desiccated ground.

Ganymede and Europa, respectively the largest and fourth-largest moons of Jupiter, may have liquid oceans deep beneath their frigid crusts, but these at best are buried beneath 2.5 miles (4 km) of ice. Saturn's Titan – a satellite larger than the planet Mercury, thickly swathed in orange smog – appears to be a world of ice and tar, according to images relayed by the European Space Agency's Huygens probe, which landed on the moon's surface in January 2005. Distant Pluto has water, but it's frozen solid; and the water on the Moon – up to 1300 million tons of it, by one NASA estimate – exists solely in the form of frosty ice crystals scattered through the lunar dust.

The collected works of the Brothers Grimm include a tale called "The Water of Life," which centers around a quest for the title liquid, a vitalizing draught of which will heal an ailing king. Unfortunately – as in all good fairytales – the wondrous drink is very hard to find. Modern scientists agree. In fact, to date, as far as we know, just one planet has it. And we're living on it.

In the myths of many cultures, water is an element of primal mystery. It's often posited to have been the first substance, the matter used to make everything else, here before creation began. In the beginning, according to the ancient Egyptians, nothing existed but an endless sea. Out of its depths rose the sun god, Atum, whose first act, sensibly enough, was to create a hill. According to the Japanese, the world began when the god and goddess Izanagi and Izanami, peering curiously down from heaven, wondered what lay beneath the endless ocean that covered the Earth; Izanagi poked his jeweled spear beneath the waves and dredged up some lumps of mud that became the islands of Japan. The Yoruba of Nigeria claim that the world began when Oduduwa, the sky god's son, shinnied down a great chain from the heavens, carrying a handful of dirt and a magic five-toed chicken. He dropped the dirt on the boundless water and set the chicken upon it to scratch and scrape, thus forming the first dry land. In the Koran, all livings things were made from water, which was around before either the heavens or the Earth. In the Old Testament's Book of Genesis, water was already here on the First Day when God, moving on the face of it, created darkness and light. He made dry land later, on Day Three.

Some modern interpreters suggest that all this primordial water simply represents chaotic nothingness, a primitive portrayal of the inconceivable emptiness in which, some 14 billion years ago, a random vacuum fluctuation set off the explosive universe-initiating event now known as the Big Bang. Or perhaps the myths hark back to real water; early myth-makers, so dependent on the tug of tides and flood of rivers, may have been unable to imagine a waterless world. The world at its inception, however, was almost certainly waterless. Most geophysicists hypothesize that water, as a substantial and separate entity, only appeared after the solidification of Earth and the accumulation of atmosphere, beginning some four billion years ago at the end of the Hadean interval.

The Hadean era comprised the first 700 million years or so of Earth's 4.5-billion-year-old history, a hot and hellish epoch during which the infant planet was relentlessly bombarded by meteors, asteroids, and planetesimals, among them a rock the size of Mars, that gouged out the Moon. Its close ushered in the comparatively peaceful Archaean era – 3.8 to 2.5 billion years ago – during which the much-mauled Earth, now solidified, cooled. As it did, water, in the form of steaming vapor, began to leach from its sizzling rocks. The steam rose, cooling and expanding as it ascended into the upper atmosphere, where it condensed into liquid water droplets and fell to earth again, torrentially, as rain. Archaean rain, scientists guess, fell for millennia – a massive panglobal monsoon, thundering across proto-continents, filling the vast bowls of future oceans. According to this traditional dogma, the seas filled fast in a single efficient multi-thousand-year downpour. An alternative theory, however, suggests that the process took a lot longer and that the water came from outer space. Colloquially, this proposal is known as "the Big Splash."

On April Fool's Day 1986, Louis Frank's theory about the creation of the oceans hit the popular press. The theory had been accepted for publication the previous month in the somewhat less popular *Geophysical Research Letters*, described in a pair of papers modestly titled "On the Influx of Small Comets Into the Earth's Upper Atmosphere." Frank, a physics professor at the University of Iowa, had concluded on the basis of evidence from satellite photographs that the Earth was being steadily bombarded by gargantuan chunks of ice – extraterrestrial snowballs the size of two-bedroom bungalows. Calculations further indicated that there were a lot of these: the icy comets were arriving at a rate of three per second – or 25,000 per day, 10 million per year. Each disintegrated as it entered the atmosphere, releasing some 100 tons of water – a collective delivery of 600 million gallons (2300 million l) a day. Spread out over the entire surface of the planet, this doesn't amount to much – a mere one ten-thousandth of an inch of water (.00025 mm) per year – but a steady patter of comet-borne water adds up. Over the 4.5-billion-year life of the Earth, Frank's cosmic water bombs could conceivably have filled the oceans to brimming.

"Man has in him the silence of the sea," writes Indian poet Rabindranath Tagore. This is a literal as well as a figurative truth; we are made of water. By all rights, we should gurgle when we walk; adult humans are about 75 per cent water by weight. Newborns, at nearly 90 per cent water, are almost as juicy as tomatoes (93.5 per cent); and much of what we eat – barring the odd desiccated cracker – is water-logged. The average celery stalk is 95 per cent water; the average cucumber, 95 per cent; the aptly named watermelon, 97 per cent; even the stolid potato is 80 per cent water. Blood is thicker than water, but not by much – plasma, the liquid part of the fluid that courses through our veins, is 92 per cent water; and even the best and brightest of human brains is three-quarters plain H_2O.

Water, as well as dominating our physical make-up, may even – possibly – have directed our evolution. The most widely accepted view of human evolution, generally known as "the savanna hypothesis," holds that we are descendants of arboreal primates who climbed down out of the trees of East Africa three million years ago to prowl across the equatorial grasslands, walking upright, looking for food, and avoiding lions. In the savanna scenario, our ancestors remained firmly on dry land. According to aquatic ape theory (AAT), however, they went into the sea.

AAT, first proposed in 1960 by oceanographer Alister Hardy and subsequently promoted by writer Elaine Morgan, argues that our pre-human ancestors evolved in water. Many of our modern features, AAT supporters point out, are characteristic of water-adapted animals. We're nearly hairless – more like the sleekly streamlined whales and dolphins than the thoroughly furry apes. And we're plump. Human beings have ten times more adipocytes than expected for a mammal of our size. Human babies are born pudgy; each is about six times fatter than the scrawny newborn chimpanzee. One guess is that our extra fat may once have insulated us against the chill of our aquatic environment; it may also, by making us buoyant, have helped us float. We have large prominent noses – not all, perhaps, of the dramatic heft of a Cyrano de Bergerac, but still sizable, the better to breathe when poking our heads out of water. Gasping – the startled gulp of air we take when confronted with the hideous, the horrifying, or the unexpected may be a holdover from our long-gone aquatic days, when the better part of valor was to take a deep breath and dive out of harm's way. Even the female breast may owe its existence to an early water environment: Morgan suggests that the bulging bosom evolved as a personal flotation device to which the plump aquatic infant could cling while nursing.

Some proponents of AAT go so far as to suggest that the development of human speech was driven by an aquatic lifestyle: we had to learn to bellow expressively to be heard and understood over the pounding surf. And then, of course, we all *like* water. Unlike most primates, who approach it with catlike distaste, we positively glory in it. We wade, paddle, swim, scuba dive, and waterski in it; we catch (impossibly large) fish in it; and we spend an inordinate amount of time messing about in boats. Perhaps, AAT supporters argue, our aquaphilia is the result of dim evolutionary memory; and everything from water polo to the string bikini owes its existence to the days when ancient hominids first hit the beach.

Chimps, being lean, sink like stones; humans, being fattish, float. This is because water, despite all appearances to the contrary, is denser than we are. The density of the average (male) human body is 0.980 kg/liter; that of the average and slightly fatter female, 0.968 kg/liter. The density of pure water, in contrast, is 1 kg/liter – which also means, incidentally, that water is heavy: a bucketful weighs about 25 pounds (11 kg). We float even better in seawater, which is denser yet (1.025 kg/liter); and in the oversalted Dead Sea – a body of water so dense that waves can't form upon it – we positively bob up like corks.

On the other hand, though people generally like to swim, it's not something that comes naturally. In 1586, when Everard Digby, Fellow of St John's College, Cambridge, published his book on the sport of swimming, *De arte natandi*, he claimed optimistically that human beings were better swimmers than fish. This was patently wrong – the book, after all, was inspired by a series of un-fishlike drowning accidents in the college pond; and studies of true water-dwellers prove that they have it all up on us in matters submarine. Fish can swim as fast as cheetahs can run, and with far less expenditure of energy. The sailfish – fastest fish in the sea – zips along at 68 mph (110 km/h); the marlin, favorite of deep-sea anglers, swims at 50 mph (80 km/h); the dolphin at 30 mph (48 km/h). Humans, comparatively, are snails: even Olympic gold medalists barely make it over 4 mph (7 km/h), and then only by dint of putting out over a thousand effortful watts of power.

Water supports us, but it also holds us back. As an object

moves through a medium – say, a swimmer through the water of a swimming pool – the fluid moving past him/her creates a friction-like resistance called drag. Streamlining – as in the sleek bullet-like shape of the bottlenose dolphin – reduces drag; the narrow head and smooth flanks create minimal disruption in flow as the dolphin speeds through the deeps. Human swimmers similarly reduce drag by lying flat and attempting to look as dolphin-like as possible, which works to a certain extent, given the limitations of the lumpy human body. Still, drag is why swimming is so effortful. Compare, for example, the energy expended in running on dry land (through air) to that expended in jogging through waist-deep water. Water is 800 times denser than air; and its retardive effect becomes worse with every increase in speed. For a swimmer to double his or her speed in the pool, power output must be *cubed*, or increased eight-fold. It's more than we can manage. We're slowpokes in water, lumbering misfits, out of our element. Our aquaphilia has physical limits. Even turtles swim five times faster.

The ancient Egyptians had a hieroglyph for swimming; and swimmers, caught in the act, appear in Egyptian bas-reliefs, Assyrian stone carvings, and mosaics from Pompeii. The sports-mad Greeks admired swimming – to describe someone as "unable to either run or swim" was a cutting Hellenic insult – but not enough apparently to include it among the competitions in the original Olympic Games. Swimming, in the Western world, was considered an essential component of military training: Julius Caesar and Charlemagne were acclaimed as swimmers, and even during the depths of the Dark Ages, when swimming was strictly avoided, knights were encouraged to master the art while wearing armor. Most didn't, which wasn't surprising: a full suit of plate armor could weigh 66 pounds (30 kg).

Natural or not, the act of learning to swim is astonishing in that it's a mastery of an element not ordinarily our own. Most of us can remember the day we discovered that water, incredibly, could hold us up, just as frustrated parents and swim instructors had been telling us all along. It's the sort of baffled thrill adolescent birds must feel when, abruptly booted from the family nest, they find themselves flying through the air. Water gives us a freedom from gravity that we can experience nowhere else – other than hang-gliding, perhaps, or in the near gravity-less corridors of the International Space Station. To swim is to soar – though inevitably with the niggling feeling that somehow, on some level, this whole thing shouldn't work. Swimming, or failing to do the same – perhaps for this very reason – is a compelling metaphor for numerous human endeavors. When things are going swimmingly, for example, our affairs are running smoothly, the economy is prosperous, and life is treating us particularly well. When we're just treading water, we're in a holding pattern, hoping for better times ahead; and when we're out of our depth or in over our heads, we're dealing with problems too difficult for our capabilities. If we've been told unsympathetically to sink or swim, we've been sent off on our own, to survive as best we can. Sinking is frequently synonymous with disaster: a sinking heart signifies misery and despair, sinking feelings are premonitions of catastrophe; and anything on its way to the financial or political bottom is traditionally compared to a sinking ship, which is why I've always thought it appropriate that the anniversary of the demise of the *Titanic* is the day Americans pay their income tax.

At some unknown point in human history, the quick – and perhaps inadvertent – dunk in the water of a neighboring lake, pond, stream, or river evolved into the sybaritic and deliberate delights of the bath. Baths are not unknown in the animal world: chickens, chinchillas, hamsters, hedgehogs, and rhinoceroses all take dust baths; and the cat, that most fastidious of animals, scrubs itself with its rough tongue. For human beings, however, the common element of cleanliness is water.

Water washes away dirt, grime, and sweat; it makes us

healthier, pleasanter people; and it makes us feel good. The ancient Egyptians showered; homes featured small rooms furnished with a drain and a limestone slab, upon which the bather sat or stood while slaves doused him or her with bowls of water. Excavations of the palace of Knossos on Crete show that by 1700 BCE the Minoans were bathing in terracotta tubs temptingly painted with water motifs. Ancient Greek gymnasiums featured hot and cold showers; and the Romans, for whom the bath was a social occasion, built luxurious heated public baths, lined with marble and fitted with mirrors, mosaics, glass ceilings, and silver spigots. The baths of Caracalla, a third-century cross between a water park and a posh hotel, covered 28 acres (11 hectares) and accommodated 1600 bathers, scattered among the dressing rooms, massage rooms, steam baths, *natatio* (outdoor swimming pool), *caldarium* (hot-water bath), *frigidarium* (cold-water bath), and *tepidarium*, whose warm waters were presumably just right. Sandwiched between these was the *laconicum*, an extremely hot bath, in which bathers spent only the briefest of times before emerging, gasping; the name derives from the province of Laconica, whose inhabitants' reputed brief and unforthcoming mode of speech gives us the modern word "laconic." Caracalla's baths were in operation for over 300 years, from 216 CE until 537 CE, when the invading and determinedly grubby Vandals destroyed the aqueducts that provided them with water.

The barbarian invasion and subsequent fall of the Roman Empire ushered in a long period of Western European bathlessness. The early Christians, who felt that the cosseting of the body was detrimental to the well-being of the soul, were for the most part reluctant and infrequent bathers. St Benedict in the detailed Rule that eventually became a near-universal guideline for monastic life, limited baths to the infirm. Queen Elizabeth I set a record for royal cleanliness by taking a bath each and every month, though this was far from common practice. When Samuel Pepys's wife took a bath in 1607, the event was notable enough for her husband to record it in his diary.

The first bathtub in the American colonies is said to have belonged to Benjamin Franklin, who acquired it in France, which country was somewhat advanced in the matter of baths. Nearly a century before Franklin's landmark purchase, Louis XIV had had six baths constructed for the royal apartments in the re-modeled Palace of Versailles, one a pink marble octagon ten feet across. (The king seldom bathed in it – the marble was cold – and the massive tub ultimately ended up as a garden fountain.) Simpler baths for the less exalted became increasingly common throughout the eighteenth century. Franklin's, a classic example of the genre, was roughly shoe-shaped – accordingly, these were known as "slipper" or "boot" baths – and made of copper with a drain in the toe. It was also equipped with a bookrest; and Franklin, ever a man ahead of his time, spent many pleasant hours in it, reading. Reactionaries, however, insisted that bathing brought on rheumatic fever and respiratory ailments; and one of the Georgian royal dukes was heard to thunder that it was not water but sweat, damn it, that kept a man clean. In some political climates, baths were positively dangerous. Jean-Paul Marat, a Jacobin leader of the French Revolution, was stabbed to death in his by Girondist Charlotte Corday.

"On December 20 there flitted past us, absolutely without public notice, one of the most important profane anniversaries in American history, to wit, the seventy-fifth anniversary of the introduction of the bathtub into These States." So wrote H. L. Mencken in the December 28, 1917, issue of the New York *Evening Mail*. The seminal tub, Mencken went on to explain, was installed in 1842 in the Cincinnati home of Adam Thompson, a grain and cotton dealer who had picked up the habit of bathing on business trips to England. The tub measured 7 feet long by 4 feet wide (2.1 m long, 1.2 m wide), was encased in polished Nicaragua mahogany and lined with sheet lead, and weighed 1750 pounds (795 kg). On its first day in service, Thompson took two baths in it, one cold in the morning, and one hot in the mid-afternoon. From Cincinnati, the bathtub spread across America: Millard Fillmore (who admired the Thompson tub on a campaign visit in 1850) installed the first bathtub in the White House; General McClellan introduced the bathtub to the Army; and public-spirited Pennsylvanians introduced the bathtub to criminals, installing one in Moyamensing Prison in Philadelphia in 1870.

Such is the gist of the famous and still much-quoted Bathtub Hoax. Not a word of it is true; Mencken made the whole thing up, Thompson, Fillmore, McClellan, Nicaragua mahogany, and all. It took him nine years to confess – "The article was a tissue of somewhat heavy absurdities," he wrote (belatedly, in 1926) – but confession came too late: by 1926 the Bathtub Hoax had taken on a life of its own. The Boston *Herald* quoted it as gospel truth in June of that year; *Scribner's* magazine followed up in October. Over the next decade, Mencken-derived bathtub facts popped up in several books, a bulletin from the Kentucky Department of Health, the New York *Times*, the Tucson *Star*, the Chicago *Daily News*, and even – in 1942, the hundredth anniversary of the bogus bathtub, in a spectacular example of what goes around, comes around – in Mencken's own paper, the Baltimore *Sun*. Often to this day the only fact popularly known about Millard Fillmore – an otherwise innocuous president – is that he (supposedly) put the first bathtub in the White House.

"There must be quite a few things that a hot bath won't cure, but I don't know many of them," wrote Sylvia Plath in her fraught and semi-autobiographical novel *The Bell Jar* in 1963. Plath was hoping to restore emotional well-being; the bath, however, has traditionally been recommended for the physically debilitated as well. Its resurgent popularity after the long and dingy hiatus of the Dark Ages was inspired less by the desire to be clean than by the hope of health. Though the medicinal bath can be traced at least to the ancient Greeks – Hippocrates refers to therapeutic bathing – the centuries following the fall of Rome were characterized by a gingerly avoidance of external water. (An exception was the washing of the hands before meals – a necessity in the days when fingers were the prime eating utensils; the medieval *Boke of Curtasye* not only insists upon a pre-prandial wash, but warns the would-be courteous against dirty fingernails or playing with the dogs during meals.) It was only at the end of the seventeenth century that the bath finally emerged again as a cure-all; and by 1830 Edward Hitchcock, a professor of chemistry at Amherst College, was vigorously touting baths as promoting "health, strength, longevity, and serenity of mind."

The bath, in its therapeutic heyday, was said to cure everything from syphilis to insomnia, including epilepsy, gout, blindness, ulcers, tetanus, toothache, earache, irregular menstruation, and infertility. Such a potent prophylactic, the contemporary literature warned, was not to be indulged in willy-nilly by slipshod amateurs, but must be administered by trained professionals. *Water, and How to Apply It, in Health and Disease*, written by R. B. D. Wells in 1885, includes several cautionary anecdotes about patients who foolishly felt competent enough to prepare their own baths. Some barely survived the experience, to be rescued in the nick of time by Wells or his assistants; though the only truly dangerous bath of the period appears to have been the "Electric Bath," in which the patient sat in warm water infused with an electric current.

The prime progenitor of the healing bath was Vincent Priessnitz, a Prussian peasant who managed in the nineteenth century to do for cold water what Bill Gates in the twentieth did for the silicon chip. As a teenager Priessnitz had been trampled by a horse, suffering injuries so severe that the local medico pronounced them incurable. However, by treating himself solely with applications of water-saturated cloth bandages, Priessnitz effected a dramatic recovery, the fame of which soon spread throughout the neighborhood and beyond. By 1829, he was the proprietor of his own "Water Cure Establishment;" by the early 1840s, he had acquired a devoted clientele of 1500 patients and a hefty bank balance.

The "Water Cure" soon acquired a pseudoscientific name – hydropathy – and a vast number of remedial offshoots, among them the Rain Bath (a cold shower, sometimes so powerful that partakers were compelled to wear protective headgear), the Sponge Bath, the Sitz Bath, the Steam Bath, the Bran Bath, and the Dripping Sheet Bath. Specific parts of the body were treated to their own special baths – one hydropathic clinic offered Arm, Leg, Nose, and Foot Baths (a subset of which was the squishy-sounding Wet Socks Bath). By the 1850s, the United States boasted sixty-two water-cure establishments, a professional organization for water-cure specialists (the American Hydropathic Society), a monthly *Water-Cure Journal*, and even a water-cure medical school. Not everyone was equally sanguine about the medical benefits of plain water: the *Boston Medical and Surgical Journal* damned the burgeoning water-cure college as a "quack institution;" and the miracle fluid in some cases did considerable harm. The death of John Roebling, builder of the Brooklyn Bridge, in 1869 was the fault of medicinal water: when his foot was crushed in a construction accident, Roebling – an impassioned advocate of hydropathy – refused any treatment other than prolonged soaking in cold water. The water cure failed, tetanus set in, and he died three weeks later.

The remedial bath was almost universally icy cold. Sylvester Graham, inventor of the eponymous Graham cracker, whose crankish health reforms swept the United States in the early 1800s, was one of the most vocal advocates of the daily cold bath, to be taken immediately upon arising at five a.m. (This was followed by a pure vegetarian breakfast of gruel and wheat-bran tea; the lot of the devout Grahamite was not a happy one.) The Duke of Wellington staunchly immersed himself in a cold tub each morning, which manly (British) behavior contrasted sharply to the decadent (French) habits of Napoleon, who favored a leisurely soak in very hot water. (The Duke's bath practice, popular rumor had it, contributed heavily to his subsequent triumph at the Battle of Waterloo.) Cold baths – said to be particularly efficacious for those afflicted with "worldly thoughts" – were prescribed for luckless schoolboys until the turn of the twentieth century. This aspect of the water cure persists to present day; the sexually inflamed are still unsympathetically advised to take cold showers. The hot bath, traditionally, was associated with debauchery – one cavorted in it, in mixed company; it was also (somewhat contradictorily) said to be enfeebling and likely to bring on confusion of thought and depression. The Roman Empire was often cited as a case in point by the cold bathers: the Romans were notoriously addicted to hot-water bathing and everybody knew what ultimately happened to them.

Luckily for the comfort of future generations, bath proponent Edward Hitchcock favored the warm bath, which, in Hitchcock's view, posed no threat to king and country provided it was taken in moderation. (The fall of Rome, he explained, was the result of much too hot baths, taken much too often.) By the turn of the nineteenth century, the order of the day was the now-familiar Soap Bath, an experience so agreeable, one bath tome of 1880 boasted, that children could be induced to take it as a treat.

"Soap and water and common sense are the best disinfectants," wrote Canadian physician William Osler at the end of the nineteenth century. Though water never lived up to its sensational billing as master therapeutic, water and a dollop of soap were indeed lifesavers, as evidenced by the drop in patient deaths when doctors were at last convinced to wash their hands. Cleanliness as standard medical technique was formally introduced in the mid-nineteenth century under the aegis of Joseph Lister, who favored carbolic acid, and Florence Nightingale, who favored elbow grease. Their innovations, tragically, were only sketchily adopted by the time of the American Civil War: of the 620,000 men who died during that conflict, nearly 400,000 died of disease, largely brought on by lack of sanitation.

Water, wrote the eighth-century theologian John of Damascus, "is the most beautiful element and rich in usefulness, and purifies from all filth, and not only from the filth of the body but from that of the soul, if it should have received the grace of the Spirit." Water not only cleanses the grubby body; by analogy, it symbolically washes away sin. Purifying baths are a feature of many of the world's great religions. This practice, formally known as ablution, covers all forms of spiritually prescribed washings, from the wholesale dunk to the token sprinkle. Devout Muslims wash hands, feet, and face before prayers; Brahmins bathe before daily morning worship; Roman Catholic priests perform a ceremonial washing of hands before celebrating mass. Jews restore themselves to a state of grace through a bath in a sanctified pool or *mikveh*; the Hindus wash their sins away in rivers; and the Christians dispense with theirs through baptism – originally from the Greek *baptismos*, meaning assimilation, but now synonymous with immersion in water. Great Britain's Most Honourable Order of the Bath – now awarded for outstanding military service with a gold medal in the shape of a Maltese cross – recalls the ritual bath taken in medieval times by young aspirants to the knighthood. The Old Testament abounds with washing imagery, as in "Wash me thoroughly from my iniquity and cleanse me from my sin" (Psalms 51:2). The New, on the other hand, features what may be the most famous washing scene in history, when Pontius Pilate publicly washed his hands to dissociate himself from the condemnation and sentencing of Jesus of Nazareth. It's a powerful image: people ever since have been washing their hands of blame, responsibilities, commitments, poor past decisions, and each other. Lady Macbeth, muttering "Out, out, damned spot!," struggled in vain to scrub away the guilt of Duncan's murder. Nellie Forbush in Rodgers and Hammerstein's *South Pacific* tried to wash that man right out of her hair; and generations of irate mothers have washed forbidden words out of their children's mouths with soap and water. A quick flip through the present-day news still finds electorates washing their hands of unpopular politicians, stockbrokers washing their hands of profitless corporations, teachers washing their hands of recalcitrant pupils, and reviewers washing their hands of the latest from Hollywood.

Plain water cleanses on many levels, but some waters are reputed to have even more extravagant powers. Holy water – formally consecrated by the Catholic Church – is said to have the power to cure fevers, cast out devils, and ensure a good crop when sprinkled upon fields or gardens. Some entire bodies of water have miraculous reputations by virtue of association with a sacred person. Muslims believe that the holiest water in the world is that from the spring of Zamzam in Mecca, to which the Angel Gabriel led Abraham's abandoned wife Hagar and her son Ishmael. The River Jordan – especially at the juncture of the river and the Sea of Galilee where John the Baptist baptized Jesus – is considered holy by many Christians. (A Tel Aviv marketing firm profits from this belief by selling ampoules of authenticated Jordan water over the internet for $18.00 apiece.) The spring in the grotto at Lourdes where the young Bernadette had her vision of the Virgin Mary is littered with discarded crutches as tribute to its healing powers. Innumerable wells, springs, and pools – many now respectably associated with Christian saints – have been held to have magical curative properties since ancient times.

Faith in consecrated water has its roots in the distant past, in ancient beliefs in magical water sources. Greek mythology tells of the miraculous fountain of Kanathos in which the goddess Hera took an annual bath to restore her virginity; Norse legend describes the fountain of wisdom that springs from the roots of Yggdrasil, the World Tree. A few mouthfuls of the water are enough to make the drinker infinitely clever and cunning; unfortunately it's hard to get at, being located in the land of the frost giants. Chinese writings of the third century CE mention the fabulous fountain of Pon Lai capable of conferring "a thousand lives" upon those who drank from it; Japanese legends tell of a marvelous rejuvenating fountain hidden on top of Mount Fuji. Alexander the Great's peripatetic career is said to have been inspired as much by the hope of discovering a youth-preserving fountain as by the quest for world domination; and Juan Ponce de Leon was certainly in search of it in 1513 when he discovered Florida – which now, in a tortuous sort of demographic irony, boasts the oldest population in the United States. A sixteenth-century painting by Lucas Cranach the Elder depicts the elusive fountain in the middle of a swimming pool: the elderly plunge in at one end and emerge at the other as dewy teenagers.

Water, as well as having magical powers in and of itself, was also believed to have a resident population of spirits who – if properly honored, flattered, or appeased – would grant good health, good fortune, and other desirables; or, alternatively, would at least be persuaded to stay out of sight and leave passers-by alone. We recall such beliefs today when we throw pennies in pools and make wishes. (Our ancestors tossed in pins.) Water traditionally sheltered a range of demons, sprites, and wraiths, from the generous Lady of the Lake in Arthurian legend who gave the young king his sword to the Welsh water-leaper (said to tangle fishing lines and eat sheep); the Irish merrow, who has green teeth, green hair, and little pig-like eyes and is only seen before a storm; and the Scottish kelpie, who sometimes appears in the form of a magical pony – those foolish enough to mount it find themselves firmly stuck to its back, at which point the animal gallops into the sea. Water spirits could be generous, but many were capricious or cruel. The cow-sized swamp-dwelling Australian bunyip and the Japanese kappa – a notably intelligent sprite who lives on blood and cucumbers – are both notorious drowners; and the Russian water nymph Rusalka is known for grabbing and dispatching unwary swimmers.

In Greek mythology, the natural environment was peopled with spirits. Trees, rivers, mountains, caves, springs, and pools were alive, each having as its essence a resident nymph. The name comes from the Greek for "young woman" and is a general designation; nymphs came in a wide range of environmental specificities and personality types (though all have legs, which distinguishes them from mermaids, who are fish below the belly button). Nereids and Oceanids, for example, were the nymphs of the sea, while Naiads were the nymphs of fresh water. Naiads included the Crinaeae (nymphs of fountains), Pegaeae (nymphs of springs), Eleionomae (swamps and marshes), Potameides (rivers), and Limnades (lakes); and Roman mythology includes a dismal group native to the dark rivers of the Underworld, known as the Nymphae Infernae Paludis, or Avernales. In most cases the nymph is intimately connected to her water source and will die if her stream, spring, swamp, or underground river dries up. In at least one instance, however, the situation was reversed: the nymph Egeria is said to have sobbed so inconsolably over the loss of her lover that she produced and/or turned into a spring. Water presided over or produced by a nymph was held to have special powers, and could – depending – heal the sick, cure infertility, tell the future, or provide inspiration to those afflicted with writer's block.

We tend to think of water as above-board and obvious – as in rain, rivers, lakes, and surf – but water is a secretive element; immense amounts of it are hidden. Springs come from water sequestered underground. The Earth's subterranean stock-piles of water are enormous. Experts figure that some 2 million cubic miles (8 million cubic km) of water slosh beneath our feet, sixty times more than the water in all the world's fresh-water lakes and rivers put together. Strictly speaking, the uppermost layer of this hidden water doesn't slosh, but seeps, creeps, trickles, and oozes. The Earth is like nothing so much as a fat sponge: water, falling upon it in the form of rain and snow, gradually works its way downward through pores, cracks, and crevices in the surface, insidiously inserting itself between soil and rock particles. This layer of soil-suspended water feeds the thirsty roots of growing plants; and it's these spongy sediments – freezing and expanding – that buckle the ground into the mogul-like frost heaves that play such havoc each winter with northern roads. Such suspended water is also responsible for the tilt in the Leaning Tower of Pisa: the weight of the building squeezed suspended water out of the underlying ground, forcing soil and rock deposits to compact and subside, throwing the whole structure out of kilter.

Eventually the perennial trickle of underground water reaches a watertight layer of bedrock, at which point it stops moving downward and begins to accumulate, forming vast rock-bottomed subterranean reservoirs. These are called aquifers, from the Latin for "water-bearing." The top of the aquifer is known as the water table; and the place where an aquifer intersects with the ground surface and spills out on to it is a spring. Springs range from a gentle surface dribble – the sort that soaks the socks of the unwary walker – to wildly foaming cataracts. Roaring Spring on the north rim of Colorado's Grand Canyon roars – literally – from the limestone cliffs in thunderous waterfalls; Florida's frothing Silver Spring spews out 800 million gallons (3 billion liters) of water a day from 150 different outlets; and the French spring of Les Bouillens ("boiling water") produces a daily 500,000 gallons (2 million liters) of naturally carbonated mineral water.

Les Bouillens is the source of France's famous Perrier water, whose official designation as a mineral water indicates that it contains more than 500 parts per million (ppm) of dissolved salts. In spring waters, mineral salts are collected as water percolates its way through the ground – generally at a snail-like pace; most groundwater travels at most a few feet a year, and can take a century or more to cover a mile (1.6 km). Long contact with the surrounding ground allows the water to absorb a battery of substances: calcium, magnesium, iron, manganese, chlorides, sulfates, and nitrates – even, occasionally, gold, silver, platinum, copper, chromium, lithium, and tin. Ideally these combine in some unspecified ratio to give water a delicious and distinctive taste. Opinions as to what constitutes the most flavorful water vary, though most agree that mineral-less – as in distilled – water tastes flat and awful, and is generally unpalatable. Too much of a good thing, on the other hand, can be equally off-putting: water high in iron tastes vaguely of rusty nails; water with a high hydrogen sulfide content smells and tastes of rotten eggs; and water with a lot of magnesium in it tastes bitter. Chlorine – commonly used as a disinfectant in public water supplies – is a no-no from a connoisseur's standpoint, since it makes water smell and taste like a YMCA swimming pool.

Carp and catfish taste the surrounding water to direct themselves toward food; salmon follow the taste and smell of water to find their birth rivers, which become their spawning grounds. Eels may use a similar mechanism to find their way back to their breeding territory in the Atlantic's Sargasso Sea. In a sense, we're all eels: flavor experts claim that most people prefer the taste of the water they grew up with. Food writer Jeffrey Steingarten argues that this isn't necessarily true, citing awful-tasting California tap water; and my husband, raised on water from a spring into which rabbits periodically fell and drowned, says his natal water isn't his cup of tea. Many of us, however, bond with our first water; and I'm still convinced that the best water comes from home. Just as Marcel Proust's childhood memories fell out of a fragrant cup of lime tea, mine can come from water: all of small-town New England in one cold mouthful. Memory is intimately linked to the senses; and smell and taste are powerful ties to our pasts. We remember our home waters, and can pick them out of a liquid line-up, discriminating among them like connoisseurs of fine wines. I could find mine anywhere. It smells of fall on Vermont's Lake Champlain, and tastes of granite, snow, and freshwater clam.

The most mineral-laden of all waters are those from thermal springs – that is, those whose water temperature is at least 13.7°F (16.5°C) higher than the average annual temperature of the surrounding air. In Greek mythology, hot springs were usually associated with a superstar of the classical pantheon: not a lowly nymph, but a god, goddess, or hero. The bubbling spring of Thermopylae, for example, was sacred to Hercules, who reputedly went to bathe there while recovering from his labors. It was near this steamy stream – the name means "Hot

Gates" – that the epic battle took place in 480 BCE, in which 7000
Spartans heroically held their ground against 200,000 Persians.
(Some sources claim 300 Spartans and 2,000,000 Persians.) The
Persians won, but nobody remembers them; the moral victors
were the intransigent Spartans, who bravely perished to the
last man.

Thermal springs range from pleasantly warm to scalding.
The hot springs of the succinctly named Bath, England, hover
around a jacuzzi-like 120°F (50°C); the pools of Warm Springs,
Georgia, where Franklin Delano Roosevelt went to alleviate
the crippling effects of polio, maintain a soothing 88°F (31°C).
The bubbling pools of Yellowstone National Park, on the other
hand, can top boiling point, as can the steam-spouting Lard-
arello hot springs in Tuscany, which are said to have inspired
Dante's vision of the Inferno. It seems to be a common impression: the hottest springs in Japan are known as *jigokus* – hells.

Hot springs get their thermal oomph from the internal
heat of the Earth itself. The deeper you go, the warmer it gets:
on average, the temperature of the Earth's crust increases
about 1.4°F for every 100 feet (0.7°C for every 30 m) in depth.
Warm Springs and Bath, natural spas, are fed by water slowly
simmered far beneath the surface by the natural heat of the
Earth's interior. The roiling pools of Yellowstone, however,
are cooked by plutons, intrusions of magma from the mantle
thrust up like red-hot pokers into the Earth's crust. These
searing blobs of molten rock, typically located 4–10 miles (6–16
km) beneath the surface, rapidly boost the crustal temperature
gradient: dig a hole in Yellowstone and the temperature will
rise 123°F every 100 feet (59°C every 30 m). With such a fiery
furnace lurking just below, groundwater becomes superheated
to temperatures far above the normal boiling point. As long
as such water is maintained at high pressures, as exist deep
beneath the crust, it remains liquid. This pressure slackens off,
however, as groundwater gets closer to the surface, and at some
point drops enough such that the water suddenly bursts into a
boil. Vaporization is accompanied by an explosive increase in
water volume, as superheated liquid converts itself to steam

– and this conversion, if constricted by narrow openings in surface rock formations, can lead to breathtaking eruptions. This is the mechanism underlying the spouting of geysers, such as Yellowstone's Old Faithful, which has been fountaining at relatively predictable intervals since the recorded history of the park began in the late nineteenth century. The name "geyser" means "gusher," and is said to come from *the* Geysir in Iceland, a geyser of sensational performance, famed throughout Europe during the Middle Ages.

Life, though Charles Darwin envisioned it beginning in a tepid little pond, may in fact have begun in just such overheated water. A trip to Yellowstone's seething stewpots, the steaming streams of Iceland, or the bubbling mudholes of New Zealand may be, in an evolutionary sense, a sentimental journey, a visit to our shared ancestral home. Ecological hell holes where temperatures often top the boiling point shelter what may be some of the oldest life on Earth: the hyperthermophilic bacteria, near-indestructible little creatures who flourish in what biologists tactfully term "extreme" environments. Like the legendary medieval salamander, these microbes like it not only hot, but incendiary, which is why so many have genus names smacking of arson. *Pyrolobus*, for example – the *pyro* comes from the Greek for "fire" – thrives at temperatures up to 235°F (113°C), well above the boiling point of water; *Pyrodictium* at temperatures of 230°F (110°C); and *Sulfolobus*, a pyro in preferences if not in name, at 194°F (90°C). (*Sulfolobus* freezes to death when the temperature hits 131°F (55°C), which is the sort of heat level humans approach with potholders.)

Some evidence suggests that our putative many-times-great-grandcreatures originated at the bottom of the sea. In 1977, a full eight years after Neil Armstrong took his famous small step on the Moon, startled scientists on board ALVIN, the Woods Hole Oceanographic Institute's VW-Beetle-sized research submersible, spotted a fuming spout of smoke 8000 feet (2400 m) down along the East Pacific Rise, not far from the Galapagos Islands. The spout – which looked like nothing so much as an infernal tea kettle boiling away in the pitch black depths of the sea – proved to be the first of many hydrothermal vents, fissures on the ocean floor that mark the doorway to the hyperheated interior of the Earth. The spouts are submarine geysers, formed where cracks in the bedrock allow seawater to seep into the Earth's volcanic interior, there to be heated and then frenziedly expelled, now carrying a load of dissolved minerals – lead, cobalt, zinc, copper, silver, iron, and hydrogen sulfide. This toxic soup, hot enough to crisp a French fry, is, scientists guess, a good approximation of the environment of early Earth, and it's a paradise for hyperthermophiles. Tough, simple, self-sufficient, and happy to live at toe-frizzling temperatures, these microbial pyros may be the best bet for life's beginnings.

The phrase "to be in hot water" has had negative connotations since the sixteenth century: if you're in it, chances are you're also in a nasty jam, sticky predicament, or other significant trouble. In a biological sense, however, to be in hot water may mean quite the opposite. In fact, those in hot water may experience an unparalleled life-support system and receive an unexpected and serendipitous evolutionary headstart.

Life on Earth may have begun in water. If so, most of it certainly stayed there. Water, as environments go, is prime real estate, the pick of the biosphere. Over 90 per cent of the Earth's biomass lives in the sea. Most of this, admittedly, is plankton; still the sea also harbors representatives of every known phylum, in an impressive range of diverse forms from flatworm to flatfish to bottle-nosed dolphin. Even that 10 per cent of us who eventually tried our luck on land never got too far away from our primordial mother. Terrestrial evolution has been an exercise in how to live out of water without quite letting go of it. Structures as diverse as the pine needle, the eggshell, and the camel's hump were invented with one thought in mind: hanging on to water. Plant leaves, stems, and the skins of fruit are coated with a waxy layer called the cuticle, designed to hold in water;

it's the cuticle that gives the polished apple its sycophantic shine. The human skin – all 15–20 square feet (1.4–1.9 sq m) of it per average person – is waterproof.

We need water. Complete withdrawal from it, as young Nathan Zohner pointed out, results in death. People, to keep themselves going, require an average of 2.5 liters of water a day – more if exercising vigorously in the hot sun – and deprived of this, few of us manage to stay alive longer than a few days.

On April 28, 1789, Fletcher Christian and his fellow mutineers seized the *H.M.S. Bounty* and set sail in it for Tubuai, leaving behind their captain, William Bligh, and eighteen loyal crewmen in a 23-foot launch. The launch was equipped with 150 pounds of bread, 16 chunks of salt pork, 6 quarts of rum, 6 bottles of wine, and 28 gallons of water. Bligh divided it all with scrupulous precision, using a scale fabricated from coconut shells, and then divided it again based on the probable number of days it would take (48) to reach the closest island, Dutch-owned Timor in the Indian Ocean, 3600 miles (5700 km) away. On an ounce of bread a day, the odd mouthful of pork, and an occasional captured seagull, they could have made it. On a quarter of a pint of water a day, they were doomed. The weather saved them. It rained almost the whole time they were adrift – miserable for men at sea in an open boat, but a godsend for replenishing their inadequate water supplies. They made landfall on June 14 with all nineteen alive, though David Nelson, the ship's botanist, died of fever several weeks later.

Bligh's unhappy party, though suffering from numerous other trials and tribulations, had been most at risk of death by dehydration. The body needs continual refilling to maintain its internal liquid status quo. Human beings, despite our protective casings, leak. Water is lost with every breath we take – hence the mist of condensation that appears when we exhale on to a cold window-pane – as well as through tears, sweat, urine, and feces. We can't afford to lose much of it either. Once 15 per cent of the body's water is gone, the physiological situation becomes critical. Blood, depleted of water, becomes sludgy and difficult to pump; this, with accompanying salt imbalances, leads to heart and kidney failure. Too dry and we're dead.

Plants similarly spend their lives in a quest for drink, often going to great lengths to battle desiccation. Consider, for example, the Herculean hydraulics of tall trees. The California redwood – an arboreal giraffe – holds the world record for tallness. The current champion, known as the Mendocino Tree, 367 feet, six inches (112 m) tall – 62 feet (19 m) taller than the Statue of Liberty – stands modestly unlabeled in the Montgomery Woods State Reserve near Ukiah, California. It owes its anonymity to the fate of its towering predecessor, the Tall Tree (367 feet, 10 inches or 112.1 m), identified in Redwood National Park by the National Geographic Society in 1963. Fame, as is so often the case, proved fatal: so many spectators arrived to gaze upon (or up at) the tree that the soil around its base became packed as hard as concrete. The tree sickened and by 1990, ten feet (3 m) of its top had withered and fallen, simultaneously reducing its height to 357 feet (109 m) and removing it from the botanical limelight. Other contenders for world's tallest include the Australian eucalyptus (runner-up at 322 feet or 98 m) and the Douglas fir, a fallen specimen of which clocked in at 435 feet (132.5 m) in 1872.

Tall trees are miracles of internal plumbing. They owe their existence to their phenomenal ability to pump water. A waterless plant, as any gardener knows, is a pitiful sight. Leaves collapse; blossoms fold; stems droop, limp as over-cooked spaghetti. Forget to water the tomatoes and they slump dismally to the ground, shriveled manifestos to human and heavenly neglect. Plants, properly provided for, are perkily upright or turgid – which word, in the innocent milieu of the garden, means pumped full of water. The turgid plant has cells as swollen as miniature water balloons, their membranes pressing tightly against the surrounding cell walls. Turgidity is the prime feature of the salad bar: it's what keeps lettuce crisp and cucumbers crunchy. Its opposite number is listless floppiness: wilt.

Turgidity, however, is temporary. Water in plants, like water in humans, is always on the move, here today and gone almost immediately, merely passing through. It moves steadily from ground to roots, then upward to stems, flowers, and leaves – and then continues onward and outward, evaporating into the atmosphere. Essentially plants are biological fountains, pulling water from the depths and spraying it out again, molecule by molecule, into the upper air. This evaporative process – transpiration – takes place through tiny openings called stomata thickly scattered over the surfaces of leaves and stems. In microscopic close-up, stomata look like bright green clown lips, mouths wide open in a startled O when puffed with water, pursed severely shut when water levels fall.

Evaporation is the key to the entire water-moving process. Plants may get their water from the bottom, but the governing factor – as is so often the case in life – is pull from the top. The drying power of air creates a continuous upward tension that literally yanks water upward from the roots, dragging it leafward at a rate of 4 feet per hour (1.2 m/h) through a system of minuscule hollow pipelines – two-hundredths of an inch (half a millimeter) in diameter – collectively known as xylem. Over 90 per cent of the water slurped up by the roots is eventually spat out by the stomata. Over a given growing season, this adds up to astounding quantities of water. An acre of maple trees, in the four lush months from May to September, transpires 830,000 gallons (3 million liters) of water; the average acre of cornfield, by the time the ears are ripe for picking, has transpired 350,000 gallons (1.3 million liters). The water-drenched air above the rainforest canopy is made by thirsty trees.

The very efficiency of the transpiration process makes leaves a liability in dry climates: a leaf in the desert is the equivalent of a chronically leaky faucet. Cacti, accordingly, sport spines, which – though officially leaves – are notably unleaflike in structure and function, possessing neither chlorophyll nor anything much in the way of surface area. Similarly leaves are no advantage in frigid weather, when water for the most part is in the unusable form of solid ice. New England's glorious scarlet and orange flare of fall foliage, followed by the shameless jettisoning of leaves by deciduous trees, is a cagey exercise in water conservation. In periods when water is essentially unobtainable, no plant in its right mind wants a population of pancake-sized leaves around, each wastefully oozing precious vapor from ten million tiny pores. The better part of valor, clearly, is to chuck the lot at the onset of winter and wait for spring.

Water, in the idiosyncratic environment of planet Earth, is physically versatile, existing simultaneously in nature in the form of a solid, liquid, or gas. Which state it assumes at any given moment has a lot to do with how fast its molecules move. Though you'd never guess it to look at it stolidly sitting there, all matter, at the atomic level, is frenetic stuff. In liquid state, as in bathtubs, teacups, and dripping faucets, water molecules are in continual nervous motion. Water wiggles, jitters, and jumps, and – when conditions are energetically right – jettisons itself into the atmosphere. Molecular motion is temperature-dependent: as things heat up, activity intensifies. Increased energy and enhanced motion cause molecules to take up more space – just as an extravagantly leaping ballerina, for example, requires more room than a sedentary reader, sitting placidly in a chair – which is why substances at higher temperatures expand. Eventually, at temperatures near the boiling point, activity levels become so great that molecules tear themselves loose from their fellows and go zapping off into space. Technically, this is called vaporization and water molecules that undergo it cease to be liquid and instead become steam. Chill things down and the opposite occurs. As the temperature drops, activity slows until – at 32°F (0°C) – sluggish water molecules click solidly together into a fixed framework. Scientists refer to this event as crystallization; the rest of us call it freezing; and, meteorologists, faced with it, start warning people off the roads.

The frozen portion of the globe is known collectively as the cryosphere. This includes all of the Earth's snow and ice, variously packaged in ice sheets, sea ice, glaciers, and – seasonally – a vast tract of snow cover. Ice covers just over 10 per cent of the Earth's surface, a total of 8 million square miles (21 million sq km); seasonal snow (at peak, in the dead of winter) covers an additional 18 million (47 million sq km). Together this frozen fraction comprises over 85 per cent of the world's fresh water, which means that the lion's share of our potential drinking supply is solid.

Water, solidified, has historically caused a lot of people a lot of trouble. It made the lives of Washington's troops miserable at Valley Forge and kept Napoleon out of Moscow. Hannibal, slogging through the snowy Alps with his thirty-seven shivering battle elephants, en route to conquer Rome in 221 BCE, called it "the white enemy." Robert Scott froze to death in it at the South Pole in 1912, along with four companions and some luckless Shetland ponies; and the eighty-three members of the Donner Party – en route by wagon train to a better life in California in the winter of 1847 – were forced to eat each other in it, while trapped in the Sierra Nevadas. In Japanese mythology, the pale Snow Maiden is the spirit of death; and Hans Christian Andersen's Snow Queen was cold, cruel, and terrible, even though she did offer little Kay the whole world and a new pair of skates. On the homely side, 8 inches (20 cm) of new-fallen snow blanketing the average household sidewalk weighs about 440 pounds (200 kg), which is no fun – even positively hazardous – for the average household shoveler. The repeated lift and heave of a loaded snow shovel can quickly boost heart rate to a potentially dangerous 175 beats per minute. In the United States, an estimated 1200 people die annually of heart attacks brought on by shoveling snow.

We routinely compare anything notably white to snow: snowy owls, snowy egrets, and snowdrops are all snow-white; and the eponymous Snow White was named for her beautiful, but colorless, complexion. Some of us admire this: Thoreau gloried in being an inspector of snowstorms; and Percy Bysshe Shelley loved the snow "and all the forms of the radiant frost." In general, though, our metaphors for snow and ice indicate that we don't like being cold. Just as spring represents joy, youth, and birth, winter traditionally represents misery, old age, and impending death. An icy glance is daunting; a cold shoulder is unfriendly; and an icy heart belongs to the sort of person who kicks puppies. For Robert Frost, hate was ice.

On the other hand, frozen water, in the nineteenth century, was financially hot stuff. Ice, before the age of mechanical refrigeration, was never available when you most wanted it, in the sultry days of summer. Lord Byron, in a poem written in 1809, paired "ice in June" with "roses in December" as examples of the absolutely impossible. In quantity and properly insulated, however, ice can be stored through long hot summers, and its value to those in sweltering climates in the days before air conditioning was beyond price. The Chinese were stockpiling ice by 1100 BCE: the ancient *Book of Odes* describes the harvesting of ice in the mountains and its preservation in the royal ice houses for the delectation of the emperor and his court. Alexander the Great, while besieging the city of Petra in the fourth century BCE, had ice brought from the mountains and buried in insulating pits for the refreshment of his overheated officers. The Emperor Nero had snow transported to Rome by relays of runners for the making of his favorite frozen desserts. At least once the icy cargo melted en route, for which the apoplectic emperor had the general in charge of the operation executed.

Wealthy Italians of the Renaissance period maintained snow- and ice-cooled wine cellars. James I ordered two brick-lined snow pits constructed for his palace at Greenwich; Louis XIV had two ice houses built for his palace at Versailles. American colonists, appalled by the sauna-bath-like New World summers – in which people sweated and suffered, milk soured, meat spoiled, and butter dissolved into slimy puddles – struggled from the seventeenth century on to keep ice. Early attempts were not always successful: even when buried in deep pits and covered protectively with straw, ice melted when confronted with brutal temperatures; and food optimistically stored upon it warmed up. Ill-considered attempts at colonial food preservation led to frequent deaths from "summer complaint" – generally a deadly diarrhea brought on by enteric bacteria, which begin to multiply rapidly when temperatures creep above 5°C (41°F). With this in mind, Amelia Simmons, the self-styled "American orphan" who wrote the first American cookbook in 1796, advised readers to kill a table-designated chicken no more than four hours before dinner and to leave milk, until needed, safely inside the cow.

Ice houses only became truly effective when their design was perfected in the early nineteenth century by the Shakers, a technologically inventive religious sect devoted to holy dancing (hence the name), vegetarianism, temperance, and celibacy. Shaker ice houses were exercises in creative insulation: each was enclosed with double wooden walls and topped by a triple roof, and gaps between layers were tightly packed with sawdust. Ice, thus protected, kept – melting losses dropped from 66 per cent to less than 8 per cent – and Shaker storage principles, incorporated into the holds of ships, made fortunes for a number of enterprising New England merchants. Foremost among these was Frederick Tudor of Boston – later to be known as the Ice King – whose first cargo of insulated ice, shipped to Martinique in the French West Indies in the 1820s, arrived profitably intact. Within twenty years, New England ice was traveling as far afield as India, China, Australia, and the Philippines, and the trade had attracted the attention of a scattering of theologians who held indignantly that ice in hot climates represented a hubristic meddling with the natural order of God's universe and was thus a sin.

The boom in ice was made possible by a physical peculiarity of frozen water. In the usual scheme of things, molecules in solids are more tightly packed than those of their parent liquids, such that, volume for volume, solids are heavier. Most substances, as they crystallize, sink. For example, dry ice – frozen CO_2 – plummets to the bottom in liquid carbon dioxide. Cold water, up to a point, conforms to the norm; for the most part, cold water is heavier than warm. Because of this, lakes are layered, their waters divided between a warm lightweight surface layer – the epilimnion – and a cold dense bottom layer, the hypolimnion. In the temperate zone, with its seasonal temperature changes, upper and lower layers periodically intermix and change places. In the crisp days of autumn, season of frosts and harvest moons, the epilimnion gradually chills, eventually dropping to 39°F (4°C), the temperature at which water is at its most dense. This leaden stuff sinks, displacing the now comparatively warmer and lighter waters of the hypolimnion and pushing them to the surface. The reverse happens in the balmy days of spring, as the epilimnion warms to 39°F (4°C). The exchange is more than an equitable matter of turns at the top: the sinking epilimnion carries with it to the depths a rich mix of dissolved oxygen; the rising hypolimnion boosts to the surface a stew of organic nutrients. The resultant intermixing supports a diverse population of lake top- and bottom-feeders. Lakes deprived of such temperature-driven turnover – such as Africa's Lake Tanganyika – have cold dreary bottoms; the only inhabitants of the oxygen-depleted depths are anaerobic bacteria.

As temperatures drop below 39°F (4°C) and approach freezing point, water reveals its eccentric nature. In the process of freezing, water paradoxically loosens up. Individual water molecules, as they fall into place in the growing ice-crystal lattice, form hydrogen bonds with their four nearest neighbors. In order to comfortably perform this acrobatic act, all the participants must move slightly apart. Imagine a roomful of dancing couples coming to a halt and – in affronted fashion – holding each other off at arm's length. As each dancer extends his or her arms, neighbors are compelled to move back and outward, until the entire standoffish throng takes up a great deal more than its original space. Such generous spacing among molecules in an ice crystal makes ice about 10 per cent less dense than water. It also causes water, unlike most substances, to expand rather than contract upon freezing, which in turn is why freezing causes pipes to burst in winter, with occasional awful consequences.

Solid ice, being less dense than liquid water, floats on top of it. This seemingly small physical quirk, attributable solely to the all-important hydrogen bond, has wide-ranging effects: because of it, ice fishermen and hockey players ply their trades; Eliza, with the bloodhounds in pursuit, leaped across the ice floes to freedom in *Uncle Tom's Cabin*; Hans Brinker skated his way to victory on the frozen canals of Holland and thereby saved the family fortunes; and the *Titanic* – colliding with a floating iceberg – sank. On a biological level, the fact that ice floats has immense implications for life on Earth. Because ice is less dense than water, ponds and lakes, as they freeze, do so from the top down. The thick layer of ice and snow on the surface, once formed and solidified, acts as an insulating blanket, preserving residual heat, preventing the lake from freezing solid all the way to the bottom, and providing protection for the lake's resident population of fish, frogs, zebra mussels, and water lilies. If ice, like any other self-respecting crystal, were heavier than water, freezing would proceed from bottom to top; lakes and ponds would completely solidify each winter, with fatal results for their inhabitants; and the North Pole, Santa Claus and all, would be lying on the bottom of the Arctic Ocean.

Ice, as found in the average pond, polar sea, or ice-cube tray, is technically known as Ice Ih. Ice Ih – sometimes called "hexagonal ice," from the shape of its crystal lattice – is all most of us will ever run into; however, it is merely one of many ices, the tip of a large and varied crystalline iceberg. With changing temperature and pressure, solids can undergo phase transitions, seguing from one form to another of the same substance. Collectively such alternative crystalline structures are known to chemists as polymorphs. Cocoa butter, for example – the main solid fat in chocolate – can crystallize into six different polymorphic forms, of which form V is the most desirable, being the glossiest, the tastiest, and the smoothest textured, as well as possessing a body-temperature melting point that allows it to dissolve delectably in the mouth. Most compounds possess two – at most, three – different polymorphic configurations. Ice has at least fourteen.

The ancient Greeks hypothesized that the quartz crystals so favored by crystal healers today were ultra-frozen forms of water, ice so adamantly congealed that it had become resistant to thaw. In practice, ultra-ice is far stranger stuff. The study of ice polymorphs was pioneered by physicist Percy Bridgman of Cambridge, Massachusetts, who entered Harvard as a freshman in 1900 and never left, barring a brief trip to Sweden in 1946 to receive the Nobel Prize. The prize was awarded for his work in the field of high-pressure physics. Bridgman designed apparatus that could sustain pressures up to a hitherto unreachable 20,000 atmospheres (150 tons/sq in; 20,000 kg/sq cm), and then spent decades crushing things in it. Ice, forcefully squeezed in this manner, shifts progressively into five different crystalline states (Ices II to VI), each more compacted than the last. In Ice VI, for example, molecules are so closely intertwined that the density of the crystal is doubled. The pressurized ices also have elevated melting points: Ice VI only melts when temperatures reach 175°F (79°C) and has thus been nicknamed "hot ice," which phrase in the criminal underworld refers to pilfered diamonds. Ice VI, however, is only hot at pressures of 6500 atmospheres (50 tons/sq in; 6500 kg/sq cm) – in effect, the equivalent of the mass of a bull elephant concentrated on the eraser end of a pencil. Release the pressure and VI instantly becomes a puddle. Squeeze some more, on the other hand, and you get Ice VII, which is even hotter than VI, with a melting point well above 212°F (100°C).

Sci-fi fans are familiar with Ice IX which – as ice-nine – starred fearsomely in Kurt Vonnegut's novel *Cat's Cradle*. Ice-nine, diabolical discovery of scientist Felix Hoenikker, was a crystalline form of ice that remained frozen at high temperatures. Worse, it could induce other forms of water to do likewise: at the climax of the book, a tiny fragment, dropped into the ocean, sets off a chain reaction that converts all the water on the planet to unthawable ice. Real Ice IX is a permutation of

Ice III; it poses little risk to humanity since it cannot exist at temperatures above -148°F (-100°C). By Ice X, hydrogen bonds have been squashed almost out of recognition, shortened to the point where all hydrogens and oxygens are equidistant from each other, and individual water molecules can no longer be distinguished in the homogeneous mass. Ices XI and XII were discovered recently in cyberspace, using molecular simulations. Such ices may occur in nature, but probably not too close at hand. There may be a bit of XI and XII on Jupiter.

Ice also occasionally exists as an amorphous solid – a slovenly mush of molecules that even chemists, a group seldom noted for their attention to personal appearance, deem "disorderly." Amorphous ice was first discovered in 1936 when a pair of Canadian researchers tried condensing water vapor on to a copper pipe cooled to a bone-chilling -220°F (-140°C). The result was an icy glass, an uncoordinated frozen blob with no hint of the geometrically organized elegance of ice crystals. As amorphous ice is warmed, its sluggish molecules first form an outlandish liquid with the consistency of melted tar. This frigid sludge persists until temperatures reach -184°F (-120°C), at which point it suddenly snaps – with military precision – into the familiar hexagonal lattice of Ice Ih.

Much of the world as we know it today is a product of frozen water. Massive build-ups of Ice Ih have punctuated Earth's history, beginning some 800 to 600 million years ago when, some geologists theorize, the entire planet may have been encased in a half-mile-thick (kilometer-thick) layer of solid ice, a chilly scenario now nicknamed "Snowball Earth." Major ice ages have occurred three times since, the most recent – which continues to the present day – commencing about three million years ago. This, by far the best studied of the four, has oscillated metronomically between cold and not-so-cold episodes as the ice sheets advance for 100,000-year-long periods, then melt back for 10,000 years or so, only to advance again. The last major melt occurred some 12,000 years ago, ushering in the balmy interglacial period that we're all enjoying now.

Ice, retreating, shaped the modern landscape of the continents, scouring the sides of mountains, carving lake basins
and river valleys, crushing hills, scattering outsize boulders,
and coating the terrain with geologic junk in tumbled deposits
up to 2000 feet (600 m) thick. What we see as we stare out our
windows today is likely to be the aftermath of ice. The land
still continues to rebound from the hulking weight of glaciers;
the plains of Canada have risen some thousand feet (300 m)
since the departure of the ice, and are still on the rise, at the
rate of an inch (2.5 cm) a year.

On a human note, the last melting of the ice also coincided
with the invention of agriculture, the establishment of towns
and cities, and the rise of civilization – for which, some pale-
obiologists argue, we are intellectually well-prepared, our
brains having been honed over millennia by recurrent and catastrophic climate change. Neurobiologist William H. Calvin, for
example, points out that the phenomenal quadrupling in size
of the early hominid brain began about 2.5 million years ago
at the inception of the last ice age. He further argues that the
evolution of human intelligence may have been driven by the
need to adapt to its prolonged progression of alternating freezes
and thaws. Only the cleverest hominids – those who ultimately
acquired the mix of problem-solving capabilities, foresight,
creativity, and versatility that we define as smart – were able
to survive in an unpredictable ice-age world. Life may have
begun in hot water; and it certainly depends on liquid water.
But it's just possible that we owe our brains to ice.

*How inappropriate to call this planet Earth, when so clearly
it is Ocean.*

Arthur C. Clarke

Water surrounds us, awes us, and fascinates us. It calls to even
the most sedentary of landlubbers. The sea is *Treasure Island*
and *Captains Courageous*; Captain Nemo and the kraken; Ahab

and the white whale; Columbus sailing into the unknown west; Francis Drake setting out to defeat the Spanish Armada; John Paul Jones defiantly declaring that he had not yet begun to fight. "Why is almost every robust healthy boy with a robust healthy soul in him, at some time or other crazy to go to sea?" wrote Herman Melville. The answer is right there in *Moby Dick*: because water is the element of freedom and adventure, romance and desire. Almost no one captured this quality better than British poet laureate John Masefield – an orphan, sent to sea at the age of thirteen – who became the author of some of the most evocative verse ever written about the ocean, sagas of salt and spray, tall ships and trade winds, quinqueremes and galleons. It entices us all, that distant call of wave and saltwater. Kids in Kansas – a state so far from water that its citizens conceivably might not recognize an oar – still sing songs about the sea.

The *Oxford English Dictionary* defines *ocean* as "the vast body of water on the surface of the globe, that surrounds the land" and points out that the earliest record of the word in written English dates to 1290, when it appeared (spelled "occean") in an account of the voyage of St Brendan. The word comes from the ancient Greek, derived from Oceanus, the primeval god of the ocean, who later abdicated in favor of the more proactive Poseidon, the blue-haired god of several powerful unpredictables: earthquakes, horses, and water. In Norse mythology, the seas were the realm of the giant Aegir, whose nine red-headed daughters drove the waves; and in Chinese legend, the seas were governed by five Dragon Kings, each with five feet, beards, and yellow scales, who lived in great palaces on the ocean floor and dined upon opals and pearls. All shared temperaments: universally, across cultures, sea gods are moody, hot-tempered, and mean. Aegir, when incensed, whipped up storms and smashed ships; and the Dragon Kings, on whims, created waterspouts, tsunamis, and typhoons. The entire *Odyssey* took place because the usually clever Odysseus got on the wrong side of Poseidon, who maliciously afflicted him with twenty years of bad luck and weather on the way home from Troy.

The fearsome power and callous unreasonableness of the sea dominate the lives of those who live in close association with it. "The sea," wrote Joseph Conrad, "has never been friendly to man." Its romance is fraught with danger, as becomes clear to anyone who watches so much as five heart-stopping minutes of the movie version of Sebastian Junger's book *The Perfect Storm*. Ancient sailors, hoping to pacify the capricious gods of the ocean, made sacrifices: the Romans offered bulls, the Japanese offered rice; and the Britons offered people – usually criminals, hurled ceremonially off the tops of cliffs. The Vikings sought to protect their dragon ships by dragging the new-laid keels over the bound bodies of prisoners. Christians once had their ships baptized; and even today we set a new ship on its way by smashing a bottle of champagne over its prow.

Our contest with the sea is eternally unequal, with all the advantage on the sea's side. Our helplessness in the face of such vast and raw nature has marked us: we speak hopefully of smooth sailing, safe harbors, and the happy day when our ships come in. Troubles are storms at sea: life and love are wild oceans, battering our tiny personal boats. Thomas Jefferson spoke of the tempestuous sea of liberty; T. H. Huxley of the dangerous sea of politics. Before we took to the highways and the air, the sea was our best bet for mass transit disaster, made all the more terrifying because its victims vanished without a trace. The sea is the wreck of the *Hesperus* and the *Mary Deare*, the graveyard of ships, Anne Sexton's mother-death, and Rudyard Kipling's old gray widowmaker. Ahab's *Pequod*, at the end of *Moby Dick*, is engulfed by the "great shroud" of the sea.

Storm-wrecked Viking ships were said to plummet into the wide jaws of Aegir; and victims of more recent centuries sank into Davy Jones's locker. Davy, according to the eighteenth-century novelist Tobias Smollett in *The Adventures of Peregrine Pickle*, was "the fiend that presides over all the evil spirits of the deep, and is often seen in various shapes, perching among the rigging on the eve of hurricanes." Others variously attribute the name to a particularly merciless pirate; an unscrupulous

London publican who operated a press gang (customers were fed drugged rum and ended up in the Navy); the Biblical Jonah; or St David, patron of saint Wales. There's less dispute over the nature of the locker, which all agree to be the ocean bottom.

The bottom, depending where you look for it, ranges from deep to appallingly deep. The underwater shelves which edge the continental land masses essentially mark Davy Jones's attic: these measure anywhere from a few miles to 600 miles (1000 km) or so wide and average 200–600 feet (60–180 m) in depth, which in ocean terms is shallow. The bulk of the land beneath the waves is ocean basin, 12,000 to 15,000 feet (3700–4600 m) down; and some regions plunge further yet. The southern end of the Mariana Trench, between the Pacific islands of Guam and Yap, is the deepest spot in the ocean – 36,198 feet (11,033 m) deep; Mount Everest could disappear into it, with over 7000 feet (2100 m) to spare. No light penetrates to these abyssal depths and the only creatures who manage to survive down there are sea cucumbers, indomitable sausage-shaped echino-derms who dine on benthic muck and somehow resist water pressures capable of crushing a nuclear submarine.

Sea traditionally is a synonym for ocean, though there seems to be some chicken-or-egg-type bicker over which is biggest and/or comes first. "Though the sea is one continuous liquid mass, it has been for the sake of convenience in descrip-tion divided into different areas termed oceans," announced a geography text of 1880, thus flatly contradicting a similar tome of 1635, which stated "Any part of the ocean marked off from the general mass of water may be called a sea." The *Ency-clopedia Britannica* explains diplomatically that in geography the term "sea" is loosely used, which is undeniably the case, since the Caspian and Aral Seas are both lakes, the Caribbean, North, and Baltic Seas parts of the Atlantic Ocean, and the South China, Coral, and Tasman Seas subsets of the Pacific.

"Everybody talks about the Seven Seas, but hardly anyone can name them or tell just where one begins and the other leaves off," writes Danish explorer Peter Freuchen in his *Book of the Seven Seas.* The evocative name, he goes on to explain, is of ancient origin, long in disuse, and was only re-popularized in 1896 when Rudyard Kipling chose *The Seven Seas* as a title for a new book of poems. The somewhat arbitrary seven are the Arctic, Antarctic, North Atlantic, South Atlantic, North Pacific, South Pacific, and Indian Seas, though today no geographer refers to them as such. The bottom line, however, is that a sea, though ordinarily defined by its adjacent coast – as in Arabian Sea, Irish Sea, and Sea of Japan – none the less remains a part of the greater whole, inextricably entwined with the world ocean. This differentiates seas, no matter how carelessly named, from lakes, which are independent entities, landlocked bodies of water firmly trapped within continental boundaries.

Most of the world's lakes are holes chipped out by passing glaciers. Roving ice hollowed North America's Great Lakes, New York's Finger Lakes, and the scattered mass of water-filled cavities that makes Minnesota the "Land of 10,000 Lakes." Some lakes created in the craters left by volcanic explosions – a classic of its kind is Oregon's Crater Lake, now filling the cauldron-shaped basin created in the wake of the eruption of Mount Mazama 7700 years ago. Arizona's Lake Powell and Egypt's Lake Nasser are both manmade lakes, the offshoots of the Glen Canyon and Aswan High Dams respectively; and Switzerland's 45-mile (72-km)-long Lake Geneva is the product of a natural dam, formed by a rocky blockage of the Rhone River. Africa's massive Rift lakes – Tanganyika, Malawi, and Victoria – fill the gaps where the African and Somali Plates are pulling apart, separating at the agonizingly sluggish rate of one-twentieth of an inch (1 mm) per year; and Russia's Lake Baikal – both the oldest and deepest lake in the world – also fills a tectonic crack in the Earth's crust.

Depth, according to some criteria, is what distinguishes a lake from its next-smaller relative, the pond. The record-setting Lake Baikal is 5315 feet (1620 m) deep at its deepest point; the Caspian Sea – despite its name, a lake, and the world's largest – is 3104 feet (946 m) deep; and Lake Superior – largest of the Great Lakes – bottoms out at 1333 feet (406 m). Ponds, by contrast, are at most 100–200 feet (30–60 m) deep, shallow enough that sunlight reaches all the way to the bottom. Illumination is what gives a pond its characteristic ecology: photosynthesis can occur throughout it, which means that ponds are lush with vegetation and accompanying wildlife. Walden Pond – measured by Thoreau himself in winter, with a stone tied to a cod line – is 102 feet (31 m) deep in the middle, right on for a pond. Walden's substantial 58-acre (23-hectare) surface area and suspicious possession of a beach, however, make its pond status a bit shaky. A pond, classically, is a body of fresh water so small that waves cannot form upon it routinely enough to create a sandy shore; Thoreau's pond – half a mile long, with beach-like stretches – may arguably be called Walden Lake.

The modern word *wave* comes from the old Teutonic *waw*, meaning to undulate or move up and down; and we light-heartedly apply *waw* in its later incarnations to practically anything that moves intermittently or occurs repeatedly, as in waves of love, longing, fear, faintness, and charging enemy soldiers, or that coordinated series of hand flaps that fans perform at football games. The word was first and best applied, however, to water. Wave, in its wettest sense, describes everything from the peacefully hypnotic ripple – the sort of wave one watches lazily from beneath a beach umbrella – to thundering monstrosities capable of toppling lighthouses, eradicating villages, and throwing ten-ton boulders 20 feet (6 m) into the air. Most waves are the cooperative products of two of the Greeks' primal elements, offsprings of water and wind.

Waves for the most part are initiated by the frictional drag of wind across the surface of the sea. Wave size is ultimately determined both by the strength or speed of the wind, and by wind duration and *fetch* – that is, the distance the wind travels unimpeded across open water. The ideal recipe for waves calls for a stormy wind blowing for a long time over many miles of featureless sea surface, which results in the chaotic condition generally known as chop or sea waves. Such wind-driven waves are usually a confused mixture of waves, wavelets, and whitecaps, all busily responding to the rapidly shifting pressures and gusts of stormy weather. Waves that have outrun the wind, however, generally settle into a more predictable and peaceable pattern. These – the smoothly symmetrical humps and dips known as swell waves – are also among the ocean's longest waves, often over 1000 feet (300 m) from crest to crest, rolling serenely along at speeds of 20–30 miles per hour (30–50 km/h).

Waves figure in some of the earliest representations of water. The Egyptian hieroglyph for water was a zigzag line, suggesting the crests and troughs of waves; and undulating lines, indicative of water, appear on pots from Ur, urns from ancient Greece, the Bayeux Tapestry (stitched beneath the boats of the invading Norman fleet), and medieval coats of arms. For many of us, it's an unpleasant image: bobbing up and down between repetitive trains of crests and troughs brings on the combination of symptoms known as seasickness, which afflicts almost all upon the unstable sea to some extent – and has, apparently, for time immemorial, since our word *nausea* comes from the ancient Greek *naus*, meaning ship. Seasickness results from a contradictory mix of visual cues from the eye and positional cues from the inner ear, an unsettling sensory mismatch that promptly induces us to turn green and rush for the rail. Even Britain's foremost naval hero, Admiral Lord Nelson, suffered from it ("I am so dreadfully seasick that I cannot hold up my head!" he wrote pitifully to Emma Hamilton) and is said to have planned his battles to take place at least four days after he set to sea to ensure that he would be over the worst. Little was available to aid the nauseous of Nelson's time – the remedies of choice were rum and absinthe; and little progress seems to have been made since. The 1899 edition of the *Merck Manual of Medical Information* recklessly urged the nautically unwell to try strychnine, nitroglycerine, or iced champagne; and modern-day palliatives include Dramamine, ginger, and crackers. If all else fails, sufferers can always beg for relief from St Erasmus, the patron saint of seasickness. St Erasmus – also known as St Elmo, of St Elmo's fire – is noted for his patronage of a somewhat mixed bag of maritime and gastric distresses, from storms at sea to labor pains and appendicitis. He was messily martyred by disembowelment in 303 CE.

Waves, like icebergs, extend well below the surface of the water. A wave, in effect, reaches downward a distance equal to about half its length. This has little effect out on the open ocean, but causes trouble as the wave approaches shore. As the lower edge of the advancing wave hits the sea bottom, its motion is retarded. It begins to drag, moving noticeably slower than the careless crest, which continues to race exuberantly forward. The effect – dramatically depicted in Katsushika Hokusai's famous woodblock print "The Great Wave Off Kanagawa" – is rather like a runner overbalancing as he races downhill, head and torso moving too far ahead of frantically pattering feet. As the wave rushes closer and closer to the beach, its underpinnings lag behind it, until finally the unsupported crest teeters, topples forward, and smashes in a burst of spray. This self-destructive header into the sand is what gives such waves the name *breakers*. Poseidon, the ancient Greek god of the sea, is said to have created horses in the shape of breaking waves.

Despite all appearances to the contrary, while waves move through water, the water itself stays in the same place. The wave is simply energy passing through; as it rolls forward, it churns the water beneath it into a whirling circle that rotates dizzily for a moment and then returns to the status quo as the wave abandons it and moves on. Barring paddles, wind, or hidden currents, a rubber raft set down in the middle of the ocean will stay essentially in the same place, lurching forward a bit with the passing crest, and back a bit with the passing trough. This peculiar behavior of water in waves was first described by German researcher Franz Gerstner who, after carefully observing the motions of wave-stirred seaweeds, determined that the water in the crest moves forward in the direction that the wave is traveling, while the water in the trough heads backward in the opposite direction. The real shape of a wave, Gerstner explained, is a cycloid – that is, the rounded curve traced by a point on the rim of a circle as the circle rolls along a straight line. Imagine the path traced by a stone stuck in the tread of a bicycle tire and you've pictured the real shape of a wave. Gerstner's observations were subsequently confirmed by fellow Germans Ernst and Wilhelm Weber, using the first experimental wave tank, which they variously filled with water, mercury, and brandy. Their only interest in this last seems to have been to create waves in it. "Their persistence in the face of temptation," writes Willard Bascom in "Ocean Waves," "has been an inspiration to all subsequent investigators."

The biggest wave of all is the vast global swell of the tide, which *does* move water, dragging the ocean up and down the beach an average of 3 feet (1 m) daily, alternating between high and low tides every 12 hours and 25 minutes. The daily ebb and flow of the ocean's waters have been fraught with superstition and symbolism since ancient times. A prevailing belief holds that deaths come with the ebb tide. The Haida of the Pacific Northwest believed that dead souls went out with the tide, to be carried away to the spirit land in a supernatural canoe. Shakespeare's Falstaff died "at the turning o' the tide;" and Mr Pegotty in Charles Dickens's *David Copperfield* holds forth on the subject: "People can't die along the coast . . . except when the tide's pretty nigh out. They can't be born, unless it's pretty nigh in – not properly born till flood."

The rhythmic turn of the tide is a persistent metaphor for change, as in the shifting tides of battle, business interests, or public opinion. Anything that flows is also commonly compared to the tides. People who throw in their lot with the majority are said to go with the tide; those who go against the tide, conversely, are dissenters and rebels. We generally prefer high tide to low: high water is a euphemism for success and maximum accomplishment; and Shakespeare's tide in the affairs of men had to be taken at the flood to lead to fortune. Low water, on the other hand, indicates failure and slump. Biology prefers neither high water nor low, but concentrates on the fertile space between the two. The stretch of territory between high and low tidelines is a unique ecological niche called the littoral or intertidal zone, whose sands, rocks, and pools are home to a lush population of organisms, including mussels, limpets, barnacles, crabs, sea urchins, and sea anemones. It's where you go with bucket and spade if you want to dig clams.

The ancient Greeks believed that the tides were caused by the rhythmic in- and exhalations of the nostrils of the world, which lay somewhere at the bottom of the sea; a Filipino legend claimed that the tides were caused by the stirrings of a supernatural crab. The definitive cause, however, remained a mystery until 1687, when Sir Isaac Newton, fresh from devising the Law of Universal Gravitation, attributed the tides to the gravitational pull of the Moon. As the Moon circles the Earth, the world's oceans are hauled in its wake. The tide, in effect, is the longest wave in the world, circling the entire planet and moving west to east at a rate of 2000 miles per hour (3200 km/ h). The effect of the lunar gravitational pull is greatest on the Earth directly underneath the Moon, which creates high or flood tides, and weakest on the side of the Earth that is farthest away, which creates ebb or low tides. Just *how* high or low depends to a great extent on local geography. The Mediterranean Sea, for example, has relatively feeble tides due to the restrictive bottleneck at the Strait of Gibraltar. High tide on the eastern (Caribbean) side of the Panama Canal rises about 1 foot (30 cm), while high tide on the Pacific side – some 10 miles (16 km) away – rises 14 feet (4 m). Canada's Bay of Fundy, with its narrow funnel-shaped coast, boasts the highest tides in the world, over 40 feet (12 m) at peak. Though most impressive at the edge of the liquid oceans, tides also occur on solid land. The crust beneath our feet heaves up and down an average of 8 inches (20 cm) a day, rising and falling with the tide; and the city of Moscow – positioned on top of a particularly responsive piece of Earth – rises and falls 20 inches (50 cm) daily.

Time as we know it is an artifact of the tides. Time and tide, in fact, according to the *OED*, are the same word, both from the same Old English root (*ti*), meaning "to extend." It's a perspicacious definition. Water indeed is the element of time; the 24-hour day that so many of us would like to cram more minutes into is a product of water and the Moon. The drag of water across the sea floor in response to the Moon's gravitational pull acts as a leisurely cosmic brake, slowing the rate of the Earth's rotation by one second every 50,000 years. Four hundred million years ago, in the heyday of the dinosaurs, the Earth spun faster and a day was a mere 22 hours long; four billion years from now, the day will have stretched to 48 hours. Eventually – if the planet survives that long – the tidal brake will slow the Earth to the point where it spins in precise synchrony with the Moon. When this occurs, the day will be 1200 hours long and the tidal bulge will be anchored in place, no longer able to circle the Earth.

So-called "tidal" waves have nothing to do with timely tides. These – more accurately known as tsunamis – are seismic waves, the monstrous progeny of submarine earthquakes or volcanic eruptions. The abrupt crustal shifts that accompany earthquakes, when they take place at the bottom of the sea, result in a massive displacement of water. The appearance of a nascent tsunami on the sea surface is deceptively mild. The wave, seen from the top, may be an unremarkable few feet high; below, however, it extends downward to meet the ocean floor, a waterborne package of energy, moving fast and looking for trouble. Tsunamis may have wavelengths of several hundred miles (kilometers) and can travel at 500 miles per hour (800 km/h) or more – as fast as the average commercial jet flies. Their rapid passage barely bobbles a ship in midocean; the wave's true nature is only revealed as it approaches shore. As the lower edge of the immense wave begins to drag along the sea bottom, wavelength shortens and the tsunami's tremendous energy is translated into height. The wave rears up into a looming wall of water, a hundred or more feet (30 m) tall – over four times higher than the Great Wall of China – and then crashes cataclysmically to the beach, destroying everything in its path. Water, if enough of it is stacked up in one place, can be as destructive as an atomic bomb.

The tsunami generated by the quake on the floor of the Indian Ocean on December 26, 2004, was the worst in recorded history, a horrific inundation that devastated the coasts of Indonesia, southeast Asia, Sri Lanka, India, and eastern Africa, resulting in the deaths of an estimated 235,000 people. It began, seismologists believe, with an abrupt 16-foot (5-meter) vertical rise of the seabed northwest of Sumatra, a geologic jolt that propelled the 3 miles (5 km) of water overhead into ominous action. The massive wave's total energy was equivalent to that of five megatons of TNT – that is, more than twice the explosive energy generated by all the weaponry fired, ignited, dropped, or detonated during World War II, including the Hiroshima and Nagasaki atomic bombs. In some places, once the tsunami reached a landmass, it roared inland for a mile or more (2 km).

Water, the humble stuff of baths and buckets, is mighty. The tsunami that followed the eruption of Krakatoa in 1883 left 36,000 dead. "Before our eyes … all the houses of the town were swept away in one blow like a castle of cards," wrote one appalled survivor. "There, where a few moments ago lived the town of Telok Betong, was nothing but the open sea."

Most of the world's water is salty: seawater is synonymous with salt, and "briny" is a traditional modifier for "deep." According to Scandinavian legend, however, the original oceans were filled with fresh water. The salt accumulated later, the output of a misplaced magical salt mill, ceaselessly grinding away at the bottom of the sea. In some versions of the tale, the mill grinds all by itself; in others it is turned by a pair of disgruntled giantesses. A later take on this tale, titled "Why the Sea is Salt," appears in Andrew Lang's *Blue Fairy Book*, and involves two brothers (one rich, one poor), a hand mill, and a Christmas ham. As in all moral nineteenth-century fairy tales, the selfish rich brother gets his comeuppance and the kindly poor brother lives happily ever after; the magic hand mill, in this case, ends up in the hands of a sea captain who sets it to grinding salt, but then is unable to turn it off. Ultimately the insanely churning mill grinds so much salt that it sinks the ship and itself along with it.

In Norse mythology, sea salt comes from the briny blood of the first frost giant, Ymir, killed by Odin, Hoenir, and Lodur, and thrown into the great pit of Ginungagap. If so, frost giants must have been saltier than humans: seawater contains about four times as much salt as human blood plasma. This amounts to about 3.5 per cent dissolved solids – of which the lion's share is sodium chloride or table salt, the stuff we sprinkle on our baked potatoes. Minor components include calcium, magnesium, potassium, sulfate, bicarbonate, bromide, strontium, boron, silicon, and fluorine; trace elements – present in concentrations of less than one part per million – include practically everything else. (Among these is gold: every cubic kilometer of ocean contains about a million dollars' worth, for a grand total of $300 trillion.) This cocktail of added salts packs on pounds: seawater weighs about 3 per cent more than fresh-water.

All our salt ultimately comes from the water of the sea. Rock salt, found in massive underground deposits worldwide, is the remains of ancient seas, evaporated hundreds of millions of years ago. Salt mines – a name now synonymous with slave-like working conditions – have been in operation for thousands of years. The oldest known, the salt caverns of Hallstein and Hallstatt in Austria, date to the Neolithic Age; these are located near modern-day Salzburg, which name, appropriately, means "Salt City." Today, collectively, the world's salt mines and salt works produce over 200 million tons of salt each year. Salt is much in demand by the chemical industry (which uses it to make everything from glass and soap to paper, plastic, and

antifreeze) and by highway departments (for sprinkling upon
icy roads). Only about 5 per cent of the annual salt crop – a
small but essential fraction – makes it into the human diet.

Table salt – which is what most of us insouciantly define as
"salt" – is almost entirely sodium chloride. To a chemist, "salt"
is a far broader term, referring not only to sodium chloride
but to a raft of other ionic compounds: bromides, iodides,
sulfates, sulfides, phosphates, nitrates, acetates, carbon-
ates, and permanganates. From a self-centered viewpoint,
however, sodium chloride is the Queen of Salts: after all, it
helps keep us alive. The average adult body contains about
90 grams of sodium, all essential for the functioning of nerves
and muscles, the maintenance of proper fluid balance, and
nutrient transport. Without our internal 18 teaspoonfuls of the
sea's salt, we're wrecks. Salt is continually lost through urine,
tears, and sweat and must therefore be constantly replenished.
Athletes and active workers lose 15 grams of salt or more daily
through sweat.

We *need* salt, which is why seawater's defining chemical
has played such a prominent part in human history. Wars
have been fought over it; taxes levied upon it; and the ancient
Chinese used coins made out of it. In ancient Rome, soldiers
were paid in salt – hence *salarium* or "salt money," the source
of our word salary; and the pejorative "not worth his salt,"
which describes the sort of overpaid slacker who spends the
bulk of the working day loitering around the water cooler or
making personal phone calls. The sharing of salt symbolizes a
bond of friendship in the Middle East; and the Bible refers to
"covenants of salt," which were particularly binding vows or
commitments. People referred to as "the salt of the earth" – the
Bible again – are those of particular worth or value. Many of
the earliest caravan routes were salt – not silk – roads. Medieval
Venice rose to power through its domination of the Mediterranean salt trade; and New York's Erie Canal was built in large
part to provide the northeastern United States with salt, from
the salt works of Syracuse.

Generally we like salt, too: one-quarter of our taste buds

are devoted to things that taste salty; and a sprinkle of salt adds savor to otherwise bland foods and enhances the flavor of other seasonings. Salt, according to Pliny the Elder, was even an antidote to poison – the modern expression "to take with a grain of salt" refers to the ancient belief that a bit of salt makes the toxic safely palatable. (It doesn't.) A grain, however, in modern medical estimation, is about all the salt we really need to take. Present-day salt consumption, according to both the European Commission's Scientific Committee for Food and the U.S. National Research Council, is outrageously high: the average American gulps down about 6 grams of salt daily; the average Briton, 8–10 grams. (About 20 per cent of this is so-called "discretionary salt" – we load it on for ourselves; the remaining 80 per cent is either a natural component of food or is added during cooking or processing.) Though our word salt comes from Salus, the Roman god of blooming health, high salt consumption can play havoc with the average body, promoting or exacerbating such evils as hypertension, gastric cancer, and osteoporosis. Most adults, physicians admonish, don't need to consume much in the way of salt: essential sodium levels can usually be nicely maintained with a salt intake of as little as 0.07–0.5 grams of salt per day, which last is about the salt content of a medium-sized handful of potato chips.

Salt is the reason we can't survive by drinking seawater: our kidneys simply aren't designed to cope with a steady diet of NaCl. The crucial process here is osmosis – from the Greek "to push" – which refers to the movement of water across semi-permeable membranes. Osmosis explains why bartenders scatter pretzels and salted peanuts around with such a generous hand: the salt makes consumers thirsty. Thirst begins at the cellular level, with the movement of water from areas with low concentrations of solutes toward areas with high concentrations, in a desperate attempt to establish equilibrium. The influx of salt that accompanies the consumption of pretzels results in a rush of water out of the body's cells, heading across cell membranes in the direction of the salt; cellular water depletion demands replenishment, which results in a call for another beer.

Extreme salt consumption can have dire consequences for the kidneys, whose reason for being is to filter toxins, waste products, and other undesirables out of the blood and eliminate them from the body via urine. Osmotically challenged kidneys are compelled to pour out immense amounts of water in an attempt to equalize salt concentrations. Eventually kidney cells – literally drained of water – shrink and shrivel; and the kidneys, effectively pickled, cease to function. Marine plants and animals, who have to deal with this threat on a daily basis, have evolved a range of physiological strategies to fight the forces of salt. Sharks, for example, possess specialized rectal glands that concentrate and excrete salt; and marine teleosts have salt-excreting chloride cells in their gills. Sea birds – such as pelicans, gulls, and the Ancient Mariner's fatal albatross – have salt glands with which they concentrate and eliminate salt through their nostrils; and the emotionally spurious tears of the saltwater crocodile are wept to eliminate excess salt. Without such salt-defeating mechanisms, creatures who live in or on seawater couldn't; freshwater fish, malevolently transposed to a marine environment, die of dehydration. The ability of salt to yank water out of cells also explains salt's preservative properties – contaminating microorganisms, inundated with it, desiccate – and its helpful role in mummification. The ancient Egyptians prepared bodies for the afterlife by first drying them for a month or more in *natron*, a mix of salts gathered from deposits along the shores of the Nile.

Enclosed seas – especially those having little freshwater input – are generally saltier than the ocean proper. While the average salinity of the ocean is 35 parts per thousand, the sunny Mediterranean Sea hovers around 38 parts per thousand and the Red Sea 42 parts per thousand. Saltiest of all are the salt lakes. The Dead Sea, 1300 feet (400 m) below sea level and thus the lowest body of water on Earth, is also the saltiest, about ten times more saline than the open ocean. The salt concentration, in fact, is so high that nothing whatsoever can survive in it;

hence its depressing but descriptive name. Utah's Great Salt Lake, home to a lush population of microscopic brine shrimp, is about five times saltier than the ocean, as is Tanzania's Lake Natron, home to spectacular flocks of shrimp-pink flamingos. At the opposite end of the scale, oceanic salt can be diluted by inundations of freshwater from rainfall and rivers. Salinity in the Indian Ocean dips seasonally with the onslaught of the monsoon; and seas and coasts with major river outlets are consistently low in salt. The river-drenched Black Sea, for example, has an average salinity of 16 parts per thousand; the Baltic Sea, 10 parts per thousand. Strictly speaking, such seas are not saline at all, but brackish, the term applied by oceanographers to water with saline concentrations between 0.5 and 17 parts per thousand.

Even freshwater is not totally salt-free. "Fresh" water contains minimal, but measurable, salt, in concentrations under 0.5 parts per thousand. This salt – primarily the result of weathering, as rainwater slowly dissolves the minerals in rocks – eventually ends up in the ocean, as do the dissolved salts spewed from undersea hydrothermal vents and submarine volcanoes. The (average) salinity of the ocean, scientists believe, has stayed reasonably constant for the last two billion years. Salt in the Earth's crust and salt in the world oceans have reached a comfortable equilibrium: new salt arrives at about the same rate that old salt departs, packaged into sea floor sediments and removed from solution.

The salt lakes, on the other hand, are growing steadily saltier. These bodies of water are slowly vanishing puddles, inexorably drying from the edges. Salt lakes are terminal lakes – that is, they're the end of the road as far as water goes; rivers may flow into them, but nothing flows out, since such lakes lack surface outlets. Utah's Great Salt Lake was once far greater; the modern body of water is a pitiful remnant of ancient Lake Bonneville, a massive freshwater lake that covered 20,000 square miles (52,000 sq km) of Utah, Nevada, and southern Idaho about 30,000 years ago. Now reduced to one-tenth its original size, the dwindling lake increasingly concentrates its salt content. The Dead Sea, trapped in an overheated basin on the border between Israel and Jordan, is shrinking at a rate of nearly 12 inches (30 cm) per year, and has long since become as salty as water can get. Dead Sea water is a saturated solution of salt: that is, not so much as a pinch more can possibly dissolve in it. As the salt concentration increases with continued evaporation, the salt simply crystallizes and sinks to the bottom and rims the shore in thick salty sediments. The effect, for those of a romantic turn of mind, is much like that of a magic salt mill, endlessly grinding at the bottom of the sea.

In the 1940s, author Holling C. Holling wrote a book about a small boy in Michigan who carves a wooden Indian in a canoe and drops it into the cold water of Lake Superior. The canoe has a message whittled on its side: "Put me back in water. I am Paddle-to-the-Sea." Young readers ever since have followed the carved Indian's storybook journey: through the Great Lakes, down the St Lawrence River, and into the Atlantic Ocean. Eventually, carried by the Gulf Stream and the North Atlantic Current, the little Indian – slightly battered, but unbowed – reaches the Grand Banks of Newfoundland, hitches a ride on a fishing ship, and comes to rest on the coast of France.

Ocean water, luckily for us, doesn't stay still. The moving thoroughfare of water that carried Paddle on the latter half of his long journey is the essential current that keeps Europe warm. Rome, city of Mediterranean sun and pigeons, shares a latitude with Chicago, the wintry "Windy City;" Paris is in line with frigid Quebec; and London with even colder Newfoundland. But for the warming insulation of the circulating ocean, the well-equipped Londoner would sport, in lieu of the traditional umbrella, mukluks, mittens, and a ski mask.

The ocean currents, diagrammed on the world map, take the form of great revolving circles that swirl around the edges of the continents, as if the waters of the globe were a witch's cauldron, stirred by a series of immense invisible spoons. These huge circulating loops of water – "rivers in the sea" – are called gyres, from the Greek *gyrus*, meaning ring or circle. Most of the stirring in the upper 300 feet (90 m) or so of ocean water is effected by wind, with a little help from planetary rotation. The Earth's 1000-mph (1600-km/h) west-to-east spin accentuates the circulation of the wind-driven gyres through an inertial force called the Coriolis effect, a phenomenon first described by French physicist Gustave-Gaspard de Coriolis in 1835. In practice, the Coriolis effect means that anything attempting to go in a straight line on top of something that's already moving in a circle isn't going to end up quite where expected. Imagine, for example, walking across the moving platform of a merry-go-round, from wooden horse to outer edge, aiming for the exit gate on the blessedly immobile ground outside. To stay in line with the gate, you're compelled to lurch dizzily sideways to compensate for the rotatory motion of the ride. If the merry-go-round is spinning clockwise, you'll have to deflect your path to the left to keep the stationary gate in view; if the ride is moving counterclockwise, you'll have to angle your path to the right. Similarly, on the interminably whirling merry-go-round of the Earth's surface, moving objects are nudged from the straight and narrow by the Coriolis effect: a current heading northward, from equator toward North Pole, will be deflected to the east; while a current running from the North Pole south will be channeled to the west.

Winds, named for their pasts, are called after the direction from which they come; water, which looks to the future, is named for where it's going. A north wind howls out of the north, causing the weathervane to point south; if you're stubbornly facing into it, you're staring toward the North Pole. A northerly current, on the other hand, flows out of the south and heads northward. The (northerly) Kuroshio Current, which speeds northward up the east coast of Asia, skimming the islands of Japan, is the world's fastest current – up to ten times faster than anything else going, moving at an average rate of 3 feet (1 m) per second; the currents of the Indian Ocean are the most schizophrenic, reversing themselves annually and running backwards during the months of the monsoon. The world's only circumglobal current is native to the southern hemisphere: the Antarctic Circumpolar Current, or the West Wind Drift, which – like a cat chasing its tail – continually whips around the continent of Antarctica. The best-known of the world's ocean currents, however, is the Kuroshio's Atlantic counterpart, the Gulf Stream, which originates in the bathtub-like waters of the Caribbean and travels up the eastern coast of North America, dispensing tropical warmth as it goes. It runs fastest as it passes North Carolina, thus contributing to the shipwreck-strewn reputations of Carolina's Cape Hatteras and Cape Fear; then heads east past the Grand Banks of Newfoundland. There it subdivides, splitting into the northerly North Atlantic Current, which keeps western Europe nine to eighteen degrees warmer in winter than anything at that latitude has any right to be, and the Azores or Canary Current, which turns south past the Canary Islands and the jutting Horn of Africa. From there, propelled by the prevailing winds and shoved by the Coriolis force, it pours into the North Equatorial Current, heading west once more toward the turquoise warmth of the Caribbean Sea.

While the cold water of the Atlantic is predominately gray or grayish-green, the Gulf Stream is blue. Descriptions of it are found in the logbooks of early whaling ships, noting that whales are never found in the blue water, preferring instead the temperature and food supplies of the green. Even earlier, Columbus remarked upon it ("vehement and furious") and Ponce de Leon, sailing north from Puerto Rico in 1513, lost a ship to it ("carried away by the current and lost from sight although it was a clear day").

The first scientific chart of the Gulf Stream was compiled in 1769 by Benjamin Franklin and was an attempt to facilitate the delivery of the mail. Franklin, then Postmaster General of the American colonies, had received a complaint from local businessmen that British mail packets traveling from Falmouth to New York took two weeks longer to cross the Atlantic than did the London-to-Rhode-Island merchant ships, which took a more southerly route. Franklin consulted New England sea captains – all familiar with the problem – who explained the nature of the easterly current in the North Atlantic, a band of swift-running water that provided an extra burst of speed on voyages east, but was best avoided on the westward trip home. Based on the sailors' information, Franklin drew up a detailed "Chart of the Gulf Stream," duly engraved in Boston and ornamented with a portrait of a bare-chested Neptune with crown and trident. It was ignored by its intended beneficiaries: the British, according to one disgruntled Nantucket skipper, "were too wise to be councelled by simple American fishermen" and persisted in sailing head-on against the current, with attendant time-wasting and delay. Franklin never pressed the point; and his chart was subsequently pulled out of circulation and suppressed during the years of the Revolutionary War, when it was to the Americans' advantage for the British to spend as long a time as possible in transit.

Seawater not only circulates; it sinks. Surface currents churn only the upper 300 feet (90 m) or so of ocean water; subsurface currents perambulate through the depths, ½–3 miles (1–5 km) down. This hidden underwater flow is the tie that binds the Seven Seas into a single unified whole. It's also the principal means by which heat is distributed and redistributed about the globe. Water warms us through the ocean's deep currents, the hot-water pipes in the radiator of the world.

The ocean is layered like a cake, and a crucial difference among layers – as in the aforementioned lakes – is density. Cold water is denser than warm water; salty water is denser than not-quite-so-salty water; and realignments between the two drive what is known as the thermohaline circulation – that is, water movements governed by temperature and salinity. Warm water tooling northward in the Atlantic, for example, not only cools as it reaches the higher latitudes, but it becomes saltier, as evaporation – helped along by steady winds from

Canada – removes fresh water from the surface. Somewhere around Labrador, this frigid, salty, and altogether leaden combination becomes too heavy to float, and plummets to the bottom to form the North Atlantic Deep Water, simultaneously boosting the underlying layer of relatively warm water to the surface.

Salt-sinking was first proposed as a mechanism of ocean-warming in 1797 by Benjamin Thompson, a physicist from Woburn, Massachusetts, whose Loyalist sympathies compelled him to leave America abruptly in the year of the provocative Declaration of Independence. He ended up in Bavaria, where he acquired the title of Count Rumford, and eventually settled in Munich, where he conducted his reputation-establishing experiments, most centering around the nature of heat. Ultra-salty water sinks in the north, Rumford explained, and then heads south, crawling along the ocean floor. This cold flow is then balanced by a compensatory warm current heading north.

This proposal proved essentially to be the case. Salt sinks in the subpolar Atlantic, notably in the region of the Norwegian Sea, a submergence so substantial that it results in a prodi-gious warming of the northern ocean, a 30 per cent heat boost beyond that provided by the warming sun. This stupendous salt dunk – known as the Nordic Seas heat pump – shields northern Europe from the nasty winters it might otherwise have. Salt also sinks at the Strait of Gibraltar, where the extra-salty waters of the Mediterranean flow into the Atlantic. Just past the Pillars of Hercules, Mediterranean surface water meets the less dense water of the ocean and plunges heavily to the bottom in a submarine waterfall of salt. The most impressive salt sink of all may be that of Antarctica's Weddell Sea, where winter sea ice formation pulls fresh water out of circulation, leaving behind the salt. The salt-rich leftovers promptly sink like a stone, creating the Antarctic Bottom Water, the densest seawater in the world.

Salt sinking is an Atlantic, rather than Pacific, phenom-enon. The Pacific, over twice as big as the Atlantic (64,190,000 square miles (166,300,000 sq km), as opposed to the Atlantic's 31,810,000 (82,400,00 sq km), is less salty, since there's more water there to dilute the incoming salt. Atlantic salt today reaches the Pacific via a long loop around the horn of Africa. Three to four million years ago it had an alternative escape hatch – straight through the isthmus of Panama, which was then open water. When the pass clicked slowly shut with the convergence of the North and South American tectonic plates, the intransigent 10-mile (16-km) strip of Panamanian jungle effectively re-ordered the globe. Atlantic water, balked of its accustomed outlet, swirled frustratedly around the Gulf of Mexico and headed north, thus joining the warm flow of the Caribbean Current to the Gulf Stream. Perversely, this enhanced warmer current may have provided the impetus for the last Ice Age, carrying enough moisture to the far north to generate mountainous falls of snow and ice. It also provided the path by which the South American armadillo headed north, eventually to populate the southwestern United States and become roadkill in Texas; threw up a peak for Balboa to stand upon for his first startled view of the Pacific Ocean; and kept Christopher Columbus from reaching China.

Salt sinks are not to be tampered with lightly. In 1961, oceanographer Henry Stommel – who just three years previously had published the first map of the global ocean's deep water circulation – pointed out that enormous dollops of freshwater have the potential to lethally disrupt salt sinks. Freshwater, unexpectedly released into the North Atlantic, can dilute the circulating seawater, decreasing salt concentration and density to the point where nothing sinks at all. Just such a catastrophic scenario may be the result of global warming. Freshwater floods of meltwater from Greenland's warming glaciers pouring into the north Atlantic could disrupt the great global conveyor belt that – among other things – keeps the temperate zones temperate. Scientists with a weather eye on the oceans now speak of climate flip-flop: in this case, global warming with attendant ice melt could trigger a disruption in ocean heat transfer, abruptly nudging us into a new Ice Age.

The world's water keeps us warm. It might also conceivably freeze us to death.

"Rain is grace," wrote John Updike in his memoirs in 1989. Rain is sustenance for the thirsty earth. "April showers bring May flowers" is a platitude that dates at least to 1560, when it was delivered as "Aprell sylver showers so sweet/Can make May flowers to sprynge." The *Book of Common Prayer* gives thanks for rain; and since 1894, the phrase "right as rain" has meant absolutely splendid. Anything beneficent is commonly said to fall like rain (mercy, blessings, pennies from heaven). When Sir John Falstaff in *The Merry Wives of Windsor* shouted, "Let the sky rain potatoes!," he was hoping for sweet potatoes, believed in Shakespeare's day to be aphrodisiacs.

Rain is water's middle child. It constitutes the intermediate step in the endless global recycling of water, in which water evaporates from the Earth's surface, accumulates in the atmosphere, and falls as rain (hail, sleet, or snow), only to evaporate again. In general, about 250 cubic miles (1040 cubic km) of water evaporate from the planet's surface daily, 210 cubic miles' (875 cubic kms') worth from the oceans, and the rest from inland water: lakes, rivers, streams, brooks, ponds, puddles, and backyard swimming pools. Also included is water transpired by plants, which adds up to substantial amounts. A field of corn, for example, transpires 3000–4000 gallons (11,400–15,000 liters) of water each day; a single oak tree, 200 to 300 gallons (750–1100 liters). Earth's thrifty and continual re-use of water ensures that the molecules contained in any raindrops falling on your head are likely to have been around for a long time. Chances are that in the distant past they fell – repeatedly – upon the dinosaurs, the mastodons, and the Neanderthals, and will be falling still in the equally distant future, when we are long gone. Traditional dogma holds that the Earth's water is essentially a closed system: that is, we've still got almost every drop we started with and we never use it up.

The idea of the water cycle, however, is comparatively new. Plato claimed that all surface water came from caverns beneath the sea, from which ocean water filtered upwards through the Earth, divesting itself of salt as it went; and this belief, or slight variations upon it, persisted for the next 2000 years. Leonardo da Vinci took a jab at the problem in the late 1400s, suggesting in his notebooks that water, rather than burbling up from a mysterious hole in the middle of the Earth, continually circulated between seas and rivers, with clouds bridging gaps in between. His proposal was deemed outlandish and largely ignored and Leonardo, who had plenty of other projects on his plate, neglected to pursue it. The description of the hydrological cycle, as we now know it, was thus left to the flamboyant sixteenth-century Italian mathematician Girolamo (also known as Hieronymus, Geronimo, or Jerome) Cardano, a brilliant but somewhat erratic character whose scientific studies were periodically interrupted by prison terms. (He was repeatedly indicted for nonpayment of gambling debts and for heresy, once serving time for impiously casting the horoscope of Jesus.) Between incarcerations, Cardano wrote a seminal treatise on the mathematics of probability theory, invented the combination lock and the universal joint, cured the Archbishop of Scotland of asthma, and devised (or stole) the solution to cubic equations, one or the other so effectively that the crucial technique is known to this day as "Cardano's Rule." His water cycle hypothesis, published in 1550, seems to have been largely original and hit all the high points: water, according to Cardano, evaporated from the seas, accumulated in the atmosphere as clouds, fell to Earth as rain, and then was carried by rivers to the sea to repeat the process all over again.

The sticking point in the acceptance of the process was the presumed inadequacy of rain. There simply wasn't enough of it, critics argued, certainly not enough to account for all the water in lakes, streams, and rivers. Subsequent calculations, however, showed definitively that indeed there was: studies instigated in the mid-1600s by French scientists Pierre Perrault and Edme Mariotte demonstrated that the amount of rain falling upon the Seine drainage basin pretty much equaled the amount of water dumped by the river into the English Channel. The Church, which balked at Arabic (infidel) numerals, the heliocentric solar system, and the theory of evolution, found the hydrological cycle quite in line with Christian doctrine and, as an exemplar of divine order, the concept passed smoothly into the popular domain.

"Man," wrote some anonymous but insightful soul, " – despite his artistic pretensions, his sophistication, and his many accomplishments – owes his existence to a six-inch layer of topsoil and the fact that it rains." It's a perverse reflection on the human condition that we tend to appreciate neither dirt nor drizzle. Many rain homilies are foolishly fraught with resentment and bitterness. Henry Wadsworth Longfellow's "The Rainy Day," penned in 1842, gave us "Into each life some rain must fall" which is clearly the nineteenth-century version of "Shit happens." "It never rains but it pours" indicates that troubles inevitably arrive en masse; and everybody knows what kind of wet blankets roam around raining on people's parades. Most modern musicians see rain as a metaphor for general drear, though there are exceptions: Gene Kelly's "Singin' in the Rain," for example, is positively gleeful and country singer Eddie Rabbitt loves a rainy night. The rest of us usually just want to get rid of it. "Rain, rain, go away/Come again another day" is a ditty of ancient lineage. The seventeenth-century British, still annoyed about the Armada, knew it as "Raine, raine, goe to Spaine."

I am fond of rain. It's a restful weather, suitable for lying on the sofa with a book, free of any guilty compulsions – after all, it's raining – to be out and about, doing something effortful and useful. Too much rain, on the other hand, can be deleterious to the mental health. Repetitive dark gray days can edge the susceptible into seasonal affective disorder (SAD), a syndrome commonly known as "winter blues." Symptoms, believed to be induced by a hormonal response to chronic low-light levels, include lethargy, fatigue, depression, insomnia, plummeting sex drive, and either appetite loss or, in the particularly unlucky, an insatiable craving for carbohydrates. About 20 per cent of the population suffers from mild SAD, primarily during the dark cold days of December, January, and February; and in 1–3 per cent, the condition can be brought on by any period of meteorological gloom and is severe. The recommended treatment is bright light, as in a lengthy vacation in sunny Bermuda.

Despite the risks of lethargy and doleful moods, humans, traditionally, have put more effort into making the rain arrive than convincing it to go away. For farmers and agrarian societies, rain is a life-or-death affair. The tribes of the Old Testament (a desert people, struggling to raise goats) deemed rainfall a sign of divine providence. Drought, by extrapolation, was thus a sign of divine displeasure, ill-temper, or at very best neglect, to be dealt with through prayer, sacrifice, and vows of future good behavior. Cultures worldwide have devised rituals intended to call down rain. The Diaguita people of Chile, who occupy one of the driest regions on Earth, traditionally made rainsticks from dried cactus stems which were then filled with pebbles: the rattle of the tumbling pebbles as the sticks were upended sounds like a rainstorm and was intended to get the rain spirits' attention, reminding them to deliver. The American Indian tribes of the dry southwest and the Australian aborigines performed elaborate rain dances, which, in the former case, involved the handling of live rattlesnakes. Rain charms in Europe ranged from sprinkling stones with water and killing frogs to pouring buckets of water over a saint's image, the village priest, or a passing stranger. Church-sanctioned rain rituals centered around prayer and penance rather than wet priests, though when conventional approaches failed, disappointed congregations occasionally rebelled. Frazer's *Golden Bough* describes a drought in Italy in 1893 during which outraged peasants, having been pointedly ignored by heaven, turned saints' statues to the wall, threw holy images into duck ponds, and tore the gold wings off the Archangel Michael. Chinese villagers, hoping for rain, constructed huge paper dragons to represent the rain god and carried them ceremonially through the streets; if no rain followed, the dragon was shouted at, spat upon, kicked, and ripped to pieces.

Rainmaking techniques today commonly involve sprinkling salt from airplanes. This effectively "seeds" clouds with dispersed particles which act as condensation nuclei, attracting atmospheric water molecules and encouraging the formation of raindrops. Modern rainmaking research, however, dates to the mid-nineteenth century – though the impetus behind it may date to the first century CE when Plutarch, in his *Lives*, claimed that extraordinary rains commonly follow great battles. Rainstorms, observers of the 1860s insisted, also frequently followed Civil War artillery bombardments. The conclusion was that explosions cause rain; and ensuing decades featured a host of heavily armed rainmakers who detonated cannons, rockets, dynamite, and explosive balloons into the upper atmosphere. Though results were at best mediocre, practitioners continued hopeful. Nebraskans, beset by drought in 1894, formed the Rain God Association and set off gunpowder at measured intervals across half the state; the northeast stayed stubbornly dry, but the southwest (rumored to have superior gunpowder) got a little rain.

Most famous of the turn-of-the-century rainmakers was almost certainly Charles Hatfield – popularly known as *the* Rainmaker with a capital R, though he modestly referred to himself as merely the Moisture Accelerator. Hatfield was born in Fort Scott, Kansas, in 1875, and embarked on his rainmaking career in 1905, using in lieu of explosives, boiling vats of "secret chemicals" set on towers. He initially had an impressive run of successes, which inflated his reputation out of all proportion, despite the protests of the US Weather Bureau, who proclaimed to anyone who would listen that Hatfield was a fraud. Few did; rain, after all, was rain; and Hatfield and his chemical tubs, in multitudinous instances, had demonstrably been there prior to and during it. In December 1915, desperate citizens of San Diego hired Hatfield to fill their nearly empty city reservoir, agreeing to pay a fee of $10,000, an astronomical amount at a time when the average yearly salary was $750. Hatfield set up his towers and tubs in January; and the outcome was worthy of "The Sorcerer's Apprentice." Rain thundered down for a week, first closing the Panama–California Exposition in Balboa Park, then flooding the railroads (passengers had to be rescued by launch), inundating fields and farmland, washing out bridges and houses, and ultimately rupturing San Diego's Sweetwater Dam. San Diego sued Hatfield for damages (claiming too much rain, too soon); Hatfield sued San Diego for his money (claiming that it wasn't his fault that the city had failed to take proper precautions). Nobody in the end paid anybody. The courts attributed the deluge to an act of God; and Hatfield gave up the rainmaking business, moved to Glendale, and became a sewing-machine salesman. His secret rainmaking formula died with him in 1958.

As Hatfield's experience proved, too much of a good thing is not necessarily wonderful. Forty days and forty nights of rain, according to the Old Testament, were enough to flood the known world, drowning everything in it except Noah, his family, and an Ark full of paired animals. Flood legends crop up in mythologies worldwide, from China to Brazil. In the 3000-year-old Mesopotamian *Epic of Gilgamesh*, King Utnapishtim is warned of a great flood by the god Ea, who directs him to build a distressingly awkward seven-story cube-shaped ship on which to load his family, servants, and animals. Once all were on board, the gods sent seven days and nights of ferocious rain which covered the world, leaving only the distant peak of a mountain poking up above the water.

The Noah and Utnapishtim floods, taken as described, are what meteorologists call "traditional floods" – that is, floods brought on by days of heavy rains or melting snows, causing lakes and rivers to overflow their banks. Such floods are reasonably easy to predict: given the amount of rain, observers can usually make a good guess as to how much and when the creek is going to rise. The annual flooding of the Nile, upon which ancient Egyptian civilization depended, could be counted upon to take place in August; the pharaoh generally got credit for it, though the seasonal inundations were actually run-off from the mountains of Uganda and Ethiopia.

Most floods are more destructive and less welcome. The Great Flood of the Mississippi River in 1993 covered an area twice the size of Massachusetts, swamping fifty-six towns and costing over $15 billion in property and crop damage. Periodic flooding of China's Huang He (Yellow) and Yangtze Rivers – triggered by a combination of heavy rains and snow melt from the Himalayas – has been historically devastating. In 1887, the Yellow River flooded 50,000 square miles (130,000 sq km) of countryside and drowned an estimated six million people; the Central China Flood of 1931 drowned 140,000 and submerged 70,000 square miles (180,000 sq km) – an area equal to that of New York, New Jersey, and Connecticut combined. Over the centuries a complex array of levees, waterways, and dikes were constructed in attempts to thwart the floods; and in 1778 – a doomed last-ditch effort – the Emperor Qianlong attempted to hold back the rising water by sinking nine life-sized iron oxen in the Yangtze. To date, however, no human tinkering has been sufficient to subdue China's rivers. In 1994, still determined, the People's Republic embarked upon the largest water control project in the world – the Three Gorges Dam on the Yangtze. The dam as planned is a structure of mind-boggling immensity: 600 feet (180 m) high and over a mile (1.6 km) long – five times longer than the Colorado River's massive Hoover Dam – containing 26 million tons of concrete and a quarter of a million tons of steel. The river, the still undefeated champion, drains an area of 700,000 square miles (1,800,000 sq km) and extends 3915 miles (6264 km), from the Plateau of Tibet to the East China Sea.

Noah's Flood, according to recent archaeological evidence, may have been less traditional than previously assumed. It may, in fact, have been a flash flood, caused by the sudden collapse of the natural dam at the junction of the Mediterranean and the Black Seas. In 2000, Robert Ballard, famed for his discoveries of the sunken *Titanic* and *Bismarck*, spotted the remains of a Stone Age settlement 12 miles (20 km) off the Turkish coast, 300 feet (90 m) beneath the surface of the Black Sea. The find reinforces the theories of geologists Walter Pitman and William Ryan, who propose that stories of a Great Flood may date back 7000 years, when a massive influx of melting glacial water caused the Mediterranean to smash through the narrow barrier separating it from the Black Sea – then a smaller body of freshwater known to archaeologists as Euxine Lake. The break created a monumental waterfall many times the size of modern Niagara which carved out the Bosphorus, enlarged Euxine Lake's boundaries by as much as a mile (1.6 km) a day, and filled it with salt, thus creating the Black Sea. Tales of this cataclysm, passed down from generation to generation, may have slowly transmogrified into flood mythology, a fearsome pan-cultural story of loss, the still-living memory of a time when the known world ended in an awesome torrent of water.

I've known rivers; I've known rivers ancient as the world and older than the flow of human blood in human veins. My soul has grown deep like the rivers.

Langston Hughes

The world's first civilizations all grew up along the banks of rivers. Mesopotamia, the land between the rivers, was born beside the waters of the Tigris and Euphrates, Egypt on the banks of the Nile, China along the Yangtze and Huang He, and the Harrapan culture beside the Indus in India and Pakistan. The Persian word for water, *ab*, is the root of the word abode. The connection is obvious: home, since time immemorial, has always been where the water is.

Rivers are boundaries. Running water delineates our limits, once separating tribe from tribe, now dividing France from

Germany, the United States from Mexico, Bolivia from Brazil. The river also represents the divide in rites of passage: we metaphorically cross rivers in passing from child- to adulthood, or from life to death. In Greek mythology, souls entered the Underworld by crossing the River Styx, a stream that reputedly dissolved all materials dunked in it except horses' hooves; and Christians sing of crossing the River Jordan to paradise. The boundary can also be a barrier; in the absence of boats or bridges, a river can be impassable. Sometimes they're impassable anyway: folk tradition holds that demons and witches are incapable of crossing running water, as in Robert Burns's poem "Tam O'Shanter," in which Tam escapes his fiendish pursuers in the nick of time by dashing across a river bridge.

Running water, sliding swiftly past us, never to return, is a powerful image of time. "Time is a sort of river of passing events, and strong is its current; no sooner is a thing brought to sight than it is swept by and another takes its place, and this too will be swept away," wrote Marcus Aurelius in his *Meditations*; and Thoreau placidly referred to time as the stream he went a-fishing in. Rivers are magnets for the imagination, writes Tim Palmer; and rivers are lures to adventure, highways and wilderness paths, homes to bass and trout, builders of deltas and carvers of canyons. Rivers are the blue arteries of the Earth, the chords, Barry Lopez writes, that bind the Earth together. The river has a life of its own, running gloriously through Kenneth Grahame's *The Wind in the Willows*: a "sleek, sinuous, full-bodied animal . . . all a-shake and a-shiver – glints and gleams and sparkles, rustle and swirl, chatter and bubble." The river is T. S. Eliot's strong brown god.

In Greek mythology, the world's rivers were all the children of Oceanus and his wife Tethys; in Chinese mythology, rivers were born from the blood and tears of the god P'an Ku; and according to the myths of the Australian aborigines, rivers came from the urine of the ancestral Rainbow Serpent. The truth is almost as peculiar: most rivers, traced to the source, simply bubble out of a hole in the ground. Not all: some, among them the Nile, the St Lawrence, and the Mississippi, originate in lakes, and others, including the Amazon, the Ganges, and the Rhine, are born in glacial meltwater. Most, however, come from springs fed by groundwater, emerging from the earth as unexpectedly as rabbits popping out of hats, geological feats of sleight-of-hand.

Whatever their source, rivers – encouraged by gravity – inevitably run downhill. An Inuit legend holds that this was not always the case: long ago at the beginning of the world, the story goes, the rivers ran uphill as well as down. However, this comfortable two-lane highway version of rivers made life so easy for people that Raven, the curmudgeonly creator of the world, decided to do away with it – thus nowadays those who are up a creek without a paddle have no choice but to go downstream with the flow.

The strength of that flow depends on a number of factors, among them the gradient of the river channel, the volume of water in the river, and the temperature of the water. The steeper the slope of the hill, the faster the river runs down it. Every river current thus has a mountain, hill, or incline somewhere behind it; and rivers flow in one direction – even when crawling across country as flat as a pancake – because they're propelled from a distance by water headed *down*.

Generally, the greater the amount of water in a river, the faster it flows – which means that as a river picks up tributaries or accumulates water from ground seepage or rain, it also picks up speed. Small rivers travel at an average of 3 miles per hour (5 km/h) – brisk walking speed; large rivers, about twice as fast. The deepest water in any given river moves the fastest: this is generally the water in the very center of the channel, where the effect of friction – which slows the current along the banks – is the least. Furthermore, for greatest speed, the river should be warm. Warm water is less viscous than cold water: that is, as temperature increases, hydrogen bonds between water molecules become weaker and less dependable, and water loses some of its internal cohesiveness, becoming in effect more watery. Warmer rivers thus move faster, picking up a 0.5 per cent increase in velocity for every degree increase in temperature between 39 and 68°F (4 and 20°C). All in all, Huckleberry Finn had the best of the river world, rafting downstream in the middle of the large, warm Mississippi.

The Mississippi is one of the world's great rivers, cutting a swathe 3710 miles (5936 km) long through the center of North America to empty into the Gulf of Mexico. It is superseded in length only by the Nile (4180 miles/6688 km) and the Amazon (3912 miles/6259 km), and is closely followed by the Yangtze (3602 miles/5763 km). The longest river in Europe is the 2300-mile (3680-km)-long Volga – the "Mother Volga" of Russian folklore; second longest is the legendarily blue Danube, 1700 miles (2720 km) long from Black Forest to Black Sea. We all tend to know the world's broadest and biggest, but sheer size is not the only road to fame: many comparatively insignificant rivers, by all rights riparian nonentities, have historical associations that have made them household names. Caesar's Rubicon, for example – now Italy's Rubicone River – is less than 100 miles (160 km) long; the *Encyclopedia Britannica* grudgingly refers to it as a "small stream." It was this river that Julius Caesar crossed from Cisalpine Gaul into Rome proper in 49 BCE, thus initiating the civil war that would leave him ruler of the Roman world. He is said, as he strode purposefully toward the water, to have growled "*Alea iacta est*" – "The die is cast" – thus immortalizing the heretofore obscure river; ever since "to cross the Rubicon" has meant to take a step that irrevocably commits the stepper to a dangerous undertaking.

While any flowing body of water, from the tiniest trickle to the mighty Mississippi, qualifies – according to the *OED* – as a stream, a term generously defined as a "course of water flowing continuously along a bed on the earth, forming a river, rivulet, or brook." The distinctions among rivers, rivulets, and brooks are fuzzy. The *OED* defines a brook as "a small stream or rivulet." Annoyingly, it says exactly the same thing about creeks; and if you read ahead a bit, you find that there's also a creeklet, defined as "a small creek," which is presumably the same as a brooklet or small brook, though the *OED* isn't quite brave enough to say so. All, in any case, are smaller than rivers. Scientists often classify streams by size: thus rivulets, lowest on the riparian totem pole, are smaller than brooks, which are smaller than creeks, which are smaller than rivers. Multitudinous rivulets, brooks, and creeks can combine to feed a single river – the Danube, for example, has over 300 tributaries; and such interconnected populations of streams are known as river systems. The component brooks, creeks, and rivers of a system usually flow together in predictable patterns, roughly governed by the surrounding terrain. Most river systems are dendritic: that is, their interlocking branches join to form multiple Ys, like the twigs and branches of a tree. Others are braided, forming a complex network of interwoven channels separated by islands, spits, and sandbars.

Braided systems generally form from rivers heavy in sediment, which brings us to a defining characteristic of rivers as a group: they carry stuff. Rivers are the pack animals of the water world; collectively they redistribute some 16 billion tons of sediment each year. Sediment can be anything from mud and leaf fragments to rocks bigger than bowling balls; when carted about by a river, such detritus has a name all its own: *alluvium*, from the Latin "to wash." The Yangtze and Huang He (Yellow) Rivers are so loaded with lemon-colored dust – loess – from China's central plains that observers compare their water to syrup; and North America's sediment-stuffed Mississippi is commonly known as the "Big Muddy." Even the clearest of rivers carries a burden of sediment; and it's this eclectic assortment of grit, pebbles, and twigs that scrapes interminably against riverbanks and bottoms, scouring out ditches, channels, gullies, gulches, riverbeds, river valleys, and – given five or six-odd million years – the Grand Canyon.

Though rivers run in one direction, they seldom do so in a straight line. Rivers are wafflers: they turn one way and then the other, looping back and forth in an apparently purposeless series of curves called meanders. The name comes from the Menderes River in Turkey, known in ancient times as the Maiandros – a river so convoluted that it was said to reverse itself and run backwards. An American equivalent of the tangled Maiandros appears in the tales of the larger-than-life lumberjack Paul Bunyan, who once tangled with the Powder River – "the orneriest, worst-behaved river in the world" – which twisted, corkscrewed, and doubled back upon itself to the point where travelers upon it often met themselves coming from the opposite direction. Paul straightened it out with the help of Babe, his fabled Blue Ox, and then challenged it to a wrestling match.

Applied to people, to meander means to lollygag or to wander about aimlessly; rivers, however, have a method to their curvaceous madness. Meanders are work-savers; they allow rivers to overcome obstacles with the least expenditure of energy. Any number of natural obstructions can throw a river out of kilter – boulders, bulges, depressions, and hummocks, for example, can all nudge running water out of alignment, causing it to deviate from the straight and narrow. Once a bend begins, erosion and sediment deposition take care of the rest, with a little help from centrifugal force. As the current moves around the nascent bend, the water – thrust outward by centrifugal force – moves faster along the outer edge of the curve, with accompanying increased friction and enhanced erosion of the outer bank. At the same time – think of horses racing around a circular track – water slows on the inner edge of the curve, and this decrease in flow rate causes the current to dump some of its load of sediment. Continued erosion on one side and deposition on the other causes the bend to curve further, looping back toward its original starting point and further, into a compensatory curve on the opposite side. Meanders reflect a river's attempt to maintain energetic equilibrium as it wriggles its way toward the sea.

To the Australian aborigines and North American Indian tribes, the Milky Way is a river in the sky. In Chinese legend, it is called the Silver River: it separates two lovers, represented by the stars Vega in the constellation Lyra and Altair in the constellation Aquila, who are allowed once each year in summer to cross the river and meet. The constellation Eridanus – longest of the eighty-eight recognized constellations, meandering across an impressive 60-degree stretch of sky – also represents a river. The Greeks alternately identified it as Ouroboros, the great world-circling river, or as the (smaller) river that flowed into the Euxine Sea at the point where Jason and his Argonauts found the Golden Fleece. The Egyptians claimed it was a stellar version of the Nile; the Chinese, the Huang He; and sixteenth- and seventeenth-century Christian astronomers, the Jordan. Phaethon is said to have drowned in it, after falling out of the stolen chariot of the Sun. Eridanus originates near Rigel in the constellation Orion and ends at a somewhat nondescript star named Achenar, or "River's End," near the celestial south pole. Eridanus is unique in this respect: earthly rivers never end. All are part of the planet's steadily turning water cycle, liquid links between sky and sea. Weary rivers don't end in the ocean. They use it to begin again.

"Water, like religion and ideology, has the power to move millions of people. Since the very birth of human civilization, people have moved to settle closer to it. People move when there is too little of it. People move when there is too much of it. People journey down it. People write, sing and dance about it. People fight over it. And all people, everywhere and every day, need it." The words are Mikhail Gorbachev's; the sentiment is universal. We need water.

We also use a lot of it. Today, in the United States, total water consumption amounts to about 341,000 million gallons (1.3 trillion liters) daily. Of this, domestic use accounts for about 3400 million gallons (12,900 million liters) – a piddling 1 per cent of the total; and industry laps up 19,300 million gallons (73,000 million liters) (6 per cent). The lion's share is used for irrigation or the generation of hydroelectric power – 266,000 million gallons (1 million liters) a day, fairly divided at 133,000 million gallons (500,000 liters) each. Worldwide, about 70 per cent of all water used by humans is used for agriculture. Food is gluttonous when it comes to water. It takes 1000 tons of water to produce a single ton of grain; 1500 tons to produce a ton of potatoes. Move a step up the food chain and water guzzling gets worse: it takes 15,000–70,000 tons of water to produce a ton of beef – enough to make 8000 hamburgers– and 3500–5700 for a ton of chicken.

Our planet is rich in water but, as we've seen, most of it is salty. By one calculation, the amount of freshwater available to humans is less than 0.1 per cent of all the water on Earth – that is, if all of the world's water were reduced to a size that would fit in a gallon jug, usable freshwater would amount to about half a teaspoon. Today over a billion people have no access to clean drinking water. The demands of our exploding global population – nearly six billion strong, and growing – are depleting even the vast stores of water sequestered in under-ground aquifers. We're taking out more than nature puts back in, and any bank manager can tell you where that leads. North America's Ogallala Aquifer is one of the world's largest – a vast subterranean lake covering 225,000 square miles (580,000 sq km) beneath the Great Plains, extending from Texas to South Dakota. The water drained from it to nourish one-fifth of America's irrigated cropland is fossil water, 25,000 years old. At the present rate of pumping, the Ogallala will be empty in another century and a half. India, China, north Africa, and the Arabian peninsula are all operating under chronic water deficits. Across the farmlands of northern China, the water table is dropping at a rate of 5 feet (1.5 m) a year.

I have some small experience of running dry. Before we drilled our well, our water here came from a natural spring, the output of which was stored in a concrete-bottomed spring-house located – due to some thoughtless nineteenth-century real-estate machinations – in the middle of our next-door neigh-bor's woods. From the springhouse, the water was pumped uphill to a pair of holding tanks, housed in a rickety shed along with a manic population of red squirrels, and then channeled through some 700 yards of century-old plumbing to emerge, in sluggish and exhausted fashion, in our kitchen, bathroom, garden hose, and washing machine. One year a chunk of this subterranean piping broke and emptied the entire springhouse into the trees. For the next five days, we were waterless; and our hitherto cavalier attitude toward water evaporated in less time than it takes to say "H_2O." If you don't see water as a blessing and a miracle, try – just try – doing without it for a bit. Some things – like true love and gasoline, writes poet Leroy V. Quintana – we only appreciate when they run out.

"When the well's dry," Ben Franklin wrote, "we know the worth of water."

Sometimes, if you stand on the bottom rail of a bridge and lean over to watch the river slipping slowly away beneath you, you will suddenly know everything there is to be known.

A. A. Milne
Winnie-the-Pooh

"Take almost any path you please, and ten to one it carries you down in a dale, and leaves you there by a pool in the stream," wrote Herman Melville, that grand old man of water, in 1851. "There is magic in it. Let the most absent-minded of men be plunged in his deepest reveries – stand that man on his legs, set his feet a-going, and he will infallibly lead you to water, if water there be in all that region. Should you ever be athirst in the great American desert, try this experiment, if your caravan happen to be supplied with a metaphysical professor. Yes, as everyone knows, meditation and water are wedded forever." Watching water, in rivers, pools, or ocean waves, mesmerizes us and calms even the most stressed-out spirit. There's a reason that a popular computer screensaver pictures a soothing melange of rippling waves and fish; and that Zen facilitators urge would-be meditators to contemplate bowls of water. Thoreau claimed that anyone who listened to the rippling of rivers could never utterly despair; and Anne Morrow Lindbergh's 1955 autobiography speaks poignantly of gifts from the sea: the healing power of water, and the ability to draw from it strength, patience, hope, faith, and inner peace.

Greek mythology tells the story of the gorgeous Narcissus, who fell in love with his own reflection in a pool. Eventually, enraptured by his own image, he pined away and died, to be transmogrified into a flower. Usually Narcissus's sad end is seen as a cautionary tale: too much self-love is fatal, so stop peacocking around staring at yourself in mirrors. Narcissistic Personality Disorder, named for that beautiful but misguided boy, is described in the American Psychiatric Association's *Diagnostic and Statistical Manual of Mental Disorders* as a malignant cluster of self-centered properties, characterized by grandiose fantasies and an excessive demand for adulation. The Narcissus story, however, is not all about Narcissus. It's about the revelatory nature of pools.

Still water is almost certainly the first medium through which we got a look at ourselves, when some ancestral hominid paused for a moment in bending over a drinking pool. It must have been a startling moment, the sudden recognition that the watcher on the bank and the watcher in the water were one and the same. Reflection is the only way in which we can see our own faces, look into our own eyes. Like water itself, however, reflection is a word with many levels. The *OED* gives it thirteen separate meanings, most telling of which is number twelve: to reflect is to turn one's thoughts upon, to meditate, or to ponder. It's a discipline best suited to water.

Water is the element of revelation. "Consider them both, the sea and the land," writes Melville, "and do you not find a strange analogy to something in yourself?" Water, still or moving, pulls mysteriously on the mind, moving us to contemplate, plumb our inner depths, examine our lives, dream. Even the most rational among us feels it. Perhaps it's the distant call of the sea in our bones and blood, an ancient resonance that lets us find meaning there. In water, we see ourselves not through a glass, darkly, but clearly and face to face. Hermann Hesse's Siddhartha gazed into a running river and saw his destiny. Narcissus found more than his face in that fatal pond.

Water shows us our secrets. We find ourselves in water.

Part III

Air

Inebriate of Air – am I...

Emily Dickinson

Air is a contradictory element. Dumpy, clay-footed creatures that we are, we equate the lightness of air with joy and freedom. Those who are in the blue sky or walking on air are bursting with delight; a light heart is carefree; and "free as the air," a condition that occurs briefly after the completion of final exams or divorce proceedings, is an expression of unfettered liberty. Air is also an element to be reckoned with. Air, moving, is wind, a universal symbol of irresistible force and uncontrollable power. "Madame, bear in mind," cautioned Victor Hugo, "that princes govern all things – except the wind." The winds of war flatten everything before them; the winds of change sweep the past away whether we want them to or not. "There is no good in arguing with the inevitable," wrote James Russell Lowell, adding that the only possible response to an argument with an east wind is to put on an overcoat.

On the other hand, although we can't change the wind, the wind itself is a perennial waffler. Sometimes it's on our side, but it can never be depended upon: winds of fame and fortune are notoriously fickle, prone to turn around suddenly and blow the other way. Still, a contrary wind can build character. To stand against the wind denotes independence and defiance, while to turn with the wind, though it can indicate a sensible adaptability, can also describe the sort of spineless stool pigeon who denounces his neighbors to the thought police.

Air fascinates us in part because it's invisible. Of the four elements, only air is nowhere to be seen, which is why, however unfairly, it is often used as a metaphor for absolutely nothing. An airhead, for example – such as those of us who have yet again forgotten a lunch appointment, the location of the car keys, or the due date of the mortgage payment – are accused of having nothing between the ears; and anyone who's running on air has a dead-empty gas tank and isn't going to get far. Things that vanish, like Shakespeare's sprites, are said to melt into thin air; and Ben Jonson's moan – "Alas, all the castles I have are built with air" – is a common plight of poets: an insubstantial wealth of imagination rather than a solid fortune in real estate. Air reveals itself only through action. We can feel it and hear it; we know it's there because we see it work. Grass ripples, trees sway, clouds flit across the sky, sails swell, waves form on water, and the Earth hums.

The hum is a recent discovery, teased out of a mass of concealing seismic data in 1998 by geophysicists Naoki Suda and Kazunari Nawa of Japan's Nagoya University. Analyzed, it's a low grumbly sort of noise, a confusion of some fifty different notes, all with pitches in the range of 2–7 millihertz. The result is a boring and cacaphonous little melody conducted about sixteen octaves below middle C, a sort of infra-bass planetary rendition of "100 Bottles of Beer on the Wall." Our ears can't detect it, which is probably just as well; while there's a chance we might find our global mantra spiritually soothing, it strikes me as the sort of annoying sound that might equally well drive us all mad.

The hum consists of free oscillations – that is, vibrations that continue on for a limited period after an initiating event, like the drawn-out twang of a plucked guitar string or the prolonged bong of a piano key with the loud pedal pressed to the floor. Its source is a matter of debate, but the best guess to date is that it's the result of global atmospheric pressure fluctuations. When air pressure increases, cool dense air shoves down heavily on the land or sea beneath it; when air pressure falls, as warmed air rises, the surface – released – rebounds. The collective play of worldwide pushes and releases creates a responsive vibration from the Earth below. Air, like water, has a voice. Though we can't see it, we can hear it: wind whispers, moans, howls, shrieks, sings, and rattles our windowpanes. Far beneath the reach of our senses, however, air also plays a four-billion-year-old melody. The atmosphere pumps the Earth like an accordion and the Earth hums.

The world is wrapped in 5500 trillion tons of air, a vast sea of life-giving gas extending upward over 300 miles (480 km) above our heads. This is the atmosphere, the encircling cocoon that Shakespeare deemed a "most excellent canopy." Oxygen, inhaled from it, keeps all of us land animals alive; shielding provided by it protects us from damaging solar radiation; greenhouse gases, seeded through it, keep us warm. (Perhaps, at present, too warm.) Air, close at hand, may be invisible, but we can see its molecular signature in the sky. Our word sky comes from the Old Norse, meaning clouds, which is presumably what Scandinavians see most of in the gloomy north when they look up. The *OED*, however, poetically defines sky as the "arch or vault of heaven" and most of us associate that soaring arch with blue.

Blue, spread above our heads from horizon to horizon, is an artifact of air. The phenomenon by which celestial blue appears is known as Rayleigh scattering after the nineteenth-century British physicist who best described it. Basically, the blue is the result of the interaction of wavelengths of visible light with gaseous molecules – primarily nitrogen and oxygen – in the air. These molecules, by virtue of their size, are ten times better at scattering the short-wavelength portions of the spectrum – that is, the violets and blues – than the long-wavelength oranges and reds. Blue light, preferentially bouncing off molecules of air, scatters across the sky, which thus, to our eyes below, looks blue. The brilliant dome of overhead blue is so imposing that some early peoples worshipped it: Genghis Khan's Mongols, for example, venerated the Eternal Blue Sky; and Christopher Columbus mentioned in a rather self-serving letter to Ferdinand and Isabella in 1493 that the natives of Hispaniola believed that all power and goodness came from the sky.

Distant mountains look blue because forests and jungles ooze organic volatiles into the air, which molecules similarly scatter blue light. A. E. Housman's blue remembered hills and the majestic purple mountains of "America the Beautiful" are artifacts of air. Crimson sunsets are the result of more intensive light scattering: when the sun is low in the sky in the evening, light, reaching us at a shallower angle, travels a longer path through the atmosphere. Even more blue light is thus scattered away, leaving only the reds to reach our eyes. The effect is intensified by dust in the atmosphere, which leads to truly spectacular sunsets in the aftermaths of volcanic eruptions. Dust in the air also gives us (though very rarely) blue moons; water droplets in the air, provided the sun is in the right place, give us rainbows; ice crystals in the air give us lunar halos and sun dogs. Stars, viewed through the dense molecular soup of our atmosphere, twinkle.

Air gives some among us the ability to fly, soar, or glide: in an airless world, no bird, butterfly, or airplane would ever leave the ground; and anything, dropped, from the flimsiest feather to the frailest leaf, would fall like a rock. Our sense of smell depends on air; as does our ability to hear. Sound travels to our ears through air; without it, no symphony orchestra or brass band could be heard; and even the most raucous of rock stars and thunderous of politicians would be stricken dumb. There's no sound in the empty vacuum of space and the airless Moon is silent as the grave.

Air, in Earth's 300-mile-thick (480-km-thick) cocoon, is layered like an asymmetrical club sandwich. The bulk of the atmosphere – the lion's share of Earth's oxygen and other gases – is all within 10 miles (16 km) of home. This region is called the troposphere – from the Greek *tropos,* meaning "turning over" or "mixing" – and it's in this layer that the turbulent mixing of air takes place that generates the world's weather. Above the troposphere is the stratosphere, which extends to about 30 miles (48 km) above sea level. There's air up here, but not much; jet planes fly through the stratosphere, but with crew and passengers in safely pressurized cabins; and oxygen is so depleted that, presuming anyone should attempt to, it would be impossible to light a candle. The lower stratosphere is home to the world's much-threatened ozone layer, a band of ultra-reactive molecules that efficiently absorbs potentially harmful ultraviolet radiation from the Sun. Above the stratosphere are the mesosphere, which extends from about 30 miles to 50 miles (48–80 km) above the Earth's surface; and the thermosphere, 50–300 (80–480 km) miles up. Beyond the thermosphere is the almost-empty exosphere, the atmosphere's wispy outer fringes, reaching outward some 40,000 miles (64,000 km), gradually blending into the bleak vastness of space. The deprived mesosphere contains only one-thousandth of the atmosphere's total gas molecules; the even skimpier thermosphere, 1/100,000 of the total. In the barely existent exosphere, the air is so thin that lonely molecules can drift for 6 miles (10 km) or more without running into a neighbor. Over 99 per cent of the mass of the atmosphere is at the bottom, within 18 miles (30 km) of Earth, and 75 per cent of it is in the troposphere. It's a thin skin to protect us from the frigid void. If the Earth were a ball 8 inches (20 cm) across, our indispensable air could be represented by coating the ball in salad oil.

*If there were collected a great bulk of wool, say twenty or
thirty fathoms high, this mass would be compressed by its
own weight; the bottom layers would be far more compressed
than the middle or top layers, because they are pressed by a
greater quantity of wool.*

Blaise Pascal

The Greeks' four elements perceptively include examples
of the three basic states of matter: earth, a solid; water, a
liquid; and air, a gas. Solids – things like rocks and kitchen
chairs – have fixed volumes and shapes. Liquids – like water,
chocolate syrup, and gin – have fixed volumes but no personal
shape; instead they sycophantically conform to the shape of
their various containers. Gases have neither fixed volumes
nor shapes: instead, like housework or hobbies, they expand
indefinitely to fill the available space.

Air, with its loopy proclivity for expansion and drift, gives
the impression of being a flimsy substance, but actually it's an
element with considerable heft. Air is astonishingly heavy – at
14.7 pounds per square inch (1 kg/sq cm), we each balance 600
pounds (250 kg) on top of our heads – but we're so attuned to
its constant and inexorable pressure that we barely notice it.
In fact, we're internally pressurized in response to it, pumped
like so many humanoid balloons. In a vacuum, our internal
pressure, shoving outwards, can be lethal, causing eardrums
to explode, and lungs and blood vessels to rupture. Astronauts
wisely wear pressurized spacesuits. Our bodies are calibrated
for this atmosphere, designed to live at the bottom of Earth's
weighty sea of air.

In 1648, air came to the attention of the French scientist/
philosopher Blaise Pascal. Then a sickly but brilliant 25-year-
old, Pascal had been a mathematical prodigy, his talents egged
on inadvertently by his father, a man of unorthodox educa-
tional views, who forbade his son to study mathematics until
he reached the age of fifteen. The young Pascal promptly
became obsessed with the forbidden topic; by the age of nine,
working on his own, he is said to have discovered the first

thirty-two theorems of Euclid; at sixteen, he published a book on the geometry of conic sections; and at eighteen, he invented an early version of the digital calculator, which he epony-mously christened the Pascaline. By then, he had managed to run afoul of mathematician/philosopher René Descartes, who considered the precocious teenager an upstart. He said so; and Pascal, in retribution, expressed doubts as to the value of Descartes's analytical geometry. They also disagreed about the nature of the atmosphere, Descartes insisting that it went on for ever, filling all space with matter; Pascal countering that it ended at a finite distance from Earth, after which space was a vacuum. Descartes responded peevishly that Pascal had too much vacuum in his head.

The wrangle was settled with the help of a device invented by Italian physicist Evangelista Torricelli in 1643, a 4-foot (1.2-m) tube filled with mercury that constituted the first barometer. Torricelli, a devoted admirer of Galileo, had been inspired by the elderly astronomer to investigate the problem of why water could be pumped no higher than 33 feet (10 m) above its natural level. The answer, Torricelli eventually hypothesized, was that the water pump, in effect, was measuring the weight of the atmosphere: air presses on the water in a reservoir, forcing it up the cylinder of the pump to the level at which the weight of the water precisely balances the weight of the air. He tested his hypothesis with mercury, which poisonous metal is both liquid at room temperature and 13.5 times heavier than plain water. Air, Torricelli reasoned astutely, should only be able to balance a 30-inch (76 cm) column of mercury – that is, a column 13.5 times shorter than the column of water. The result, as they say, was history.

Back in France, Pascal still had Descartes to deal with, and the troublesome question of the vacuum in space. To this end he co-opted his athletic brother-in-law and convinced him to climb a mountain, armed with two barometers and 6 pounds (3 kg) of mercury. The air might balance 30 inches (76 cm) of mercury at sea level, Pascal reasoned, but – if his conception of the atmosphere were correct – on the top of a mountain, above which there was less air, it should balance less. The brother-in-law, who seems to have been a robust type, repeated the experiment several times, with the help of an interested company of friends. The results were conclusive: the column of mercury supported by the air on top of the 3000-foot (900-m) mountain was three inches (7.6 cm) shorter than that supported in the town below. Air, Pascal determined, had a finite weight; thus it also had a finite height. It went so far and no farther. Air was the peculiar possession of our planet, and there was nothing but emptiness beyond.

He never managed to convince Descartes.

Our word *gas* comes from the Greek *chaos*, from the chaotic and formless nature of drifting vapors. The name was first adopted around 1620 by Jan Baptista van Helmont, a wealthy Flemish aristocrat, physician, and hobby alchemist; and we have an inkling of what he sounded like because "gas" is his phonetic rendition of "chaos" in Flemish. Today Helmont is often referred to as the father of biochemistry, largely because he was the first known experimenter to attempt to answer a biological question through quantitative methods. In an attempt to demonstrate the elemental nature of water – "All vegetables," surmised Helmont, "proceed out of the element of water only" – he planted a 5-pound (2-kg) willow sapling in a pot and watered it faithfully for five years. At the end of this period, the tree, uprooted and weighed, was found to have gained 164 pounds (75 kg). Such a gain, Helmont concluded confidently, could only have come from the added water. In this he was only partially right. At least half the added weight, ironically enough, came from an atmospheric gas that Helmont himself had discovered. Since he had originally produced it from burning charcoal, he called it *spiritus sylvestris*, or spirit of wood. We call it carbon dioxide. It comprises some 0.035 per cent of the atmosphere.

Helmont discovered a gas, but never successfully isolated

any – in fact, he insisted that such an act was impossible since gas, enclosed, would cause its vessel to "dangerously leapeth asunder into broken pieces." That was left to Robert Boyle, the first scientist to call himself a chemist, now often referred to as the father of modern chemistry. (Also in competition for the paternal title are Antoine Lavoisier, Joseph Priestley, and the eighth-century Arabian alchemist Geber.) Boyle, like Helmont, was an aristocrat, and a dedicated experimenter, taking to heart the Royal Society's injunction *Nullius in verba* – that is, nothing from hearsay, but from hands-on investigations in the laboratory. A lab coat seems so suited to his personality that it is almost a shock to see a portrait in which he appears very much the earl's son, looking rather sullen, in an immense ringleted wig and lace cravat.

Among Boyle's major chemical accomplishments was not only the isolation of a gas – hydrogen – without bursting its containing vessel dangerously asunder, but the discovery that gases, unlike liquids and solids, are compressible. Further-more, his experiments – which involved trapping gas in a 17-foot-long (5 m) J-shaped tube partially filled with mercury – demonstrated that the volume of the compressed gas changed in inverse proportion to the amount of applied pressure. That is, if the pressure on a given amount of gas is doubled, the volume of the gas is halved; if pressure is tripled, volume is reduced to a third. Conversely, if pressure is reduced, gas proportionally expands – if pressure is halved, volume doubles. This elastic relationship so reminded Boyle of the behavior of stretched and compressed coiled metal springs that he referred to it as "the spring of the air." Still known today as Boyle's Law, it was the first application of mathematics to chemistry.

*I have discovered an air five or six times as good as common
air.*

Joseph Priestley

Air is not an element in the modern sense, but a mixture,
consisting of 78 per cent nitrogen, 21 per cent oxygen, 0.9 per
cent argon, tiny but constant quantities of hydrogen, helium,
neon, and xenon, and variable amounts of water vapor, carbon
dioxide, methane, nitrous oxide, ozone, and particulate junk.
As far as human beings are concerned, most of air is filler: for
us, the pièce de résistance is oxygen. Oxygen was discovered
in the eighteenth century – and nearly simultaneously by three
different scientists, whose deserved degrees of credit are still
a matter of debate today. This thorny dilemma is the theme of
the play *Oxygen* by Carl Djerassi and Roald Hoffman – possibly
the only play ever written by a pair of prize-winning chemists
– in which a committee assembles in present-day Stockholm
to award a retro Nobel Prize to one of the three claimants:
Carl Wilhelm Scheele, Joseph Priestley, and Antoine-Laurent
Lavoisier.

The winner should have been Scheele. Scheele was born
in 1742 in Pomerania – now eastern Germany, but at the
time Swedish territory, having been annexed by the Swedes
following the Thirty Years' War. The seventh of eleven children
in a poor family, Scheele was apprenticed at the age of fourteen
to an apothecary, a good choice for a boy who was to develop
into a brilliant self-taught chemist. Over the course of his short
career – he died at the age of forty-three, of what looked suspi-
ciously like mercury poisoning – he discovered or investigated a host of new chemical substances, among them copper
arsenite, which lethal arsenical compound, known to this day as
Scheele's green, was implicated in the death of Napoleon, whose
bedroom wallpaper was painted with it. He also characterized a
long list of previously unknown acids; studied the effect of light
on silver salts, a phenomenon later instrumental to the development of photography; and isolated noxious hydrogen sulfide
gas, as well as the elements chlorine, manganese, and barium.

By 1770, Scheele had ascertained that air was "composed of two fluids, differing from each other," which he called "Foul Air" and "Fire Air" – respectively, nitrogen and oxygen. Shortly thereafter, he managed to isolate Fire Air – "a colourless gas, in which a taper burned with a flame of dazzling brilliance" – by several different procedures, among them heating mercuric oxide and capturing the liberated gas in a bladder. He described his experiments in a book, *Chemical Treatise on Air and Fire*, which, though sent to the printer in 1775, was not published until 1777, nearly seven years after Scheele's initial discovery. By that time Priestley, who discovered it later, had published first.

Joseph Priestley, born in 1733 in Yorkshire, was a Unitarian minister, and a highly learned one, with grounding in philosophy and logic and an extensive background in languages. He spoke French, Italian, German, Greek, Latin, Hebrew, and Arabic, and in 1802 – at the age of seventy – he set about learning Chinese. He also has some minor fame for bestowing the name "rubber" on the sap of the South American caoutchouc tree, having discovered – while writing a treatise on the art of perspective drawing – that small pieces of the stuff could be used to rub out ill-placed pencil lines.

Priestley had no formal education in science, but clearly had a prodigious knack. He became interested in gases in 1767, initially studying carbon dioxide or "fixed air" – so called because it could be released from solid calcium carbonate, and then recombined or "fixed" into solid form again. Isolating carbon dioxide from fermenting grain, he found that the gas could be dissolved in water to produce an appealingly fizzy beverage, thus discovering soda water, for which the Royal Society gave him a medal. He described the process in 1772 in a paper titled *Directions for Impregnating Water with Fixed Air*. He first isolated oxygen in 1774, by heating mercuric oxide with a large convex lens or burning glass, soon discovering that a candle "burned in this air with a remarkably vigorous flame" and that a mouse, confined in a bottle of it, lived twice as long as a mouse confined in the same volume of "common air." By the end of that year, before Scheele had so much as approached his unhelpful printer, the excited Priestley had published the first volume of his trilogy *Experiments and Observations on Different Kinds of Air*.

Lavoisier was undeniably a latecomer to the oxygen scene, though he tried to put a good face on it, referring to oxygen in 1789 as "this air, which Mr Priestley, Mr Scheele, and I discovered about the same time." He didn't discover it. He learned how to make it in 1774, through a dinner conversation with Priestley, who visited Paris in October, and from the instructions in a letter, written at about the same time, by Scheele. He was soon producing some of his own and investigating its properties using the precise quantitative methodologies for which he is justly famed. His claim to oxygen stems not from being the first to find it, but from being the only scientist of the original three to figure out how it worked. Scheele and Priestley clung firmly to the increasingly awkward phlogiston theory [see Fire]; and Priestley, to the end of his days, referred to the trio's much-disputed gas as "dephlogisticated air." Lavoisier, however, with his exquisitely calibrated balances, clearly demonstrated the nature of oxygen's role in combustion, respiration, and oxidation, definitively laying phlogiston to rest. It thus seems only fair that Lavoisier's chosen name for the new gas – oxygen – is the one that survives.

Ironically, both Priestley and Lavoisier came to grief during the French Revolution, Priestley for supporting it too enthusiastically, and Lavoisier for not supporting it at all. Enraged at his pro-French politics and controversial religious views, a murderous mob burned down Priestley's home and laboratory in Birmingham, after which he and his family emigrated to America. He died in Pennsylvania in 1804. Lavoisier, less lucky, was sentenced to death by a Revolutionary tribunal and sent to the guillotine on May 8, 1794. "A moment was all that was necessary to strike off his head," mourned a friend, the mathematician Joseph LaGrange, "and probably a hundred years will not be sufficient to produce another like it."

Air is the most spiritual of elements. Our word spirit comes from the Latin *spirare,* meaning to breathe; and the Latin *anima* – variously meaning air, breath, or soul – gives us our verb animate, to endow with life. The Greek *aura, psyche,* and *pneuma* all are simultaneously defined as breath, soul, or spirit. *Pneuma* today survives in pneumatics, the branch of physics that deals with the mechanical properties of air; *psyche,* however, refers solely to the spirit or inner life, the mind as opposed to the biological body. It's only logical that we associate breath or air with soul; breath and life are inextricably interconnected. We can go for two weeks or more without food, for three days or so without water, but – under normal conditions – we can't last for more than a few minutes without oxygen. Ordinarily, within five minutes, brain damage sets in; and shortly thereafter convulsions, coma, and death. On average, we take between fourteen and twenty-five lung-filling breaths per minute, usually without so much as thinking about it. Breathing is an unconscious act, ordinarily left to the control of the autonomic nervous system, which is the body's way of ensuring that none of us will ever, in a moment of carelessness, forget to take in air.

On the other hand, we're also capable of over-riding the system, which means that we can hold our breath. In 1912, archaeologist Edgar James Banks of the University of Chicago, excavating the ancient Mesopotamian city of Adab, discovered conch-shell oil lamps and ornamental stone rosettes inlaid with iridescent slices of mother-of-pearl. Both substances had been harvested from the bottom of the sea, mute evidence that 6500 years ago people knew how to dive. All early coastal dwellers, archaeologists guess, probably did their best to explore the world underwater. Homer mentions diving in the *Iliad* (an admiring metaphorical reference to a notably nimble leap from a chariot); and Thucydides, Aristotle, and Livy all discuss it (respectively, for purposes of warfare, sponge-gathering, and treasure-hunting). The ancients also dived for shellfish and pearls – particularly profitably in the case of the latter, since pearls were greatly in demand among the fashionable of Rome. Julius Caesar, a maker of grand gestures, is said to have given a pearl worth six million sestertii as a gift to Servilia, the mother of Brutus.

Ancient pearls and sponges were acquired with a gulp of air and a plunge – what is now called free or breath-hold diving. Expert breath-hold divers can accomplish amazing feats: the *ama* divers of Japan, for example, routinely swim down 145 feet (44 m) in quests for coral, pearls, and seaweeds; and champions in free diving competitions have reached depths of 240 feet (73 m) or more. Such dives involve holding breath for periods of approximately two minutes. According to the diving literature, this isn't all that unheard of: human beings can generally hold their breath for one to two minutes, and, with practice, up to four. It's not easy, however, because our bodies, if deprived of oxygen, start screaming for it, and when overloaded with carbon dioxide, start shrieking to let it out.

Oxygen is so essential because it runs the body's energy supply system. A series of chemical reactions, all dependent on oxygen, lead efficiently to the formation of adenosine triphosphate – best known by the acronym ATP – whose intramolecular bonds serve as storage containers for energy. Chip a phosphate group off ATP and you release a dollop of energy; chip a lot of such phosphates, and you generate enough power to support the synthesis of proteins, the propagation of nerve impulses, the contraction of muscles, or, if you're a firefly, the rhythmic posterior blinking that tells receptive fireflies of the opposite sex just where and who you are. Our metabolism is entirely dependent on oxygen; and the more we do, the more of it we need.

Many of our anatomical and physiological complexities are clever evolutionary solutions for feeding us enough air. Primitive single-celled organisms, whose volumes are small relative to their surface areas, can obtain enough oxygen for their metabolic needs simply by absorbing it across their cell membranes. In larger creatures, however, the demand for oxygen soon outstrips the ability of surface area to supply it. Those who feared the Blob – the gelatinous and carnivorous lump from outer space that terrorized townships in the 1958 and 1988 movies of the same name – needn't have bothered; surface area to volume ratio would have ensured that the Blob would have asphyxiated before it managed to ooze across the laboratory floor. Frilled gills and multi-branched lungs are both creative systems for maximizing oxygen uptake. Human lungs, for example, are vast trees of dividing and subdividing tubules, ultimately ending in millions of minuscule sacs called alveoli where inhaled oxygen is transferred to the blood. Tidal lung volume – that is, the amount of air we ordinarily move in and out of the lungs with each inhalation and exhalation – amounts to about one pint or half a liter of air, though the lungs inflated to full capacity can hold ten times that. Total lung surface, spread out flat, amounts to about 850 square feet (80 sq m) – approximately the size of a tennis court – over which the entire blood volume of the body passes once a minute.

Or more, should we do something inordinately strenuous, such as flee a bear, benchpress a crate of cannonballs, or swim the English Channel. Chemoreceptors in the aorta and carotid arteries and in the respiratory center of the brain stem continually monitor concentrations of carbon dioxide and oxygen in the blood; if too much of the first or too little of the second, respiratory rate is elevated to restore the balance. A brisk session at the gym leaves us panting because repeated contraction of large muscles requires substantial expenditure of energy, generated by the breakdown of cellular ATP. ATP must be promptly replenished, which leads to a demand for oxygen. Textbooks like to describe this in monetary terms: ATP is the energy currency of the cell; we spend it on metabolic activity; and then we must replace it with deposits of oxygen and carbohydrate to avoid a debilitating overdraft.

In many ways, it's an apt comparison. Just as money circulates from hand to hand, so air travels companionably from nose to nose. While the average lifespan of the average dollar bill is just eighteen months, the average air molecule lasts indefinitely. All of the air we breathe has been internalized innumerable times by somebody else; molecules in our lungs at this very moment have almost certainly once resided in those of *T. rex*, Socrates, and Mother Theresa. A popular calculation given to aspiring physicists involves determining the probability of your present inhalation containing at least one air molecule from Julius Caesar's last breath – the one which the expiring emperor, bleeding from multiple stab wounds, used to gasp *"Et tu Brute?"* to the son of Servilia, recipient of that expensive pearl. The conclusion – based on an assumption of 10^{44} total molecules in the atmosphere and 2×10^{22} molecules in the average human breath – predicts a 98.2 per cent chance that at least one of the molecules presently in your lungs was one of Caesar's last. There's no such thing as fresh air. The elements are common property. All air is used.

Of the four elements, air – at least in its present form – is the newest. The atmosphere hasn't been around for ever; and scientists hypothesize, in fact, that the one we have at present is our third. Our first – an evanescent envelope of hydrogen and helium – vanished in the early days of the Earth's formation. Our second – a thick steamy froth of carbon dioxide, nitrogen, sulfur dioxide, and water vapor – accumulated about 4.4 billion years ago, at about the same time the solidifying Earth separated itself into layers, forming crust, mantle, and core. This second try at air was the product of a million-year-long volcanic belch, a process known as outgassing, in which volatiles from the planet's interior were spewed up and outward from the surface, then captured by gravity. Water fell steadily out of this second atmosphere, filling the oceans and establishing the hydrologic cycle; everything else stayed behind, waiting to be inhaled.

The second atmosphere contained hardly any oxygen at all – about 1 ppm (.0001 per cent), as opposed to 208,500 ppm (21 per cent) today. As such, it was Eden for the first primitive microbes, anaerobes which live happily in a smog of CO_2. They would doubtless be doing so yet, if it were not for the unfortuitous appearance, some 2.5 billion years ago, of cyanobacteria, aggressive greenish creatures who formed themselves into fat microbial mats and proceeded to photosynthesize. Photosynthesis is the process by which green plants, with the help of incoming sunlight, convert carbon dioxide and water into organic carbon compounds and oxygen. Organic carbon compounds, from a cyanobacterium's point of view, are desirable energy-providing food; oxygen, on the other hand, is the equivalent of industrial waste, a junk molecule, suitable only for throwing away. Preferably far far away, since oxygen, to the average innocent denizen of the late Archaean and early Proterozoic Eras, was poison gas.

We know that a good amount of oxygen was around during the Proterozoic because it left its signature in the rocks in the form of iron oxides. Pure iron is rare on our planet, found only in the Earth's inaccessible core or in fallen iron meteorites. Our iron is invariably oxidized; that is, most of it is rusty. In rocks, oxidized iron is found in the form of ferrous oxide, or hematite – the name comes from *heme*, ancient Greek for blood, since it was once believed that the red rocks were stained with the blood of warriors seeping into the ground in the aftermath of great battles. The mid-Proterozoic, 2.2–1.8 million years ago, saw the formation of an immense amount of hematite, the result of cyanobacteria-generated oxygen interacting with the iron dissolved in seawater. The hematite sank to the ocean floor, eventually to solidify and form the continental red beds, vast rust-red layers of rock that, to a geologist, are smoking guns: clear evidence that somewhere in the Proterozoic something was spewing out a lot of oxygen.

As long as iron was around to mop up the awful stuff, oxygen was relatively benign. Once the available iron was exhausted, however, oxygen – with nowhere else to go – began to accumulate ominously in the air. This was the beginning of the third atmosphere, the air of today, now primarily a mix of nitrogen and oxygen. Life on Earth has accustomed itself to this over the past two billion years; at its inception, however, the third atmosphere was catastrophic. Oxygen is a danger-ously reactive molecule; and oxidation, in effect, is a vicious chemical burn. It affects the innards of unprotected cells like a blowtorch – which is how it put the kibosh on the virulent space bugs in Michael Crichton's *The Andromeda Strain*. Set loose in ancient Earth's air, it created what biologist Lynn Margulis has dubbed an oxygen holocaust. The resident anaerobes, lacking biochemical defenses, died like flies.

The elements, though beautiful, are not necessarily our friends. All the the Greeks' classic four have terrible aspects: water, floods and tsunamis; earth, quakes and avalanches; fire, volcanoes, lightning, and wildfire. Air – innocently blue above us, the Chinese color of immortality – is an enemy; life, confronted by it, has been hard put to survive. Organisms that did were forced to devise means of dealing with the deadly new oxygen atmosphere – most cleverly (one of the greatest metabolic coups of all time, writes Margulis) by learning how to breathe it. The result was oxidative respiration; and its acquisition changed the world.

About 1.8 billion years ago, scientists believe, shortly after the formation of the final red rock, eukaryotic cells appeared: our own distant unicellular ancestors, distinguished from primitive bacteria by their possession of nuclei and internal organelles. Among these organelles were the mitochondria, the eukaryote's first line of defense against air. The name comes somewhat confusingly from a mesh of the Greek *mitos*, thread, and *chondros*, grain, which is what mitochondria variously looked like to biologists squinting through nineteenth-century microscopes. Actually, now that microscopes are better, they've turned out to look a bit like high-tech sneaker soles. Mito-chondria were and are all that stand between us and death by oxygen. Nicknamed the powerhouses of the cell, mitochon-dria carry out oxidative respiration, the essential metabolic process that simultaneously converts potentially lethal oxygen to harmless water and manufactures energy-laden packets of ATP.

Most biologists today believe that we didn't develop mitochondria on our own, but instead snatched them, ready-made, out of the surrounding environment. This theory, known as endosymbiosis, harks back to an ancient and mutually beneficial partnership in which early eukaryotes engulfed free-living bacteria in much the same way that the Biblical big fish swallowed Jonah, but to markedly better end. The relationship proved so advantageous that it became permanent. The eukaryotes supplied their newly internalized bacteria, the ancestors of our present mitochondria, with food; the bacteria, in exchange, protected their eukaryotic hosts from the ravages of oxygen. Darwin argued that survival is a matter of sheerest luck, thrust upon whatever is best fitted for a particular place and time. We don't choose it, in other words; it chooses us, because we happen to be bigger, furrier, faster, or the color of a tree trunk. In the case of our most distant of ancestors, however, it's empowering to imagine that it might have been the other way around. The first eukaryotes didn't sit about helplessly, hoping for survival, but instead went out and swallowed it.

Air may be biochemically hazardous, but it has also been considered, depending on author and era, invigorating, revitalizing, refreshing, and health-promoting. In 1856, domestic self-help book author Catharine Beecher – older sister of the inflammatory Harriet Beecher Stowe – wrote: "It is probable that there is no law of health so universally violated by all classes of persons as the one which demands that every pair of lungs should have fresh air at the rate of a hogshead an hour." Fresh air was the signature ingredient of the early nineteenth-century health movement; and its proponents variously railed against overheated air, "respired" air of the sort breathed by "persons who are fond of frequenting unwholesome crowds," and the corrupt and infectious air of cities. Unventilated bedrooms, filled with the "slow poisons" of sleepers'

exhalations, were condemned as "experimental boxes for the synthetic development of pulmonary disease;" and a volume of the *Journal of Health*, published in 1829, insisted that the bedchamber be not only ventilated, but cold, since a foolish dependency on "external warmth" rendered people susceptible to colds, coughs, and consumption. By the latter half of the century, the health-conscious slept with their windows briskly open.

Until the development of antibiotics in the mid-twentieth century, air was medicine. Fresh air – preferably mountain air, tinged with the resinous scent of balsam – was the recommended treatment for tuberculosis; and the great sanatoriums that were built across Europe and America in the late nineteenth century were not only attempts to isolate the afflicted from the general population, but to provide them with healing draughts of clean air. In the 1880s Marc Cook, a newspaper reporter from New York City, popularized the "wilderness cure" – basically camping – in which the "pure air of the tent" was touted as a remedy for a raft of ills, including weakness, fever, night sweats, and general debility. Invalid Victorians flocked to the coast, to breathe sea air. Americans with weak chests traveled to the Southwest, to breathe the arid air of the desert.

The best air of all, according to a growing group of urban enthusiasts today, is pure oxygen, delivered to the lungs through plastic nose tubes at the cost of a dollar or so per minute, and scented with bayberry, cinnamon, strawberry, or peppermint. The craze for oxygen bars, where customers order gourmet gas instead of Chardonnay, began in the 1990s in Japan, and rapidly spread worldwide, notably flourishing in cities notorious for smoke and smog, among them Tokyo, Los Angeles, Mexico City, Bangkok, Chicago, and New York. Ten minutes or so on pure oxygen, claim its proponents, does everything from induce euphoria to cure hangovers. Medical practitioners are skeptical, pointing out that a normal person's hemoglobin is already 99 per cent saturated with oxygen, which leaves little room for effects due to additional oxygen uptake. Still, as early as 1775, Joseph Priestley pointed out that breathing pure oxygen seemed to make mice unusually frisky, and when he tried it himself, he found that his "breast felt peculiarly light and easy for some time afterwards." "Who can tell," he continued, perceptively anticipating Tokyo, "but that in time this pure air may become a fashionable article in luxury?" On the other hand, he noted, "the air which Nature has provided for us is as good as we deserve."

In light of what we now know about oxygen, this last line has an ominous ring. Some scientists, experts in the study of ageing, suggest that nature's air, ultimately, is what kills us off. The culprits in this case are oxygen free radicals or reactive oxygen species (ROSs) – damaging molecules with itchy unpaired electrons, such as hydroxyl and superoxide anions and hydrogen peroxide (which last, applied to hair, produces peroxide blondes). Free radicals are inevitable off-shoots of mitochondrial respiration – over 20 billion such molecules are formed by each body cell each day in the course of normal metabolism. We also accumulate free radicals from the outside, from air pollutants and cigarette smoke. All are highly reactive charged molecules that, given their head, annihilate vital cellular components, including our cell membranes and DNA. Throughout our lives we gamely fight these molecular malfeasants off, primarily with a pair of antioxidant enzymes: catalase and superoxide dismutase, the latter known crudely to its familiars as SOD. According to the free radical theory of ageing, however, we've got limits: what catches up with us in the end is oxidative stress. The accumulation of free radicals gradually overwhelms the body's antioxidant defenses and eventually radical-engendered cellular vandalism wreaks so much destruction that it becomes lethal. Air is a necessary evil. Breathing, points out Nick Lane, author of the absorbing *Oxygen: The Molecule that Made the World*, is a slow form of poisoning. Do it long enough and you'll die.

It has been known since ancient times that some air was certainly worse for people than others. Our word *miasma* – from the Greek for pollution – means noxious and infectious air; and well into the nineteenth century, such diseases as yellow fever, cholera, and typhoid were blamed on tainted air. The name malaria comes from the Italian for "bad air," in reference to the foul-smelling swamps near Rome, thought to be the fever's source – which they were, though it was the *Anopheles* mosquitoes breeding in the stagnant waters that were the carriers of the disease, not the atmosphere.

This isn't to say that air can't be infectious. Many diseases – among them diphtheria, measles, mumps, and the ubiquitous common cold – are airborne contagions, spread primarily by coughing and sneezing. Influenza travels from host to host in this fashion with fearsome efficiency, as in the massive pandemic of 1918 which ultimately killed 20 million people worldwide. Overwhelmed public health authorities struggled to contain outbreaks by quarantining victims and limiting public gatherings, with their enhanced potential for air-mediated disease transmission. In the United States, in the wake of the flu, saloons, dance halls, and cinemas were closed; and people were urged to walk to work rather than ride the unsanitary and crowded streetcars. Switzerland banned concerts, shooting matches, and theater performances, although in Great Britain the shows went on, provided they lasted less than three consecutive hours and the theaters were thoroughly ventilated afterwards. Medical personnel – and in some cases the entire populations of cities – adopted gauze face masks. In the case of the virulent 1918 influenza strain, such measures were generally ineffective; strict enforcement of quarantines, however, did successfully halt the SARS (Severe Acute Respiratory Syndrome) epidemic, first identified in Asia in February 2003, which ultimately infected 8400 people and killed 813. In the absence of effective vaccines, heroic measures are often required; a single sneeze can propel an aerosol of 2 million virus particles a distance of 30 feet (9 m). On average, it takes just ten particles to establish the disease in a new host; in theory, one sneeze, properly disseminated, could infect a small city.

Sneezing – the formal term for this explosive act is sternutation – is the body's mechanism of removing unwanted irritants from the inner linings of the nose. It's a surprisingly complex reaction involving the muscles of the diaphragm, abdomen, chest, and throat, all interacting in rapid sequence to blast particles out of the offended nasal passages with a jet of air traveling over 100 miles per hour (150 km/h). Such particles are a problem for us because air, despite its reputation for being empty, is amazingly full. At any given moment, the air worldwide holds some 1–3 billion tons of particles: molds, spores, algae, viruses, dust, and – depending on the season – a truly stunning amount of pollen. Even the cleanest air carries an invisible baggage of junk: the air over uninhabited Antarctica contains about 200,000 particles per breath. A lungful of air from the average suburban neighborhood contains up to 25 million particles; a breath unwisely taken on the LA freeway at rush hour in the height of summer may hold 375 million.

Over half the world's population these days lives in cities and cities, by and large, have the worst of the world's air. Part of the problem is heat: buildings and pavement, steel, asphalt, and concrete all absorb solar energy during the day and release heat into the atmosphere at sunset, creating overhead heat bubbles. The air over a city can be as much as 20°F (10°C) hotter than that over the surrounding countryside. These urban heat islands are made even hotter by population growth – the swelling population of Los Angeles, for example, has upped the average temperature of the city one degree every decade since the Second World War – and by the accumulation of waste heat from factories, residential buildings, cars, trucks, and trains. City sizzle is ameliorated by reflective roofs, vegetation – New York without Central Park would be hotter yet – and by wind. However, these can't always do the trick. Growing cities stew in their own toxic juice; the urban heat island is an incubator for smog.

In the US, the Environmental Protection Agency monitors daily air quality, grading it according to its content of five major pollutants: ground-level ozone, carbon monoxide, sulfur dioxide, nitrogen dioxide, and particulate junk. Results are reported using the Air Quality Index (AQI), a color-coded scale that runs from 0 to 500. Air is considered good if rated 0–50 (green); moderate at 51–100 (yellow); unhealthy for sensitive types at 101–150 (orange); unhealthy for everybody at 151–200 (red); and very unhealthy at 201–300 (purple). Anything over 300 is said to be hazardous and is colored maroon.

The nastiest of red, purple, and maroon conditions is probably smog, the impenetrable mix of toxic smoke and natural fog through which Sherlock Holmes stalked criminals and in which Jack the Ripper lurked, awaiting his victims. Smog, in Holmes's day, was a product of burning coal, a combination of water vapor, sulfur dioxide, and particulate ash yielding a thick yellowish miasma often referred to as "pea soup." In the right proportions, such smogs can turn deadly. Ash particles encourage the reaction of sulfur dioxide with oxygen to produce sulfur trioxide, which in turn dissolves in suspended water droplets to form sulfuric acid. Inhaled, this has a disastrous effect on lung tissue; in 1952, a five-day smog, settling like a malevolent blanket over the city of London, caused 4000 deaths. Many smogs today – among them the acrid brown haze that periodically envelopes the freeway-laden city of Los Angeles – are photochemical smogs, largely the result of automobile exhausts. Encouraged by heat and sunlight, ingredients in tailpipe emissions combine to produce toxic quantities of nitric oxide, nitrogen dioxide, and ozone.

One of the many Bab Ballads – written by William S. Gilbert, of Gilbert and Sullivan fame, for the periodical *Fun* in 1865 – is called "Ozone." In it, Gilbert compares ozone to a policeman – never there when you want it. It's not one of Gilbert's best, but it's an interesting reflection on the Victorian age. Ozone, at the time, was a health fad. The elusive gas was considered an energizing tonic, thought to bolster the normal constitution and rejuvenate the invalid. The presumptive high-ozone content of the coastal atmosphere inspired restorative trips to the seaside to breathe the bracing sea air; for those forced to stay close to home, ozone was pumped helpfully into churches, hospitals, theaters, and underground railways. People must have known it was there, too, because ozone – unlike oxygen, which is colorless and odorless – is blue and smells.

The name ozone comes from the Greek *ozein*, meaning to smell, in recognition of its peculiarly piercing metallic odor, the nose-tingling scent associated with lightning and electricity. It was discovered in 1840 at the University of Basel by chemist Christian Friedrich Schonbein – later to attain notoriety as the discoverer of nitrocellulose [see Fire]. Schonbein initially believed his ozone to be a new element, akin to chlorine, but subsequent studies proved it to be a variant and highly reactive form of molecular oxygen, possessing three atoms rather than the more conventional two. It's a natural component of air in the lower atmosphere, though only in tiny quantities – for every 100 million molecules of air, for example, 20 million are oxygen molecules, and just two are ozone. In the sun-drenched days of summer, however, ultraviolet radiation interacting with pollutants from automobile exhaust can boost ground-level ozone concentrations by five-fold or more, to the point where it becomes a biological hazard. In amounts upwards of 0.1 ppm, ozone irritates or damages lung tissue, and plunges sensitive hay-fever and asthma sufferers into allergic paroxysms. The ozone-sniffing Victorians, who already had smog to contend with, weren't doing themselves any favors.

Air is a two-faced element, part beauty and part beast, part good cop and part bad. Oxygen, on the one hand, is a deadly poison; on the other, it's the indispensable breath of life. Ozone near the Earth's surface is bad news; in the stratosphere, however, it's our shield against damaging radiation from the Sun. The stratospheric ozone layer is concentrated some 12–20 miles (20–35 km) above the ground. Here the air is in continual chemical turmoil as ozone molecules repeatedly sacrifice themselves to incoming ultraviolet rays. Ozone, zapped by UV, dissociates into a diatomic oxygen molecule (O_2) and a single oxygen atom (O); the lone oxygen atom then recombines with an oxygen molecule to generate ozone (O_3) once more. In this manner, about 300 million tons of stratospheric ozone are destroyed daily; and an equal amount re-formed. The solar energy used in this seemingly pointless process is, providentially, prevented from reaching the Earth's surface.

Ultraviolet radiation – the part of the electromagnetic spectrum just to the left of visible blue, indigo, and violet – comes in three different varieties of decreasing wavelength: UV-A, B, and C. The shorter the wavelength, the greater the energy of the radiation, which means in theory that C poses the greatest danger to living things. In practice, however, UV-C isn't a problem; thoroughly absorbed by ozone in the upper reaches of the atmosphere, it never reaches the ground. A, the gentlest of ultraviolets, slips through the ozone layer, but does little harm: in fact, it's UV-A that stimulates the reaction by which our bodies manufacture Vitamin D; and in pale-skinned people, it triggers the production of melanin by skin cells, thus turning Caucasians a sun-protected tan. UV-B, however, is trouble. B is energetic enough to be damaging; and, though much of it is absorbed by stratospheric ozone, some still reaches the Earth's surface. Just how much varies, depending on weather, air conditions, and season of the year. UV-B intensity is measured with the erythemal UV index – from erythema, the medical term for sunburn, which is an early sign of UV-B overdose. Prolonged exposure to UV-B increases the likelihood of skin cancers and cataracts in people and animals; and inhibits photosynthesis and slows growth in plants.

The less ozone we've got, the greater our exposure to UV-B, which is why the environmentally conscious are jittery about the Antarctic ozone hole. In a Snoqualmie tale of the Pacific Northwest, Blue Jay tears a hole in the sky through which he,

Fox, and Beaver crawl to steal the Sun; in another version of the legend, a great eagle pecks a hole to let the Sun shine through. In such stories, making a hole in the sky was the sort of heroic deed that could only be done by magical animals in the days when the world was young. Now it seems that we've managed to do it ourselves. The hole – scientists refer to it as an "ozone depletion area" – is a vast tear in the structure of the sky, first identified in 1985. By 1988 studies had pinpointed the cause. Reactive chlorine atoms and other catalysts, derived from chlorofluorocarbons (CFCs) – the stuff of freon, styrofoam, and industrial solvents – had contaminated the stratosphere, with effects on the molecular level akin to loosing piranhas in a vast goldfish bowl. A single chlorine, it was found, could rapidly account for a million ozone molecules in a destructive chemical feeding frenzy, removing ozone from the atmosphere faster than it could be regenerated. The result, by the year 2000, was a record-breaking hole, covering an area 10.9 million miles square (28.3 million sq km) – three times larger than the continental United States – in the center of which the ozone concentration had dropped by 60 per cent. In computer-enhanced photographs, the hole appears as a vast atmospheric blot of menacing inky blue.

The seasonal Antarctic ozone hole, which peaks each winter and dissipates in the (relative) warmth of summer, is enhanced by cold; it will only get worse, meteorologists predict, as accumulating greenhouse gases ratchet up the temperatures in the troposphere below, forcing the stratosphere to cool. When the stratosphere gets cold enough, dipping to -108°F (-78°C), water squeezes out of its thin air to form glimmering clouds of ice crystals called polar stratospheric clouds, or PSCs. The ice is a death knell for ozone: the crystals provide a supportive surface for reactions that sever chemical bonds, releasing hyper-reactive chlorine. At the same time, the heavy cold air of the Antarctic winter sinks and, propelled by the Coriolis force, begins to circle, forming a vortex around the pole. The vortex, a whirling wall of frigid air, insulates the Antarctic from the rest of the atmosphere, blocking entry of ozone-laden air from the upper latitudes, and sealing the polar ozone in with the deadly PSCs.

The North Pole, which also develops a vortex and a crop of icy PSCs, tends not to develop an Antarctic-style ozone hole because the Arctic is warmer. The northern vortex is buffeted by planetary-scale atmospheric waves, generated by large-scale weather systems acting over continental land masses. Unlike its stable southern counterpart, it's unpredictable and leaky, perpetually nudged, shoved, and occasionally forced right off the pole, as planetary waves provide aggressive infusions of warmer air from the lower latitudes. In the southern hemisphere, where there are fewer land masses to create planetary waves, the frigid southern gyre spins on its merry way undisturbed, its ozone the victim of an inadvertent conspiracy among people, ice, and wind.

*Of all natural phenomena, there are, perhaps, none which
civilized man feels himself more powerless to influence than
the wind.*

James Frazer
The Golden Bough

"Wind is almost anything you want it to be," writes Lyall Watson in *Heaven's Breath*, his engaging natural history of the wind. "There are few things as steady or as changeable, as fierce or as gentle, as unstable or as undeviating, as light or as bold, as wroth, as balmy, or as protean as wind." Actually, what we want it to be has little to do with it; air, once it decides to move, has an agenda all its own. Occasionally we manage to capture a bit of it or divert it for our own purposes, but air, of all the elements, is the most independent: it goes where it wants to, and that's essentially everywhere.

The Sun, beating down over the steamy equator, is the engine that drives the world's winds. As heated air over the equator rises, bobbing skyward like a massive invisible balloon, it leaves a gap behind it, known to meteorologists as a low-pressure area. Low-lying air from the cooler upper and lower latitudes promptly rushes in to fill the hole, is warmed in turn by the relentless Sun, and also rises, pushing the air above it to either the north or the south. In the cooler upper (or lower) latitudes, the pushed air – chilled – sinks, forming a dense lump of cool high-pressure air, poised once more to rush in and fill the yawning equatorial gap. Such endlessly cycling convection cells, yanked sideways by the rotation of the Earth as the planet rockets west to east at upwards of 1000 miles per hour (1600 km/h), are the heart of the wind.

Before global physics came in to play, the winds were people or animals. To the ancient Norse, the wind was the work of Hraesvelgr, a giant in the guise of an eagle, who made the gusts and gales by flapping his great wings. In Greek mythology, the winds were controlled by Aeolus, son of the sea god Poseidon, who kept them chained in a cave beneath his bronze-walled island in the middle of the Mediterranean Sea. His sons, Boreas, Notus, Eurus, and Zephyrus, were respectively personifications of the north, south, east, and west winds. The Latin poet Ovid described each in detail in his *Metamorphoses*. Boreas, the harsh cold wind of winter – Shakespeare called him a ruffian – was the most powerful of winds, with barrel chest and wild gray hair and beard, shown blaring on a conch-shell trumpet. Notus, beardless, volatile, and younger, came wrapped in cloud, carrying a clay water jar filled with rain. Eurus was gloomy, dark, and elderly; while Zephyrus, the west wind, was a lovely long-haired boy, in a mantle filled with fruit and flowers. Cross-culturally, people seem to have been of two minds about the mincing Zephyrus: English poet John Masefield saw April in the west wind, and daffodils, but the Chinese identified the flowery west wind as a tiger; the Egyptians saw it as a serpent; and Joseph Conrad described it as "a monarch gone mad."

Most mythologies agree that the essence of wind is its unpredictability. There's either too much of it or too little, and there's always the risk of it running dangerously out of control. The wind – like a bellicose genie in a bottle – is often seen as supernaturally imprisoned, trapped by whimsical deities in anything from mountain caves to mystic coconuts. The Hawaiian wind god Laamaomao, for example, who lived on the island of Molokai, stored the winds in hollow gourds; and the Chinese wind god, Feng Po, carried the winds tied in a yellow sack on his back. In the days when sea travel was wind-powered, much depended on the gods' willingness to let the winds out or rope them in on demand. Feng Po – when he felt like it, which he didn't always – merely opened the neck of his sack and pointed, and the winds blew from that direction. This wasn't always the desired direction; and Chinese sailors once launched paper boats from their ships in hopes of distracting improperly aimed winds and leading them astray. Sometimes the wind didn't blow at all: in Homer's *Iliad*, becalmed Agamemnon, desperate for a wind to waft his ships to Troy, sacrifices his eldest daughter Iphigenia, thus setting himself up for a retributive bloodbath on his return home. Or, alternatively, it blew too much: in the *Odyssey*, Odysseus, heading back to Ithaca and the arms of his patient Penelope, is given a goatskin bag of winds by the wind god Aeolus. All goes well until he falls asleep and his curious crew, hoping for treasure, pulls the bag open and unleashes a disastrous gale.

In the Middle Ages, the wily winds were thought to be the province of witches. The *samis* – weather magicians – of Lapland were famed for capturing the winds in knots on ropes, which they then sold to sailors. Gustavus Adolphus, the Swedish king who invaded Germany during the Thirty Years' War and thus gained possession of Schonbein's Pomerania, is said to have owed his victories in part to Lapp wizards, who called up helpful winds. Shakespeare was no stranger to the wind wisdom of witches: the witches in *Macbeth*, denied chestnuts by a shortsighted sailor's wife, call up a storm to batter her husband's hapless ship; and Prospero uses his magical arts to provoke the fateful tempest. Anyone, on the other hand, is said to be able to whistle up a wind, or – by tossing coins in the water – to buy a wind; and sticking a knife in a ship's mast is said to summon a wind blowing from the direction the handle is pointing.

The creation myths of the Navajo or Dine people of the American Southwest are tales of wind. The first people – made, depending on the story, either from soil and water or from corn, turquoise, and crushed white shells – were flabby and lifeless until the wind filled them, giving them breath, the power of thought, and the ability to stand erect. When a child is conceived, it takes one wind from its mother and another from its father, and merges them into a distinctive wind of its own. That personal wind, spraying out from the top of the head and from the whorls on the tips of the fingers and toes, determines an individual's character and personality. It was also the wind, according to the Navajo, that taught people how to talk.

Navajo has no written alphabet; and, as of the twentieth century, only a handful of people other than members of the tribe understood the Navajo language. During World War II, Navajos developed a military code based on their native tongue for use in the Pacific theater, a brilliant device which – despite the best efforts of Japanese cryptographers, who cracked every other code the Americans came up with – remained impenetrable throughout the war. The Navajo code was an indecipherable mix of substitutions and transliterations – Navajo "whale," for example, for battleship; and "turkey rain" for terrain; and spoken over a field radio, it sounded, said one listener, like a cross between the call of a Tibetan monk and the gurgle of a hot-water bottle being emptied down the drain. The code played an essential part in the Allied victories at Guadalcanal and Iwo Jima, and the participating Navajos – very belatedly, in 2001 – received Congressional Gold Medals for their heroism. They were known, then and now, as windtalkers.

Wind talk, in a sense, is common to all of us, for the wind as metaphor has infiltrated human language. The wind is nothing or everything: if you know which way the wind blows, you're a canny customer, but speaking to the wind is an exercise in futility. Going down the wind, since the financially unstable days of Samuel Pepys, has meant going bankrupt, though stumbling upon a windfall means happening upon an unexpected gift of money. Windbags bore us at parties; and critical companions take the wind out of our sails, at which point we may fly to the bottle, thus rendering ourselves a lugubrious three sheets to the wind.

In Margaret Mitchell's Civil War epic *Gone with the Wind*, the wind is a metaphor for war, sweeping away the lifestyle of the antebellum South – white-pillared plantation houses, thoroughbreds, and mint juleps on the veranda, all gone with Sherman's army, the "wind that swept through Georgia" – though in black poet Langston Hughes's "A New Wind a Blowin,'" that same wind is freedom. Catullus, who must have been thwarted in love, wrote that a woman's faithless words to her lover are best written in wind and water; and folk singer Bob Dylan mourned that the unknowable answer to righting humanity's wrongs "is blowing in the wind."

Wind is a synonym for breath: to knock the wind out of somebody is to literally or figuratively render them breathless; and to get a second wind – a phenomenon familiar to exhausted runners – is to experience a revitalizing burst of energy. "Don't spit against the wind" is not only practical advice, but a warning that those who oppose prevailing opinion may suffer for it; and the Biblical "He who troubleth his own house will inherit the wind" – though subject to interpretation – is usually taken to mean that those who deliberately stir up dissent will ultimately end up with nothing. On the other hand, "It's an ill wind that blows no man good," a platitude that has been smugly repeated since the sixteenth century, refers to the optimistic belief that even the most awful of situations is of benefit to somebody.

Some winds, however, are just plain ill. We live on the western face of a mountain. Admittedly it's a small mountain

– Vermont's Green Mountains average 2000 feet (600 m) – but it still gets wild up here. To date, we've lost a storage shed; three trees; the copper roof over the living room – it peeled like a banana, popping shingles; the vinyl cover to the barbecue grill – last seen heading north, flapping like an outsized bat; and a straw hat and most of the Sunday *New York Times*, which I foolishly left on the unprotected front porch. And this, compared to what the wind *can* do, is small potatoes.

On November 7, 1940 – a date that lives in infamy in the annals of physics – the Tacoma Narrows Bridge over Washington's Puget Sound, buffeted by a 38-mile-per-hour (60-km/h) wind, began to oscillate ominously up and down. Within two hours, the oscillation had built to a series of rollercoaster-like standing waves, rising and falling to such an extent that panic-stricken drivers abandoned their cars and fled on foot to safety. The wind then began to blow harder, increasing to 42 miles per hour (70 km/h), at which point a crucial support cable snapped, setting up torsional oscillations – twisting corkscrew-like motions – with wave amplitudes of 28 feet (8.5 m) or more. A few minutes later the bridge tore itself apart and crashed into the Sound, an event that was later to become known as the "Pearl Harbor of engineering." For years afterward, physics students – including me – were taught that the Tacoma disaster was the result of resonance: that is, the frequency of the wind-induced oscillations had matched the natural frequency of the bridge structure, which cooperative pairing dramatically increased wave amplitude and ultimately ripped the bridge to pieces. Later and more precise calculations suggest that the destruction was due to something else altogether: von Karman vortex shedding, for example, or aerodynamic instability oscillations. The prime culprit, however, was wind.

Just how fast the wind blows is measured precisely by anemometer – from the Greek *anemos*, wind – or by educated guess, using the Beaufort Wind Scale. The Beaufort Wind Scale was devised in 1805 by Admiral Sir Francis Beaufort – then the 30-year-old captain of HMS *Woolwich*, assigned by the Royal Navy to conduct a hydrographic survey of the Rio de la

Plata region of South America. The scale was based on what Beaufort knew best – the British sailing frigate – and ran from 0 (dead calm) to 12 (hurricane), each ranking based on astute observations of a ship's behavior in rising wind. In the late nineteenth century, the scale was extended to 18 to accommodate the raging winds of newly explored Antarctica.

Beaufort's scale, formally adopted by the Navy in 1838, proved to be almost infinitely adaptable, as applicable to people on land as to ships at sea. At a Beaufort force of 2, as the wind fills the average frigate's sails, people become physically aware of it as a light breeze ruffling hair and blowing coolly across skin. By force 3 and 4 – what Beaufort deemed a moderate breeze – a frigate is barreling along at five knots, all twenty-two sails unfurled; while on land, twigs and leaves are flapping, dust is swirling from the ground, and kites are lifted aloft. By force 5, branches are whipping wildly and we have to struggle to open an umbrella; and by force 6 – strong breeze headed for gale – a frigate, getting nervous, is reefing in its sails; the Weather Bureau is issuing small craft warnings; and people are having a hard time walking upright. Force 6 – a wind speed of about 30 miles per hour (48 km/h) – is the biological wind threshold; people and animals attempting to navigate through it are reduced to crouching, cowering, or crawling on hands and knees.

Wind, as Lyall Watson points out, can be iffy stuff for the bipedal, many of us towering six feet (1.8 m) or more above the ground, and all of us balanced on narrow pedestals no bigger than the soles of our shoes. Wind increases in strength with altitude – that is, it's worse higher up; conversely, the closer to the ground you are, the less of it there is. Official wind velocity is measured 32 feet (10 m) above the ground – that is, higher than the rooftop of the average house; it's this figure that appears in the weather reports. The blast that hits you in the face as you edge out of the neighborhood coffee shop, however, is considerably less. A storm-force wind of 55 mph (90 km/h) blowing high above the chimney tops is slowed by a third – to 36 mph (60 km/h) – at average human head level;

and is reduced to a gentle zephyr (6 mph or 10 km/h) by the time it reaches the level of the knees. This stratification of wind velocities helps keep you upright, though still, when you're walking through a storm, "keep your head up high" is poor advice.

By the time wind speed reaches 12–18 mph (7–11 km/h) or so – force 4 on the Beaufort scale – we're also usually beginning to get cold. The wind chill index, which details the relationship between wind speed and ambient temperature of the air, was developed by Antarctic explorer Paul Siple, who began his career in 1928 when, as a teenager, he represented the Boy Scouts of America on board the *City of New York*, Commander Richard E. Byrd's Antarctic expedition flagship. His wind chill index, perfected in 1939, is based on the observation that the faster the wind, the colder a given temperature feels. In some circumstances, wind chill is dangerous; temperatures that are ordinarily cold but bearable become, when coupled with wind, fearsomely frigid, putting those exposed to it at risk of frostbite. A temperature of 5°F (-15°C) with the air standing still, for example, dips to an effective -10°F (-23°C) with a wind blowing at 10 mph (16 km/hr) and to -20°F (-30°C) at 30 mph (48 km/h).

Wind chills because it strips away our insulating blanket of air. As mammals, we're warm-blooded creatures, chugging biological engines, continually manufacturing heat at a rate of 50 kilogram-calories per hour per square meter of body surface. If we were completely insulated, this would be enough to cook us to death within a few hours. In the 1964 James Bond movie *Goldfinger*, secretary Jill Masterson dies when her evil gold obsessed boss coats her entire body in gold paint. The infallible Mr Bond attributes her demise to asphyxiation, but the true explanation is overheating. The gorgeous Jill, impenetrably encased in pigment, self-broiled.

Under unpainted circumstances, we steadily get rid of our heat excess, which is how we manage to maintain a normal body temperature of 98.6°F (37°C). We do this by direct radiation – mostly from the tops of our heads; or by the evaporation of water through our two million sweat glands. We also lose some heat through convection. Each of us is encased in a personal atmospheric haze – a thin layer of air, some 4–8 mm thick, that is constantly in motion as air closest to the body's surface warms and rises, to be replaced by cooler air from our surroundings. Wind rapidly dissipates this cozy layer; in a 4–7 mph (7–12 km/h) breeze, it's reduced to 1 mm; and at 30 mph (50 km/h) – the biological wind threshold – it's down to a mere 0.3 mm. Wind-induced loss of insulation increases our rate of heat exchange with the air, which in turn makes us cold. ·

It's not love that keeps us warm. It's air. Fur works because it's an air-grabber, keeping the insulating layer of surface air warmly intact and close to the skin. Feathers fulfil the same function; when birds, in cold weather, fluff out their feathers, they do so to trap the maximum amount of insulating air. People, who in cold climates would operate much more efficiently if furry, insulate themselves with clothes. Clothing, be it Armani, Calvin Klein, or Army–Navy surplus store, exists for a single practical purpose: to trap air. The insulating value of clothing is measured in units of clo, in which 0 clo means stark naked and 1 clo is the amount of insulation required to keep a relatively sedentary person comfortable at room temperature (70°F or 21°C). Since the clo scale was established in pre-feminist 1941, the 1-clo standard of comfort was defined as businessman's garb: a three-piece suit and a set of BVDs. The secret to effective insulation is layering, in which multiple layers of clothing trap multiple layers of protective air. This technique is maximized by the Inuits, experts in the matter of clo, whose parkas, pants, boots, and mittens total up to 4.0 clo, almost as good as the insulating value of the average sled dog's fur coat (4.1 clo), but not a patch on the average polar bear (8.0 clo).

Even if we're properly insulated, wind can cause us problems. Some studies indicate that wind boosts blood pressure, exacerbates chronic pain, triggers the synthesis of stress hormones, and makes schoolchildren quarrelsome and hyperactive. The sound of wind makes us anxious, which is why its eerie whine is such a staple of horror movies and Halloween ghost stories; and if the wind blows long enough, it can make us crazy. A mid-nineteenth-century law in windswept Wyoming ruled wind-induced insanity a viable defense for murder.

In psychological and physiological terms, some winds are worse than others. The east wind, according to Voltaire, who blamed it for his chronic ill health, was a bringer of "black melancholy," responsible for depression and suicides. (King Charles I was beheaded in an east wind, he pointed out tellingly, and King James II was deposed in one.) Shakespeare blamed the north wind for "gout, the falling evil [epilepsy], itch, and ague;" and Theophrastus accused the south wind of rendering men "weary and incapable," due to a thickening of the lubricant in their joints. Hippocrates, who arguably initiated the study of biometeorology in the fourth century BCE with *Airs, Winds, Places*, gave each wind its own disease profile: the warm wet south wind, for example, caused headaches, deafness, dimness of vision, sluggishness, and epilepsy; the cold dry north wind, coughs, sore throats, and constipation.

In north Africa and the Middle East, the sirocco – a hot dry southeasterly wind from the Sahara, most common in winter and spring – brings with it a rash of adverse reactions, among them migraines, allergies, irritability, insomnia, depression, nausea, and fatigue. This evil wind is known by a host of regional names: *sharav* in Israel, *levante* in Spain, *leveche* in Morocco, and in Egypt *khamsin*, which name comes from the Arabic for fifty, which is the number of days it is said to blow. The sirocco is featured in the Bible, where it carried blights and plagues of locusts.

The *foehn* of the Alps – like the chinook of the North American Rockies and California's Santa Ana – is a katabatic wind, formed when a moving bloc of air encounters a mountain range, loses its moisture on the mountain peak, and then rolls downhill in a hot dry wave, boosting leeward temperature by as much as thirty degrees in minutes. "Chinook" is a northwest American Indian term meaning "snow-eater;" "foehn" means fire. This parched wind – said to be hot enough to bake apples on the trees – touches off forest fires, and plays havoc with the population: people complain of aching joints, headaches, exhaustion, and respiratory problems; traffic accidents multiply; and crime rates climb. The Santa Ana winds, writes detective novelist Raymond Chandler, "come down through the mountain passes and curl your hair and make your nerves jump and your skin itch. On nights like that every booze party ends in a fight. Meek little wives feel the edge of the carving knife and study their husbands' necks. Anything can happen."

There are also truly horrific winds. Empedocles, when he first proposed his four-element model of the world, attributed each to a ruling god. Earth was the province of Hera; fire of Hades; and water of Nestis or Persephone. Air was the element of Zeus, the angry and tempestuous god of storms. From air, writes Pliny the Elder, comes trouble: "clouds, thunderclaps, thunderbolts, hail, frost, rain, storms, whirlwinds and most human misfortunes, and the struggle between the elements of Nature."

Dominating such airborne human misfortunes are the phenomenal winds known as hurricanes. Hurricanes are named for the Mayan weather god Huaracan, whose tantrums were said to whip up the horrendous storms that annually devastate the Caribbean. Such storms are anger on a truly supernatural scale, measuring hundreds of miles across, generating winds up to 250 miles per hour (400 km/h), and unloading up to 20 billion tons of water a day on the shrinking Earth below. A hurricane is the weather equivalent of a thousand simultaneously exploding 20-megaton bombs.

Egged on by the rapid rotation of the Earth, hurricanes whirl counterclockwise in the northern hemisphere and clockwise in the southern. Those born in the Indian Ocean are called cyclones, from the Greek for "circle;" those that develop west of the International Dateline are known as typhoons, from the Chinese for "great wind." In Japanese history, the typhoon is the prototypic ill wind blowing unexpected benefits; without it, Japan in the thirteenth century would almost certainly have become a resentful subsidiary of the powerful Mongol Empire, under the heavy thumb of Kublai Khan. The Khan, amassing a vast armada of over a thousand ships, attempted invasion twice, but in both instances was thwarted, his navy dispersed and destroyed by raging typhoons. The Japanese, certain that the providential storms had been sent by the gods, began referring to them as "divine winds." In World War II, young Japanese pilots willing to volunteer for suicide missions were known collectively as the Divine Wind Attack Force: *kamikaze*.

Tornados, in contrast to the gargantuan hurricanes, are mean but small. While the average hurricane is half the size of the state of Texas, most tornadoes can be measured in feet (or meters); and almost all are under a mile (1.6 km) in diameter. The United States is the world leader when it comes to tornados, spawning 800 or so per year, mostly on the plains of the Midwest. The swathe of territory stretching north from Texas through Oklahoma, Kansas, and Nebraska is commonly nicknamed "Tornado Alley;" and it was here that a tornado, roaring across the farmland of Kansas, picked up Dorothy and Toto in L. Frank Baum's classic fantasy, and transported them, farmhouse and all, to the land of Oz.

Tornados are the offspring of supercells, particularly powerful and long-lived thunderstorms in which rising air, lashed by turbulent winds, begins to rotate as it climbs, forming a mesocyclone. As the rotating mesocyclone picks up speed, air pressure plummets at its center. Warm moist air promptly roars in to fill the low-pressure gap and is yanked upward at speeds of over 200 miles per hour (320 km/h). Fed with enough air, the mesocyclone grows steadily larger, developing into an immense funnel-shaped tube of whirling wind. The low-pressure area at the mesocyclone's heart allows air to expand, which causes it to cool. As it does so, its water vapor condenses, forming a cloud, which is what makes the nascent tornado visible. At this stage it resembles a Brobdingnagian trumpet, poking downward, mouthpiece first, from the bottom of a cumulonimbus cloud. Some of these eventually drop to the ground like lethal tops, at which point they become tornados.

Once on the ground, a tornado has all the destructive unpredictability of a terrorist with a bomb. A tornado's spinning winds are the most violent on Earth, reaching velocities of 300 miles per hour (480 km/h) or more, creating a central updraft of 200 miles per hour (320 km/h), capable of sucking up trucks, trailers, and entire herds of cows. Most tornados travel at speeds of 20– 40 miles per hour (32–64 km/h), in a surface path generally about a mile wide and only a few miles long (1.5 km wide to a few km long), though some determined examples travel 50 miles (80 km) or more before dissipating. Some bounce, like outsize jumping jacks, lifting skyward, then touching down again to blast all beneath to bits.

One theory holds that the first known description of a tornado is found in the Old Testament, in the Book of Ezekiel, which dates to approximately 600 BCE. The crucial passage (Ezekiel 1:4) speaks of "a whirlwind" from the north, "a great cloud, and a fire infolding itself, and a brightness was about it, and out of the midst thereof as the colour of amber, out of the midst of the fire." The meteorological validity of the observation is somewhat weakened by subsequent verses, in which Ezekiel describes four-winged creatures emerging from the aforementioned whirlwind, each with four faces and brass calves' feet; and Josef Blumrich in *The Spaceships of Ezekiel* claims that the incident refers not to a tornado, but to a rocketship bearing visitors from outer space. The Romans were somewhat more reliable. Pliny the Elder in his *Natural History* describes a whirlwind that "snatches things up and carries them back with it to the sky." If such a thing approaches you at sea, the only remedy is to pour vinegar in its path; unfortunately, he adds regretfully, this doesn't work very well.

Wind, when there's too much of it, is a killer. A cautionary

North American Abenaki legend, on the other hand, describes what happens when there's no wind at all. A young warrior, Gluscabi, wants to hunt ducks on a lake but is unable to launch his canoe because of the wind. Angered, he travels to a distant mountain where he finds the cause of the wind – an immense eagle perched on the summit, beating its wings. Gluscabi captures the Wind Eagle, ties its wings to its sides, wedges it upside-down in a cleft in the rocks, and heads triumphantly for home. It soon becomes clear, however, that the young man has made a terrible mistake: the air is now hot and stagnant, the water filthy and motionless, the land withered and dry. Gluscabi, chastened and appalled, hurries back to the mountain, frees the Wind Eagle, and returns him to his place. Once more the air is cool and fresh, the water is clean, and the Earth washed with rain. Gluscabi, however, sagely manages to strike a bargain. He convinces the eagle that the wind doesn't have to blow all the time.

Aspiration and inspiration, like spirit, derive from the Latin for air or breath; and both lead us skyward. When striving to improve ourselves, we struggle to rise; to succeed is to fly up; to fail is to fall. We have soaring visions and flights of fancy; if we're hard-working and lucky, our dreams can take us sky high. Angels fly through the air as a mark of having attained spiritual perfection; and heaven, traditionally, is located somewhere far above our heads. The human race, according to Dante, was born to fly upward, though our sinful natures often take us quite the other way; Thoreau thought this was just as well, writing prophetically "Thank God men cannot fly, and lay waste the sky as well as the earth." Still, our myths and folktales are filled with fantasies of flight. We imagine flying horses, flying carpets, flying chariots, and whole islands that hover magically in the air. Fairies flit from flower to flower; witches scud gleefully across the sky on broomsticks. Hermes, messenger of the gods, flew on winged sandals. Iris, goddess of the rainbow, Nike, goddess of victory, and Nut, the Egyptian sky goddess, all had wings.

Eagles, bats, butterflies, and flying squirrels are all creatures of the air, able to move through it with an ease that humans – with our leaden bones and uncongenial shapes – have always envied. Traditionally, however, once we launch ourselves into the air, we're asking for disaster. In the story of Daedalus and Icarus, the quintessential cautionary flight tale, father and son attempt to escape from King Minos's prison on Crete by flying to the Greek mainland on immense feathered wings held together with wax. Icarus, an exuberant teenager, ignores his father's instructions and flies too near to the Sun; the wax melts, his wings disintegrate, and he falls fatally into the sea. Daedalus, horror-stricken, flaps sadly on alone.

In practice, of course, both father and son should have taken a dive. Wings large enough to lift a human being would be too heavy for human arms to move; anatomically, we're simply not meant to get off the ground. Birds, in contrast, are built for flight, stripped for aerial action. Avian anatomy is a masterpiece of parsimonious engineering. All excess has been eliminated: birds lack a urinary bladder, and their reproductive systems – except during breeding season, when a female's single ovary may increase 1500-fold in size – are reduced to the bare minimum. Heavy tail, teeth, and jawbones, and assorted finger and leg bones have been done away with altogether; and many bones, such as those of the skull and pelvic girdle, have been economically fused, thus eliminating extra weight due to joints. In mammals, bones are broader and thicker at the joints, the better to absorb the force of life's activities, such as running, walking, and jumping up and down. The jogging knee routinely absorbs a force up to seven times body weight; it's necessarily a heavy-duty construct. The trade-off for birds, having markedly fewer joints than we do, is inflexibility. We're earthbound, but we can do yoga. Birds are stiff.

Bird bones, unlike the solid marrow-packed bones of mammals, are famed for being hollow, which gives the erroneous impression that they're also as light and flimsy as soda straws. The opposite is, in fact, the case; the hollow bones are strongly reinforced with multiple internal crosspieces or struts, and are often equal in weight to those of an equivalently sized mammal. A duck's skeleton, for example, weighs about the same as a rabbit's, though weight is apportioned differently. A rabbit's skull, with all those weighty carrot-chomping teeth, is about three times heavier than that of a duck. Ducks, on the other hand, while light-headed, have truly impressive breastbones. While the mammalian sternum is flat, the avian sternum is equipped for flight with a large and prominent "keel" – so named for its resemblance to the keel on the bottom of a ship – to which the massive muscles of the chest attach. These muscles are further braced by the furcula – a Y-shaped fused collarbone positioned in front of the keeled sternum. (This, commonly known as the wishbone, is the bit of the skeleton that is traditionally saved from the holiday turkey.) The reason for such elaborate chest structures is that flight – no place for sissies – requires immense expenditures of muscle power. Chest muscles in the average human account for a mere 1 per cent of body weight; in the average bird, a whopping 15 per cent. Every chickadee, chest muscle for chest muscle, is a biologically enhanced Arnold Schwarzenegger; birds, in effect, are muscular barrels on toothpicks. There's no way Daedalus had the thoracic wherewithal to flap. A human, to support the chest muscles necessary for flight, would need a breastbone six feet (1.8 m) long.

We're also barred from flight by our sluggish metabolisms. The immense demands of taking to the air are reflected in the nature of avian respiration. The average mammal dedicates about 5 per cent of its body volume to its respiratory system; the average bird – a veritable balloon – dedicates 20 per cent. A bird's lungs connect to a system of large thin-walled air sacs, which in turn connect to the air spaces in the hollow bones. This elaborate interconnecting network makes for a highly efficient rate of oxygen exchange, which in turn is helped along by the bird's powerfully pumping heart. The avian heart is proportionately four to five times larger than ours – to keep up, the fist-sized human heart would have to swell to the size of a cantaloupe – and it beats at a rate of up to 400 beats per minute at rest, 1000 or more per minute in flight. (The average adult human heart, in contrast, beats about 70 times per minute when its owner is lolling about in the hammock; double that after a hard bout of exercise at the gym.) Birds also breathe faster than we do. Our average 14–25 breaths per minute when not doing much of anything increases to some 80–90 breaths per minute after a fast run; birds, just pottering about the feeder, take about 40–50 breaths per minute, and 450 per minute or so when airborne.

A congenial anatomy and physiology help, but the bottom line in flight is lift, a phenomenon generated in birds by the shape of the wing. A bird wing, seen in cross-section, is roughly shaped like a lopsided teardrop, with a flattish lower surface and a longer convex upper surface, rounded at the leading edge and tapering to a point in the rear – a shape known to the aeronautical industry as an airfoil. When a bird launches itself forward, the air separates before it, flowing both over and under the wing. Since the upper surface of the wing is longer than the lower, air is forced to travel faster to cross the wing's top. Rapidly moving air exerts less air pressure than slowly moving air; thus higher pressure beneath the wing lifts the bird up. This crucial effect is known as Bernoulli's Principle, after Swiss mathematician Daniel Bernoulli, who published an elegant description of it in 1738.

Once you've got wings, flying looks deceptively easy, which is much of the allure of air. Who wouldn't prefer swooping effortlessly across the sky to trudging around on the ground, hauling things? In some cases, swooping is indeed effort-less: king-size birds such as the condor and albatross, with wingspans of 10 feet (3 m) or more, tend to ride dreamily on likely wind currents or uprising thermals, or to glide, which is a matter of spreading out as flat as possible in the air and entering into a slow graceful fall. Smaller birds routinely flap – essentially rowing themselves through the air, with deft twists of wings and feathers designed to maximize the wing-crossing flow of air. Smaller wings, which cover smaller areas of air, generate less lift; thus in general the smaller the bird, the faster the flap. Tiny hummingbirds, in order to keep aloft, beat their wings at a frenzied rate of 80 beats per second. Hummingbirds, in fact, can't glide – their wings are too small – and their exhausting lifestyle explains why 30 per cent of their body mass goes into chest muscle.

A butterfly flutters at the gentle rate of 8–12 wingbeats per second, but insects by and large are frenetic flappers. The vibration of their wings is so fast that it generates a whirring buzz, which increases in pitch with beat rate: thus the low ominous buzz of the bumblebee (130 wingbeats per second) versus the ear-piercing whine of the hungry mosquito (600 wingbeats per second). Insects, however, are puzzling exceptions to the airfoil rule; their wings are as flat as sheets of paper, which means that they are not subject to classic Bernoullian lift. Still, insects do fly, and well, despite the much-touted legend that the bumblebee is aerodynamically impossible, since its wings are too small to generate sufficient lift to hoist its fuzzy bulk out of the hive. This apiaran slander dates to the 1930s and – though no one likes to name names – results from an erroneous calculation in which the bumblebee wing was equated to the wing of a conventional airplane. Insects, however, function neither like airplanes, nor – for the most part – like birds. Some, like butterflies and moths, can glide, skilfully riding air currents like the seabreeze-riding albatross; most airborne bugs, however, depend on sheer wingbeat, employing a battery of adroit claps, flaps, twists, and rotations to keep themselves aloft. Some entomologists hypothesize that insects do not so much fly as swim – air, to something the size of a gnat, is thick viscous stuff, through which insects forcibly row themselves, as if paddling through a vat of molasses.

From earliest times, primitive people doubtless watched all this enthralling flapping, soaring, and gliding with envy and calculation. Lacking the means to launch themselves, early humans first flew by proxy, via handcrafted airfoils. The oldest human-made flying device is probably the boomerang. The name is aboriginal Australian, but variations of it – a two-armed throwing stick, with a curved upper surface and flat bottom – were in use in India, Africa, Europe, and North America by at least 6000 BCE. Boomerangs, brandished with lethal intent, are featured in European cave paintings and Australian petroglyphs. The natives of the American Southwest used them for downing ducks and rabbits. The ancient Egyptians used them for bird-hunting – a particularly elegant model carved from ivory was found in Tutankhamen's tomb – and the ancient Greeks had a boomerang-like weapon known as a *lagobolon*, or "hare thrower."

Old, but not quite as old, is the tethered airfoil or kite, which seems to have originated in China, perhaps as early as 2600 BCE, though the first written record of Chinese kite-building dates to the fourth century BCE. The described kite, which reportedly took three years to construct, was an eagle made of silk and bamboo. From China, kites spread to Japan, where they were called *kami tobi* – "paper hawks" – suggesting that early Japanese kites were also in the shape of birds, and then to southeast Asia and Indonesia. Eventually they even reached the Pacific Islands, where the Samoans cleverly harnessed them to pull their canoes.

By the time Marco Polo reached China in 1282, kites were routinely being used as aerial workhorses to carry bricks and tiles to workmen on pagoda towers; and some were capable of lifting men (fools or drunkards, according to Polo, who found the process unnerving). In Japan, which acquired the kite in the seventh century CE from Buddhist missionaries, the building of man-lifting kites was strictly forbidden for fear of aerial invasion. None the less, in a famous Japanese legend of the twelfth century, the samurai warrior Minamoto-no-Tametomo, exiled with his son to the island of Hachijo in the aftermath of a clan war, builds a giant kite on which his son escapes to the mainland. Less edifyingly, a man-lifting kite was used by an enterprising robber with an eye on the solid gold dolphins that adorned the roof of Nagoya Castle near Osaka, the seventeenth-century residence of the Tokugawa shogunate. He managed to make off with the fins, but couldn't resist boasting of his accomplishment, and so was apprehended and boiled in oil.

The kite only reached Europe in the sixteenth century, most likely via trade with the East Indies. Marco Polo's earlier kite descriptions seem to have made little impact; his *Travels*, published in 1300, some eight years after his return from the Far East, were referred to by a skeptical public as *Il Milione* – "The Million Lies" – which may have had something to do with such inclusions as the story about the monster birds who picked up elephants, dropped them on rocks, and then devoured their battered remains. Once introduced, however, the kite rapidly became popular, especially among children; Isaac Newton as a schoolboy used his to hoist lighted lanterns at night, thus terrifying the neighbors, who thought they were comets.

The first European to engineer a person-lifting kite appears to have been an English schoolteacher from Bristol named George Pocock, a hobby inventor who had already in his spare time devised an automated caning machine for the effortless discipline of unruly boys. In the early 1820s, Pocock managed to send his 12-year-old daughter Martha 300 feet (90 m) in the air in an armchair suspended from an immense kite. He then went on to invent the *char-volant*, or "flying carriage," a light-weight kite-drawn vehicle capable of carrying four passengers. The original carriage, which looks quite fun in the illustrations, was propelled by two kites, respectively 10 and 12 feet (3 and 3.7 m) across; and could reach the – for the day – lightning-like speed of 20 miles per hour (30 km/h). Pocock, driving one, managed to outrun the Marlborough stagecoach and (impertinently) the coach belonging to the Duke of Gloucester. He also made a triumphant appearance at the Ascot races in 1828, where his creation is said to have much impressed King George IV.

The most productive use of the kite, however, was undoubt-edly that of a pair of bicycle-shop owners from Dayton, Indiana, named Wilbur and Orville Wright. The Wright brothers tested their early aviation theories on specially designed kites, perfecting their designs until – on the cold morning of December 17, 1903, at Kitty Hawk, North Carolina – the *Flyer* made its maiden flight: a distance of 120 feet (36.5 m), a total of 12 seconds in the air. The first heavier-than-air engine-powered machine had successfully gotten off the ground; and man – to what would have been the vast dismay of Thoreau – had at last conquered the elusive element air.

"Sail away from the safe harbor. Catch the trade winds in your sails. Explore. Dream. Discover." The quote is attributed to Mark Twain, but the desire is universal. Most of us have longed at one time or another to throw caution to the wind and go. Air, like water, is the element of travel and adventure; the song of the wind is the song of the open road.

As a host of proverbs and metaphors attest, we can't control the wayward wind. Sometimes, however – if and when a likely breeze shows up – we can catch a ride. Our first device for effectively doing so was almost certainly the sail. Sails have been around for at least 5000 years. An Egyptian pot from 3000 BCE pictures a boat equipped with one; and a lovely model of a fully rigged sailing ship was found in Tutankhamen's tomb. By 1000 BCE the Phoenicians, through a masterful exploitation of the device, had become the major naval power of the ancient world, dominating the Mediterranean Sea, founding the city of Carthage in north Africa, and establishing commercial ports as far away as Spain – which name comes from the Phoenician meaning "rabbit coast," possibly in honor of ancient sailors' suppers. Phoenician ships ventured past the Strait of Gibraltar, traveling as far as the British Isles for Cornish tin, and in the seventh century BCE, under the aegis of the Pharaoh Necho II, a particularly daring Phoenician crew managed to circumnavigate Africa.

The Greeks were sailors. Odysseus, canny when it came to wind and water, managed to escape from Calypso's island by building a raft and rigging a linen sail; and the Greek sail is commemorated in the night sky in the constellation Argo Navis, the Ship, said to be the vessel in which Jason and his Argonauts set off in search of the Golden Fleece. The Ship is visible low on the southern horizon, though not much of it, unfortunately, from my particular latitude. Its Sail, Vela, billows in the celestial wind, a somewhat raggedy rectangle of stars projecting into the Milky Way.

Wind drove much of the exploration of the globe. The Polynesians, in double-hulled outrigger canoes with sails of woven pandanus leaves, settled ten million square miles (26 million sq km) of the Pacific; the Vikings, in longships topped with striped wool sails, established colonies as far afield as Russia and Newfoundland. The wind-powered sail was the hallmark of the European age of exploration (or invasion, if you happened to be on the receiving end), blowing Vasco da Gama to India, Columbus to the Caribbean, Cortez to Mexico, Magellan around the world, and Captain James Cook, explorer extraordinaire, from Tahiti to Antarctica to the Bering Sea.

Wind allowed Portugal, Spain, Holland, France, and ultimately Britannia to rule the waves, as well as – for a brief moment in the fifteenth century – the Chinese. In the heyday of the Chinese navy, a period stretching from 1405 to 1433, a treasure fleet of over 300 immense nine-masted junks set sail periodically from Nanjing, eventually visiting thirty-seven countries, from southeast Asia to the Malay archipelago, India, the Persian Gulf, and Africa. The fleet was commanded by Zheng He, a eunuch, master navigator, and trusted confidant of the emperor – and a giant of a man, said to tower over eight feet (two and a half meters) tall. His ships collectively carried over 27,000 men and were crammed with cargo – porcelains, silk, lacquerware, gold and silver vessels, mercury, umbrellas, and straw mats – for trade with the barbarians, with whom the commander was expected to establish friendly relations, while awing them with Chinese might. The expeditions were highly successful. Seventeen of the countries contacted subsequently sent emissaries to China; and Zheng He's various voyages swelled the emperor's coffers with pearls, coral, ivory, gemstones, and rare woods, and added to his private zoo lions, leopards, ostriches, zebras, an oryx, and a giraffe (misidentified by the Chinese as a unicorn). Retired British naval officer Gavin Menzies, in his controversial book *1421: The Year the Chinese Discovered America*, argues that Zheng He and his colossal fleet may even have reached Australia, rounded the Straits of Magellan, and explored the western coast of North America. In lieu of establishing an empire upon which the Sun never set, however, China furled her sails. A return to conservative Confucianism brought with it isolationism; the barbarians of the outer world were deemed to have nothing to offer the elite and sophisticated citizens of the Middle Kingdom. By the beginning of the next century, Chinese ships and shipyards had been deliberately destroyed. The Chinese never set sail again; and the wind and seas were left to the smaller and shabbier ships of the West, who set out in them to colonize, convert, and exploit the globe.

Many of us live where we live today because of wind. The millions of Americans descended from the 102 *Mayflower* Pilgrims, for example, are on the far side of the Atlantic because their forbears set sail. I live in New England because one morning in 1643 Stephen Bradley, born in Cornwall, and his wife Elizabeth, stepped on board a sailing ship bound for the New World.

Air, for all its seeming fragility, is power; and some time in the sixth or seventh century CE the Persians inveigled it into operating windmills. The first wind-powered mills, dotting the windswept plains of what is now Iran, were squat square towers topped with a horizontal bladed wheel that spun in circles rather like a carousel or Christmas glockenspiel. The wheel was attached to a shaft that ended in a grindstone; turning, it ground grain into flour, a task that to date had been performed by hand by women or, at best, donkeys. The windmill arrived in Europe in the eleventh century, either brought back from the Middle East by the Crusaders or perhaps invented independently – some sources credit a nameless bright soul in East Anglia. The European version was an improvement on the Persian model, the horizontal blade flipped upright to create the post windmill with its four broad vertical sails – the charming windmill of Dutch-patterned plates and children's picture books. It was a cluster of post windmills that Don Quixote battled on the plains of Spain.

The giant-armed vertical windmills were three to five times more powerful than their horizontal predecessors; and by the twelfth century, they had become so popular and numerous that they attracted the attention of the pope, who proceeded to tax them. The ubiquity of wind- (and the even earlier water-) mills, points out Lyall Watson, is reflected by modern surnames: the plethora of Millers, Mills, Milnes, Milners, Millmans, and the like – check any telephone directory – stems from ancestral job descriptions. By the 1800s, arguably the peak of the wind age, there were 100,000 working windmills in Europe. Germany boasted 18,000 mills, England and Holland 10,000 each. (The windmill, valiantly draining the lowlands, had given the medieval Dutch a 40 per cent increase in farmable land.) Collectively the mills ground grain, pumped water, drove saws and stamping apparatus, crushed olives for oil and pigments for paints, and pulped fiber for papermaking. Over the course of the windmill's prolonged heyday, wind was the world's number two major energy supplier, superseded only by wood. The age of wind lasted for 700 years, and only ended with the invention of James Watt's steam engine and the start of the Industrial Revolution.

Wind as a prime power source lasted longer in vast and sparsely populated North America. The American Midwest in the latter half of the nineteenth century had some six million windmills pumping water for the parched farms and cattle ranches of the Great Plains. The repetitive clunk and squawk of the working mill was a hallmark of turn-of-the-century American farm life. Illinois poet Carl Sandburg, whose poems are filled with midwestern imagery, wrote in 1918 of "Slow-circling windmill arms turning north or west/Turning to talk to the swaggering winds . . ." and farm pictures without a signature windmill look oddly blank in the middle.

The traditional American farm windmill resembles an ungainly mechanical sunflower: 20 feet (6 m) or more of scaffolding topped by a broad circular fan of feathered metal blades. Modern windmills – more accurately, wind turbines – are altogether glitzier propositions, sleek and sculpted as ballistic missiles, topped by aerodynamically tapered propellers. Gathered together in massive arrays, these look like gleaming high-tech forests – the sort of thing that might sprout if you planted titanium pellets – or eerie armies straight out of George Lucas's *Star Wars*. The largest such wind farm, 6000 turbines strong, is in Altamont Pass northeast of San Francisco; there each spinning propeller – its blades the size of jumbo jet wings – uses a rotating shaft and a generator to convert wind energy to electricity. Altogether the Altamont turbines have a 550 MW (megawatt) capacity – enough to comfortably electrify some 55,000 California homes.

About 2 per cent of the energy that falls on us as sunlight is converted into kinetic energy in the form of wind. This adds up to a gargantuan amount of energy – researchers estimate that the power in the world's winds at any given moment is sufficient to supply all global energy needs. That power, however, isn't necessarily available where and when we want it – a lot of it, for example, is blowing across the Pacific Ocean or roaring over the ice fields of Antarctica – and we have yet to devise efficient methods for storing excess wind energy for use during lean periods of windless calm. Still, collectively to date, the world's wind farms – concentrated in Germany, Denmark, Spain, India, and the United States – have a combined capacity of 14,000 MW. There is hope for the future: once a few nagging technological snags are removed, the world may indeed eventually run on air.

"Eating the air," for Shakespeare, was synonymous with star-vation; but some of life's creatures, quite literally, are fed by air. In the 1950s, biologist Lawrence Swan – born and raised in Darjeeling, India, and a lifelong lover of the Himalayas – became interested in life above the snowline. Scientific dogma to date had held that there wasn't any: the snowline marked the uppermost boundary of the alpine or polar zone, beyond which temperatures never rise above freezing, snow never melts, and all plant life – even the hardiest of lichens – perishes. Swan, however, identified a small but thriving community of snow-dwellers: tiny bristletails, springtails, and jumping spiders, these last found 22,000 feet (6600 meters) up the sides of Mount Everest, making them the highest permanent residents on Earth. Their diet, Swan discovered, was supplied via airlift. Updrafts of wind brought in regular cargoes of algae and insects, dropping them on to the waiting population below in much the same manner that post-World War II Allied pilots delivered chocolate bars to the hungry children of Berlin. Swan had discovered a previously unknown ecosystem, which he dubbed the aeolian biome in honor of Aeolus, Greek god of the winds.

The air is not only a cafeteria; for many of the planet's life forms, it's a sex machine. In some mythologies, in fact, the air is a potent sexual aggressor. An Indonesian legend claims that human beings are descended from a sky woman who climbed down a giant palm tree to Earth where she was impregnated in her sleep by the south wind; and Boreas, the north wind, is said to have impregnated the mares of Erichthonius, King of Troy, after which they bore foals as light and fleet as the wind. The Algonquin hero Hiawatha, immortalized in lengthy verse by Henry Wadsworth Longfellow, was fathered by the west wind – a faithless creature, as it turned out, who caused Hiawatha's mother to die of a broken heart. The ancient Egyptians, insisting that there were no male vultures, believed that females were fertilized by the wind.

Modern biology, however, reduces air from supernatural stud muffin to prosaic mode of transport. Plants, to maximize their chances of cross-fertilization, must disseminate their pollen; and the most straightforward way of accomplishing this is to pitch it into the breeze. Such anemophilous or "wind-loving" plants – the bane of hay-fever sufferers – collectively coat every square meter of the planet each year with 100 million grains of pollen, largely concentrated in two annual peaks: one in spring, when the temperate forest trees flower; and another in mid-summer, when herbaceous plants prepare to form seeds. Only about 10 per cent of plants are wind-pollinated, but reproductively speaking, these are wildly successful, comprising over 90 per cent of the Earth's total plant population. Among the anemophiles, for example, are all the world's grasses, including such essential cereal grains as wheat, rice, barley, oats, rye, and corn (maize). About 70 per cent of world farmland is planted in grains, with annual yields of 1.8 billion tons. Collectively these supply over half of our food energy, and the bulk of feed for domestic livestock. In effect, the wind feeds us, since without the wind-loving grasses, the world goes fatally hungry. The famines of sub-Saharan Africa that now threaten the lives of millions are the result of persistent drought and accompanying failure of essential grain crops. An apocalyptic science-fiction novel by John Christopher titled No Blade of Grass takes this frightening scenario a step farther, describing a world in which all grasses are stricken with a mysterious and fatal disease. The predictable result is global famine and total collapse of the social order. The few survivors hunker down behind barricades and prepare to plant potatoes.

Pollen travels for much the same reason grown children clamor for their own apartments: to establish themselves in new habitats, seek genetic novelty, and – through creative sexual mix-and-match – further the process of evolution. "Nature," wrote Charles Darwin, "abhors perpetual self-fertilization" – though, to be fair, it uses it in a pinch. Cross-fertilization, however, is the more powerful species survival tool, a genetic

re-scrambling that broadcasts novel genes, increases population variation, and thus enhances the ability to cope with a
broad spectrum of external selection pressures. In pursuit of
all this, amenophilous plants produce phenomenal amounts of
pollen. A single birch catkin yields over 5 million pollen grains;
a typical corn tassel produces 2–5 million grains, a total of 300
pounds of pollen per corn-planted acre (330 kg per hectare). A
mere breath of wind can sweep lightweight pollen grains half
a mile (1 km) in just a couple of minutes, though once aloft
windblown pollen may travel hundreds of miles in search of a
mate. Pine and fir pollens, for example, have been found over
700 miles (1200 km) from their parent trees.

Spores and seeds must similarly be dispersed to maximize
chances of individual and species survival; and many travel by
air. Fungal spores – just twenty-five times larger than bacteria
– can ride air currents around the globe. Mushrooms, toad-
stools, and puffballs all depend on air for distribution of spores,
which they produce in enormous quantities: the average
edible mushroom, for example, produces about 16 billion
spores, shed at a rate of 100 million per hour; and the football-
sized giant puffball (*Lycoperdon giganteum*), when struck by
falling raindrops or prodded by a curious hiker, coughs out
7000 billion. (An exception to fungal wind dependence is the
desirable truffle, whose spores are spread by animals. In North
America, the job is usually done by squirrels and chipmunks;
in Australia, by bandicoots, mice, and potoroos.)

Seeds, which consist of a plant embryo, a protective
outer coat, and, in most cases, a food reserve, are larger and
more complex than spores, though the smallest, produced
by orchids, are as fine as talcum powder. These, by virtue
of their size, have almost the travel potential of spores.
Larger seeds, in their attempts to avoid a leaden plop on to
parental ground, possess a range of aerodynamic features to
help them catch the wind: botanical versions of parachutes,
propellers, and wings. Seeds of the ubiquitous dandelion
(*Taraxacum officinale*), for example, are each attached to an
umbrella-like crown of delicately branched hairs; blown by

the breeze, these can travel anywhere from 6 to 120 miles (10 to 200 km), before parachuting gently down to sprout into yet another dandelion. Cattail or bulrush seeds (*Typha latifolia*) are released to the wind in a silky tangle of cotton fluff, each seed bearing a comet tail of fine white hairs. An acre (half a hectare) of marsh can produce up to a trillion such seeds, and these, scattered to the skies, are indefatigable travelers, Magellans of the air. Wind-transported cattails and thistles grow on such remote islands as St Helena and Tristan da Cunha, respectively 1000 and 1700 miles (1600 and 2800 km) from the nearest mainland.

Some seeds sprout stiff membranous wings that allow them to glide, flutter, or twirl like tiny helicopters. The twirlers – typically one-winged seeds – are called samaras, though the maples (*Aceraceae*), among them New England's gaudy sugar maples, bear twin versions, a double samara in which two winged seeds are joined together at the base. Gliders are characterized by paired flat lateral wings, and look rather like miniature toy airplanes. The most remarkable example is the seed of *Alsomitra macrocarpa*, a climbing gourd native to tropical Asia, that sports a pair of translucent pearl-colored wings measuring five inches (13 cm) across. These, released by the hundreds from cantaloupe-sized gourds in the rainforest canopy, spiral slowly downward in wide, swooping circles to settle on the forest floor. Some airborne seeds are completely surrounded by a thin circular wing, somewhat like Lilliputian frisbees; examples include the Argentinian jacaranda tree (*Jacaranda mimosifolia*), the Chinese sumac or Tree-of-heaven (*Ailanthus altissima*), the catalpa (*Catalpa speciosa*), and the trees of the elm family (*Ulmaceae*).

In some plants – most famously the tumbling tumbleweed of the American West – the entire plant is transported by the wind. Tumbleweed, or Russian thistle (*Salsola kali* or *tragus*), a native of the plains of Russia and the steppes of Siberia, was introduced to North America in the late nineteenth century. It grows as a nearly spherical bush with prickly pointed leaves – hence the "thistle" – which, when dry, detaches from its roots and is bowled along the ground by the wind, scattering its 20,000–50,000 seeds as it goes.

Not all plants are pollinated by the wind; many – known as zoophilous plants – are pollinated by animals. Magnolias and waterlilies, for example, are pollinated by beetles; apples and primroses by bees. Moonflowers, which open at night, are pollinated by death's-head moths; baobab trees and saguaro cacti by bats. Butterflies pollinate verbena and hollyhocks; hummingbirds pollinate fuchsia and trumpet flowers. Aspidistras are pollinated by snails; and the African stapeliads – starfish-shaped succulents that smell repulsively of rotting meat — are pollinated by flies. Zoophilous plants, unlike their modest wind-loving cousins, tend to be brazen and flashy, with large bright-colored flowers specifically designed to catch the eye of passing pollinators. As added attractions, many produce sugar-laden nectars; and many exude intense and tempting odors. Even the zoophilous, in other words, depend on air.

Air, of the four elements, is the foremost servant of the senses. Air is the purveyor of odor; we need it to smell a rose or a rat, to determine the freshness of fish, or to detect lethally leaking gas. The causative agents of smells are small volatile molecules, which move by diffusion: that is, they ooze from regions of high concentration to regions of low, until equilibrium is reached – at which point the entire room is uniformly perfused with the aroma of lily-of-the-valley, the fumes of airplane glue, or the stench of decaying potato. Each travels at over a thousand miles per hour (1800 km/hr), as fast as a high-powered jet plane flies. At such speeds, in theory molecules could cross the average living room in fractions of a second, flinging smells near-instantaneously from source to sniffer. In practice, however, odors spread relatively slowly. This is due to the tangled path the average volatile molecule takes through the air. Air, as we have seen, is far from empty; and odorants therefore can't travel in unimpeded straight lines.

Instead, since air at human nose level is molecularly crowded, the odor-bearing particle is compelled to shoulder its way through a throng, continually colliding with and bouncing off its molecular neighbors. The result is a frenetic three-dimensional billiard game: a typical gas molecule at room temperature suffers about a billion collisions per second, each collision causing the participants to ricochet off in a different direction – up, down, sideways, forward, back. Volatile molecules thus travel not as the crow flies, but in convoluted and unpredictable squiggles, an inefficient path that physicists describe as a "random walk."

Once a meandering molecule finally arrives at a receptive nose, it routinely enters the nostrils at ordinary breathing speed – about 7 miles per hour (11 km/h), though a forceful and deliberate sniff can up the rate to 20 miles per hour (32 km/h), force 5 on Admiral Beaufort's wind scale. Sucked up through the nose, odorants enter the nasal cavities and, deep inside the skull at about eye level, come in contact with paired patches of yellowish tissue that constitute the olfactory epithelium. The more highly developed an animal's sense of smell, the larger these patches are. Humans, as smell goes, are unimpressive: our olfactory patches are the size of a thumbnail, each populated with approximately 5 million olfactory neurons or receptor cells. (Dogs, in contrast, whose folded olfactory epithelia are about the size of a standard mailing envelope, have over 200 million.) The receptors – each specific for a different odor molecule – are borne on hairlike cilia that poke outward from the olfactory cell surface into a thin protective layer of mucus. When receptor and scent molecule meet, an informative electrical signal is sent via the olfactory nerves to the plum-sized olfactory bulb at the base of the brain. This in turn communicates with the brain's emotion-governing limbic system and knowledgeable cerebral cortex, where the signal is matched to memory storage banks, thus allowing us to identify, for example, the smell of garlic, and to associate it hungrily with spaghetti.

Air is an encyclopedia of odor. There are hundreds of

thousands of molecules that smell, though generally these are packaged into a limited handful of categories based on perceptual or chemical similarities. Among the first to devise a taxonomy of smells was the obsessively orderly Carl Linnaeus whose *Odores Medicamentorum*, published in 1752, identified seven basic kinds of odors, ranging from fragrant to foul. Fragrant, to Linnaeus, meant flowers: his *Fragrantes* category includes such botanically sweet aromas as jasmine, saffron, and wild lime. Equally appealing are his *Ambrosiacos*, which include the rich heavy scents of geranium, sage, mint, and musk; and the spicy *Aromaticos*, such as the grandma's-kitchen aromas of cinnamon, citron, anise, and clove. Linnaeus's *Tetros* – foul – category, on the other hand, is somewhat more controversial, lumping together the scents of bedbugs, marigolds, tomatoes, and opium poppies, though few would disagree with his examples of *Nauseosos* (feces and spoiled fish), *Alliaceos* (onions and garlic), and *Hircinos* – goaty – which encompasses sweat, urine, and rancid cheese.

Many smells, however, are difficult or flatly impossible to classify – what to do, for example, with the scents of pickle brine, corrugated cardboard, or new car upholstery – and furthermore, smell has proved to be a slippery and subjective sense. Odors perceived by some as pleasant are perceived by others as utterly awful. People disagree, for example, about the appeal of gasoline; and even the mephitic aroma of skunk has its rare proponents. To some extent, judgment of smells is cultural: the cattle-raising Dassenetch of Ethiopia dress their hair with cow dung; and the Dogon of Mali favor the smell of onions, whose sulfurous fumes set most Westerners to gobbling breath mints. Smell is also intensely personal because, by virtue of its intimate connection to the limbic system of the brain, it plays a compelling role in human memory.

Air, with its lush battery of smells, is for us a sensory scrapbook. The limbic system is the brain's emotional center, governor of such grand opera passions as rage, fear, aggression, and lust. It is known as the "old mammalian brain" since in the evolution of mammals it appeared first, long before we acquired the rational cerebral cortex; and is sometimes referred to as the rhinencephalon – literally, "smell-brain" – since its organs enter around the olfactory bulb. The combination of smell and emotion are powerful memory triggers; the two combine to imprint events indelibly in the brain's long-term memory banks. For Marcel Proust, for example, the massive *A la recherche du temps perdu*, his 15-volume autobiographical novel, was inspired by the smell of lime tea. The astonishing ability of smell to trigger reminiscences is sometimes known as the Proust effect, from a famous scene *Swann's Way* in which the protagonist dunks a madeleine in a cup of lime-flower tea. The distinctive scent instantly transports him to the little French town of Combray in which he spent his childhood. Even the crusty Ebenezer Scrooge was not immune from the nostalgic effect of odor. "He was conscious of a thousand odors floating in the air," writes Charles Dickens in *A Christmas Carol*, "each one connected with a thousand thoughts, and hopes, and joys, and cares long, long forgotten." Dickens himself was affected by the smell of paste, which called up vivid memories of the blacking factory in which he had worked during a miserable period in his poverty-stricken youth.

Air is an aesthetic experience. Human beings lead a rich olfactory life, picking, choosing, and discriminating among some 10,000 detectable airborne odors. Though some smells are almost universally perceived as horrid – the odors of feces or decaying meat, for example, which probably once served to warn our primitive selves away from predators or poisons – others are delectable. Early humans must have reveled in the alluring aromas of flowers, fruits, and charbroiled mastodon; and eventually someone among them must have discovered that certain woods, tossed on to the family fire, produced sweet-smelling fumes. The first perfume – from the Latin *per fumus*, meaning "through smoke" – was incense. The practice of deliberately burning aromatic herbs and resins dates at least to ancient Mesopotamia; in the *Epic of Gilgamesh*, the title character – Gilgamesh, King of Ur – cannily burns cedarwood and myrrh as an offering to put the gods in a good mood. Our early passion for perfume made frankincense and myrrh among the most valuable commodities of the ancient world, craved by Egyptian pharaohs and Roman emperors, burned in the great Temple of Jerusalem.

In the fifteenth century BCE, Queen Hatshepsut of Egypt fielded a fabulous expedition to the distant land of Punt (modern Somalia) in search of aromatic frankincense, myrrh, and cinnamon wood. The five immense barges, depicted on the walls of her tomb, returned bulging with a cargo of gums and resins, incense trees in pots, ivory, ebony, apes, and panther skins. Frankincense (*Boswellia sacra*) and myrrh (*Commiphora myrrha*), both squat, scrubby, and altogether unprepossessing plants, flourish in the Dhofar region of the Arabian peninsula, a mountainous stretch of territory extending through modern-day Oman, Yemen, and Somalia. Their resins are harvested by making incisions in the bark and collecting the fragrant drippings, which harden upon exposure to air into sugary golden "tears." In the ancient world, the tears, packed in goatskin bags, were shipped north by camel caravan to the ports of the Mediterranean or loaded on to ships and dispatched to Egypt via the Red Sea. Collectively, frankincense tears netted Arabian merchants some 85 tons of coined silver each year. The fabled Lost City of Ubar owed its wealth to the frankincense trade; and the gift-laden Queen of Sheba, who traveled to Jerusalem to meet the Hebrew King Solomon, was a frankincense baron. Smell was big business: the trade was so lucrative that proprietors went to great lengths to protect it, concealing the location of the trees and spreading rumors that the groves were guarded by winged serpents. Frankincense processors in the port of Alexandria were strip-searched.

Smell is a social sense: the sweet-smelling are appealing; and even in the ancient world, the would-be popular worried about BO. Resins and spices, mixed with fats and oils, were used as unguents and ointments for luxuriously perfuming the body – a pot of frankincense in fat, found in Tutankhamen's tomb, still gave off a faint fragrance when opened, giving archaeologists a hint of how the young king must have smelled. Socrates frowned on personal perfumes, grousing that their use would blur the obvious distinction between slaves (who smelled of sweat) and free men (who didn't); and the Spartans, who presumably *all* smelled of sweat, deemed perfume effeminate and banned it. Pliny condemned perfumes as "the most pointless of all luxuries." (Perfumes, he pointed out, were the trademark of the Persians, who doused themselves in the stuff to hide the smell of dirt.) Most Romans, however, gloried in scent, perfuming not only themselves, but their dogs, horses, slaves, bathroom tiles, draperies, clothing, and couch cushions. Roman gladiators perfumed their bodies before battle. ("Never think of leaving perfume or wine to your heir," ran a Latin proverb. "Administer these to yourself and let him have your money.") By the first century CE, Rome was importing 2800 tons of frankincense and 550 tons of myrrh each year; and Roman endearments included such fragrant pet names as "my cinnamon" and "my little myrrh" in the same spirit that modern sweet-talk features "honey" and "sugar." In Shakespeare's day, it was still frankincense that the guilt-stricken Lady Macbeth referred to when she mourned that "All the perfumes of Arabia will not sweeten this little hand."

Women reputedly have a more sensitive sense of smell than men; and women are more likely to suffer from cacosmia – an acute intolerance for disagreeable odors or a false perception of awful smells where none apparently exist. England's Queen Elizabeth I, of whom it was said that the sharpness of her nose was only exceeded by the slyness of her tongue, may have been a victim. To cope with the often-noxious aromas of the sixteenth century, the queen routinely wore a perfume amulet filled with rose water, carried pomanders of apple, nutmeg, cloves, aloe, and ambergris, slept in sheets sprinkled with lavender, and wore perfumed gowns, cloaks, and gloves. She demanded that each of her residences and all of her courtiers be perfumed; Sir Walter Raleigh, seeking to please, obliged by drenching himself in strawberry cologne.

Perfumes not only smell good; there's some evidence that certain smells make us feel good as well. This is the basis of aromatherapy, which purports to exploit the effects of scent on the psyche. The word was coined in the 1920s by French chemist René-Maurice Gattefosse who, having scorched his hand in a laboratory accident, cured the pain by dunking the burn in an oddly available basin of lavender oil. The soothing effect of the lavender inspired him to promote the healing effects of essential oils in general and to publish a book on the subject, forthrightly titled *Aromatherapy*, which is still in print. The idea, however, was hardly new. The first to notice the psychological and physiological effects of smells is said to have been Marestheus, an ancient Greek physician, who observed that the wearing of certain kinds of flower garlands perked people up, while others made them droopy and depressed. The great Hippocrates, sounding startlingly like my mother, touted the health-promoting properties of the scented bath.

Paul Jellinek in *The Psychological Basis of Perfumery* groups odors into four basic categories according to the behavioral effects they induce in human smellers. Aphrodisiacal scents – the prime example is musk – are sexually arousing; narcotic-intoxicating scents, such as floral fragrances and balsam, are relaxing; refreshing scents, such as mint, citrus, camphor, and evergreen, are enlivening; and stimulating scents, such as the spicy or bitter odors of many seeds, woods, or roots, are intellectually enhancing. Thus musk for a hot date, jasmine for insomnia, peppermint or lemon if you're falling asleep at the wheel, and sage or cinnamon before an exam. Some studies do suggest that a whiff of spiced apple lowers blood pressure and helps avert panic attacks, a spritz of lavender promotes alertness, and a bracing sniff of peppermint enhances concentration. However, measured physiological effects are generally small; and claims for the effects of various odors are often contradictory. A study by the Smell and Taste Research

Foundation in Chicago, for example, revealed unhelpfully that the scents of pumpkin pie, cinnamon buns, licorice, doughnuts, and lavender, singly or in combination, all increased penile blood-flow, though results varied depending upon whether or not the participants' partners wore perfume and the number of times each had had sexual intercourse in the preceding month.

Air holds the key to taste as well, since nearly 75 per cent of the flavor of food is a matter of smell. Gastronome Jean-Anthelme Brillat-Savarin, author of *The Physiology of Taste* (self-published anonymously in 1825), hypothesized that smell and taste were, in fact, a single sense. "A man who eats a peach," he wrote, "is first of all agreeably impressed by the smell emanating from it; he puts it into his mouth and experiences a sensation of freshness and acidity which incites him to continue; but it is not until the moment when he swallows, and the mouthful passes beneath the nasal channel, that the perfume is revealed to him, completing the sensations which every peach should cause." On the other hand, "consider the case of the unfortunate patient forced by his doctor to drink a huge glass of black medicine." Smell, at once alerted, "warns him of the repulsive taste of the poisonous liquid; his eyes grow round as at the approach of danger; disgust shows on his lips, and his stomach begins to heave in anticipation. But the doctor urges him to be brave, he steels himself for the ordeal, rinses his throat with brandy, holds his nose, and drinks." Holding the nose, in the latter case, is a wise but not wholly effective precaution. Volatile odorant molecules reach the olfactory epithelium both through the nose and via the retronasal passageway at the back of the throat. This double entryway prevents us from altogether escaping the unpleasant effects of consuming something awful; at the same time, it puts the pizzazz in dinner. Deprived of smell, even the most creative meal tastes blandly of wallpaper paste; and people who lack the ability to smell – a condition known as anosmia – are understandably prone to depression.

Despite this, smell is reputedly the least appreciated of human senses – Helen Keller, who valued it more than most, referred to it as the "fallen angel" – and Diane Ackerman in *A Natural History of the Senses* points out that our vocabulary of smell is shamefully impoverished. We can describe the look of magenta, the sound of a piccolo, and the feel of velvet, but we have no precise words for the distinctive smells of new books, clean sheets, or crayons; and it's near impossible, verbally, to explain the subtle differences between the scents of lime and lemon, or the tangs of pine and balsam fir. Not only are we descriptively stymied; we're also mechanistically confused. The way smell works remains somewhat of a puzzle.

The Roman poet and philosopher Lucretius, whose six-volume poem *De Rerum Natura* ("On the Nature of Things") survives from the first century BCE, explains smell as a matter of odorant shape: smooth round particles entering the nose cause pleasant odors; rough barbed particles, nasty smells. He may not have been far wrong: a model of smell elaborated in the 1960s by John Amoore proposes that the members of specific odor classes have related molecular shapes. Amoore's scheme groups odors into seven different categories based on molecular conformation. Smells can thus be camphor-like (roughly spherical molecules), musky (disk-shaped), floral (kite-shaped), minty (angular axhead-like wedges), and ether-like (sausage-like rods). Two additional classes, the nose-ticklingly pungent and the off-puttingly putrid, were based not on molecular shape but electrical charge. In the olfactory epithelium of the nose, Amoore hypothesizes, odor molecules encounter corresponding olfactory receptors and the two click together like lock and key. Thus a kite-shaped odorant – attar of rose, for example – fits neatly into a kite-shaped slot on the appropriate receptor cell. The connection causes the receptor neuron to fire, sending an electrical impulse to the olfactory bulb of the brain.

As it turns out, however, the one-to-one correspondence of

odorant shape and receptor isn't quite that neat. For one thing, there simply aren't enough receptors to go around. That's not for lack of trying: a substantial portion of the human genome is devoted to genes for detecting odors in air – approximately 1000 of our total 30,000, an impressive allotment of genetic resources. Of these 1000, about two-thirds are pseudogenes – that is, genetic deadbeats, superficially present but no longer capable of producing functional proteins. Which leaves us with about 350 active receptor genes to react with some 10,000 different odorant molecules. The numbers simply don't add up; and it now appears that Amoore's molecular keys are skeleton keys, insouciantly fitting several different locks at once.

Odorants each trigger not one, but a battery of receptor types, such that a smell – say, lemon – is not so much an individual note as a complex chord. Odor is combinatorial: the number and types of receptors activated determine the sensory end result. Thus, almost identically shaped molecules are sometimes perceived as having wildly different odors. Octanol, for example, a small eight-carbon alcohol, is generally described as smelling of oranges, while the closely related eight-carbon octanoic acid smells of sweaty socks. The slight structural difference between the two molecules becomes a major difference in the olfactory epithelium. The two react with overlapping, but distinctly different, arrays of olfactory receptors.

Smell is a complex and multifaceted process because it's such an essential part of daily existence. In most animals, all social behavior is a matter of discriminating smell. Air, in this sense, mediates quality and quantity of life: based on odors, animals zero in on food, avoid predators, recognize the boundaries of their home territory, identify friends and family, and finger likely members of the opposite sex as prospective mates. Odors used to convey informational signals from individual to individual are collectively known as semiochemicals – from the Greek *semeion*, or signal – and versions of these are near-universal. Plants, attacked by bugs, secrete volatile compounds that constitute screams for help. Corn, beets, and cotton, for example, infested with leaf-munching larvae of the beet armyworm moth, put up a fight, throwing out insidious indoles and terpenes – chemicals that attract parasitic wasps. The wasps lay their eggs in armyworm caterpillars, with fatal results; the young wasp larvae eat their host as they hatch. Fava beans, beset with aphids, produce chemicals that attract aphid-eating predators.

The beans and beets, strictly speaking, are producing synomones – that is, semiochemicals that send messages between species, with benefit to both the emitter (the threatened veggie) and the receiver (wasps and aphid-eaters). Another subset of semiochemicals passes information solely from individual to individual within a species. These in-group signals are known as pheromones, from the Greek *pherein*, meaning excitement, paired with the English word hormone. The exciting pheromones appear throughout the animal kingdom, species-specific chemical semaphores that can spell out anything from "Food here!" to "Don't kill me; I'm a cousin!" to "I'm hot mate material!"

The existence of such invisible come-on chemicals was first postulated in the late nineteenth century by the great French entomologist Jean-Henri Fabre, who saw them in action. Fabre was keeping a caged female greater emperor moth (*Saturnia pyri*) – an attractive creature with a wingspan of 6 inches (15 cm), ornamented with prominent eyespots – when his laboratory was abruptly inundated by an invasive horde of greater emperor males, all flailing against the cage as if called by the flute of some invisible Pied Piper. The female, Fabre concluded, was exuding some imperceptible and aphrodisiacal odor. The crucial chemical was finally isolated eighty years later in Germany by Adolf Butenandt, noted for his studies of human sex hormones. (He won the Nobel Prize for this work in 1939, but was prevented from accepting by Hitler; he finally made it to Sweden in 1949.) Butenandt's first pheromone, identified in 1959, was collected from 500,000 female silkworm moths (*Bombyx mori*) and named bombykol. Analyzed, it turned out to be powerful stuff – a single female's output is enough to reel in a trillion sex-crazed males. An equivalent chemical secreted by the female gypsy moth, an erotic volatile known as gyplure, can, when temptingly spotted about by scientists, induce males to copulate with anything from oak leaves to filter paper. The bola spider cunningly exploits this effect by producing a faux pheromone, a chemical that mimics the sex signal of female noctuid moths. Lured males, stumbling eagerly into its parlor, get eaten.

Pheromones were subsequently found to govern a wide spectrum of insect behavior. Ants find their way to picnic crumbs by following a pheromone trail; worker bees, sensing pheromone disseminated by the queen, fail to mature reproductively and so spend their lives as efficient and supportive servants. Social insects – wasps, termites, ants, and bees – ensure the function of complex communities through batteries of pheromonal signals, variously commanding cooperative labor and defense. An enraged and stinging bee (be warned) broadcasts an alarm pheromone that summons others of her hive to come and do likewise.

Higher animals similarly are governed by a host of odoriferous chemical signals. Dogs peeing on lampposts, for example, are depositing volatile pheromones that announce occupancy of their territory. Female elephants release pheromones in urine that notify passing males of their sexual receptiveness; and the rank body odor of male giraffes – unkindly described by researchers as "fetid" – is thought to send females a message boasting of sexual prowess. Male pigs emit androstenone from their salivary glands, the scent of which elicits mating behavior in sows; and it's this steroidal odor that attracts female pigs to buried truffles (truffles smell of aroused boar). Mice, whose eyesight is so poor that they can barely see past the tips of their whiskers, depend on pheromones for information about their companions: various chemicals, secreted in urine and from skin glands, allow sniffing rodents to determine whether to fight, mate, or ignore. Knockout male mice – in whom the genes for crucial pheromone receptors have been eliminated or "knocked out" in laboratory experiments – are rendered socially clueless; deprived of pheromonal input, they fail to show aggression toward other males and attempt to mate indiscriminately with mice of both sexes.

In humans, such faux pas are easier to avoid since we rely more heavily on visual cues than pheromonal signals to differentiate among friend, foe, and potential bedmate. However, evidence indicates that we too respond to a subtle chemistry of airborne smells. Smell influences our moods, toys with our emotions, and manipulates our sex lives; we communicate with it subconsciously, no matter how determinedly we slather ourselves with soap or drench ourselves in deodorants. Our ancestors seem to have appreciated the upfront impact of natural body odors more than we do today. Medieval temptresses kept peeled apples in their armpits until the fruit was impregnated with their sweat, and then presented them to their lovers. Napoleon, from the battlefield at Marengo, wrote a famous note to Josephine: "I return in three days. Do not bathe!"

The alluring scent of pheromones is effective, however, even if we don't consciously know it's there. The crucial sensory organ is a pair of tiny pits inside the nose on either side of the nasal septum that constitute the vomeronasal organ, or VNO. In amphibians, reptiles, and most mammals, the VNO is an alternative olfactory system, communicating, via a pair of accessory olfactory bulbs, with the amygdala and hypothalamus of the brain's limbic system – those structures responsible for explosive primal emotions such as rage, lust, and fear. The VNO is a literal sixth sense, a carrier of subliminal pheromonal messages that bypass the olfactory epithelium and head straight for the emotion center. A whiff of alien male, targeted at the recipient's VNO, can turn a placid father mouse into a raging berserker and induce a pregnant female to abort her litter.

While humans have a VNO, scientists are unsure whether or not it is – like *Star Trek*'s versatile android Data – fully functional. The human VNO may be a vestigial leftover from the early days of evolution, a sort of nasal appendix, no longer actively fulfilling the role for which it was originally intended. On the other hand, an increasing number of studies indicate that people do respond to airborne signals that may slip past unnoticed by the conventional nose. Such odors are persistent features of the body's most unmentionable parts – prime producers are the armpits and crotch – and other apocrine and sebaceous glands, all sending under-the-table messages that elicit physiological and behavioral changes. Women's menstrual cycles become synchronized when women are routinely exposed to each other's underarm secretions – an evolutionary plus, researchers suggest, because simultaneous ovulation promotes genetic diversity by (theoretically) preventing a single man from impregnating every female in a given group. Pheromones may persistently maneuver us into mating with the genetically distant, with an eye toward profitably scrambling the gene pool. The evidence comes from a range of T-shirt-sniffing tests: women, for example, routinely prefer the odors of T-shirts worn by men who share fewer of their major immune-system genes and are thus less likely to be close relatives. Air, in other words, wards off incest.

Still, human pheromones clearly pack less punch than those of other species. Most effects are at best commonly described as vague feelings of well-being, which hardly puts us in a class with all those inflamed moths. Some researchers suggest that the feebleness of human pheromones is countered by our possession of color vision: we rely more heavily on visual cues than on smell to choose appropriate mates, distinguish our own children from the next-door neighbor's, and determine when a stranger is threatening enough to warrant striding into battle or shrieking for the police. On the other hand, 25 per cent of people, deprived of smell, lose their sex drive, which argues that Helen Keller's fallen angel still has an active part to play. Air not only keeps us alive; it also delivers the goods that keep the human race going.

In 1515, the bumptious Paracelsus wrote a treatise on elementals. These, he explained, were nature spirits, mysterious beings caught midway between the physical and the ephemeral and thus behaving like neither but still displaying aspects of both. Of these, the terrestrial gnome – Paracelsus seems to have coined the name – is the spirit of earth, an inhabitant of mines, burrows, and caves. The flame-dwelling salamander is the spirit of fire; the sinuous undine, denizen of lakes and streams, the spirit of water. Air, according to Paracelsus, was the province of the sylph, a willowy and ethereal creature whose name by the eighteenth century was flirtatiously applied at parties to willowy and ethereal young women. A sylph stars in the 1832 classical ballet "La Sylphide," a Romantic tragedy in which James, a Scottish peasant, jilts Effie, his solid and practical fiancée (tactlessly, on their wedding day) in favor of the beautiful winged Sylphide, an airy and unattainable sprite who ultimately brings about his doom.

He should have known better. The classical sylph is a frail and finicky creature, a Twiggy-like study in anorexia, a high-maintenance type, prone to vapors and swoons. To my mind, that's not air. Air is an element with muscle and determination. It pounds the planet, froths the oceans, carries 400-ton airplanes. At the same time, air is large and generous, a nurturer, a dispenser of warmth and personal fuel. Air encases Earth and its inhabitants in a thick protective blanket, shielding us, like frog's eggs in jelly, from the hostile big pond of the universe. "World-mothering air," poet Gerard Manley Hopkins called it.

If air has an elemental spirit, it's no flittery Sylphide. It's solid and practical Effie, who survived the treacheries of James, married somebody else, and got on with her life.

Part IV

Fire

How great a matter a little fire kindleth.

The Book of James

Fire is the element of passion: anything we get excited about is bound to be described in terms of blazes, burns, heat, and flames. Fire is the language of love and lust: the sexy are red hot, the infatuated are inflamed, and the love-stricken carry torches. (The opposite also applies: when love dies, it grows cold and dwindles to ashes.) Fire is the language of religious, as well as erotic, fervor. Zealots burn with belief. ("I set myself on fire," wrote the eighteenth-century cleric John Wesley, the fiery founder of Methodism, "and people come to watch me burn.") Political extremists are known as firebrands. Burning issues are pressing and important, matters demanding our attention and having to be dealt with fast. When infuriated, we go up in flames; anger also makes us fume, smolder, grow hot under the collar, or explode with rage. The hot-tempered reputation of redheads is the fault of their flame-colored hair – though often undeserved; Thomas Jefferson, for example, a carrot-top, was a diplomat, a devotee of reason, and a cool customer. Logic is cold, but we burn with curiosity, and genius traditionally is creative fire. Fire is said to burn in human hearts, minds, and bellies; and fire – a divine spark – is a metaphor for the human soul.

Fire is the element of danger; like passion, it can run amok, destroying everything in its path. A host of folk sayings warn the unwary about the risks of playing with real or metaphorical fire; and fire, traditionally, is a double-edged sword: a good servant, but a fearful master. Fire, in its fearful aspect, is ordeal and adversity: we pass through it and emerge transformed and improved. Fire, however, according to much of world folklore, is something we're not supposed to have. Like the famous apple in the Garden of Eden, fire is associated with forbidden knowledge, denied us by the gods in order to keep us in a state of happy innocence. According to legend, we often only got it by stealing it.

In native American legends, compassionate and clever animals often went to a great deal of trouble to pinch fire for us. According to the Apache, a fox stole fire from a firefly village and scattered it across the Earth; according to the Nez Percé, a beaver stole it from the pine trees. Many creatures were marked for life in this altruistic process: the crow's black feathers, the frog's tailless behind, and the redpoll's red-dabbed head all are said to date to prehistoric fire heists. In Western literature, the most famous theft of all was that of the Titan Prometheus, who stole fire from the Olympian gods, secreting the glowing spark in a hollow stalk of fennel. According to Aeschylus, the enraged Zeus, in retribution, had Prometheus chained to a rock in the Caucasus Mountains where, daily, an eagle tore out his liver. In an alternate version of the story, Prometheus got off relatively scot-free; while Zeus revenged himself on humankind by creating Pandora, marrying her to Prometheus's brother Epimetheus, and giving her a box of troubles. The box came with a "Do not open" stricture, and we all know where that landed us.

On the other hand, it may have been a fair trade. Fire is a prerequisite for progress, bringing with it a wealth of possibilities for new and increasingly complex technologies. With fire, amazing things were at last within our grasp: ceramics, barbecued beef, metal-bladed plows, steam locomotives. Appreciation of fire's infinite potential is echoed in mythology, in which fire is often paired with generous gifts of the finer things in life. Along with fire, Prometheus brought human beings civilization and the creative arts. Tirawa, the Pawnee sky god, taught the first people not only fire-making, but agriculture, hunting, body painting, clothing design, and tobacco smoking. Brigid, the ancient Celtic goddess of fire, was the sophisticated patron of poetry, medicine, and metalcraft. The Chinese fire god – sometimes called the Blazing Emperor – brought agriculture, the art of smelting metal, and oil lamps. "Scarcely anything," writes Pliny the Elder admiringly, "is brought to a finished state without the involvement of fire. It takes sand and melts it, as occasion offers, into glass, silver, cinnabar, lead, pigments and drugs. Ore minerals are smelted to produced copper. Fire produces iron and tempers it, purifies gold, and burns limestone to make mortar that binds blocks together in buildings." Even the leftovers of the fire are useful: a draught of fireplace ash, Pliny continues, cures gladiators of bruises; charcoal mixed with honey is a remedy for anthrax; and then there's the story of the penis that once emerged from the ashes of a hearth to impregnate a servant girl.

No one knows when humans first acquired fire, but some evidence indicates that it predates *Homo sapiens*. The oldest known campfires in the world, anthropologists suggest, come from Koobi Fora and Chesowanja in Kenya, where they left behind ovoid stains of orange earth that date back 1.6 million years. These fires may have been tended by our immediate evolutionary predecessor, *Homo erectus*, a hominid of roughly human proportions and posture – hence the upright *erectus* – with a brain about two-thirds the size of our own. Chemical analysis of the scorched earth supports the idea that the Kenyan fires were deliberately man- (or hominid-) made. While the temperature of most bush fires hovers around 100°C (212°F), the orange ground burned at 400°C (750°F), implying the continued and judicious addition of fuel; and the orange smears are smallish lens-shaped blobs, suggesting controlled fire pits.

This newfound tool was almost certainly instrumental in the movement of *H. erectus* out of Africa into the chillier climates of Europe and Asia. Bipedal locomotion, tribal social structure, and stone axes were all to the good, but it was fire that allowed our predecessors to infiltrate the greater world. Without fire, proto-man was restricted to warm climates; even the most determined hominid would have found points north just too damn cold.

By the time our own ancestors arrived on the scene about 400,000 years ago, fire was an established fact of life. Anthropologists have unearthed proper hearths from this period – rings of stone, the sort of confining circles still favored by campers today – along with charcoal and charred chewed-upon bones, the remains of ancient fire-centered feasts. At this point, anthropologists argue, fire took on its cultural and social significance, which intensified as humans, once possessed of language, gathered around the fire to eat, talk, sing, and share experiences. The fire was the unifying point of the prehistoric day, the primitive equivalent of Happy Hour. It was here, around the fire, that the first stories were born. Mythology, legend, and boastful heroic epic have their roots in fire; and it was here perhaps that the first groping toward a theory of elements began, as newly articulate man attempted to explain sun and stars, wind and rain, and the entire marvelous and utterly confusing world.

The first fire that human ancestors encountered must have seemed like a wrathful special delivery from the gods, since it almost certainly arrived in the form of a bolt of lightning. Lightning, even to the scientifically savvy, is intimidating: a massive 30,000-ampere shock of electricity, capable of turning beach sand into glass, exploding trees, and frying unwary golfers, hikers, and whole herds of cows. To early man, lightning was an awesome and unmistakable sign that whoever was up there was not a being to trifle with. Infuriated gods sent lightning: Zeus, powerful lord of the ancient Greek pantheon, hurled lightning bolts when peeved; and in Scandinavia, the irascible red-headed god Thor made thunder and lightning by flinging his lethal hammer. Many North American Indian tribes shared a vision of storms as a winged monster, the Thunderbird, who shot bolts of lightning from its beak (or eyes).

Fire from the sky is selective: in terms of lightning strikes, not all geographical regions are created equal. Cold dry climates – Siberia and Greenland, for example – are relatively lightning-free; the tropics, on the other hand, suffer more lightning strikes than anyplace else on Earth. Equatorial Africa, presumptive birthplace of *Homo erectus*, and Amazonian Brazil boast as many as 180 thunderstorm days per year. In the United States, the lightning hot spot is the southerly state of Florida, with an average of 100 annual thunderstorm days. The Great Plains experience a yearly 30–50 storm days; the New England states, 10–30; and sunny California a mere five. The annual average in Europe ranges from 10 to 25, with the exception of Britain and Scandinavia, which rate less than ten; either Thor's temper is seriously overrated, or he spends an inordinate amount of time in Kenya, Zaïre, Brazil, and Palm Beach.

Lightning is the product of thunderclouds, 7-mile-high (11-km-high) monsters that form particularly effectively in the hot muggy days of temperate summers and practically anytime in the steamy tropics. Under such conditions, hot water-drenched air rises, bubbling up geyser-like from the Earth's overheated surface at a rate of 3000 feet (900 m) per minute, faster than an airplane can climb. It cools and expands as it rises toward the chilly upper altitudes of the troposphere, eventually condensing to form a nimbostratus cloud whose flat anvil-shaped head towers 40,000 feet (12,000 m) or more above the ground.

If fair-weather cumulus clouds are placid puffy sheep, nimbostratus clouds are the quintessential big bad wolves. They look the part too, looming above the horizon like Gothic novel villains, dark, evil, and ominous. Internally, each is more ominous yet: a roiling broth of wind, rain, and hail that spells a prelude to electrical disaster. Inside a thundercloud, violent collisions of ice pellets and water droplets lead to a separation of electrical charges, as electrons are stripped from molecules by friction and redistributed into opposing camps, rather like players lining up on opposite sides of the field for a rousing game of Red Rover. The top of the cloud thus gradually accumulates a positive charge, while the electron-heavy bottom becomes negative. At the same time, since opposites attract, positive charges begin to build up on the surface of the ground below. There's a good deal of air between the bottom of a thundercloud and the surface of the ground; and air ordinarily is a good insulator – that is, under normal circumstances, electricity can't move through it. Air, however, is no match for the power of an electrically charged thundercloud. Once enough negative charges build up in a nimbostratus bottom, something has to blow.

Lightning, to the cowering observer on Earth, may look like a single blinding zap from the sky, but in practice it proceeds in a measured series of sequential steps. It begins as a 200-foot (60-m)-long leader of negative charges, extending tentatively downward from the underside of the thundercloud. The charge is powerful enough to ionize the air, converting it from unresponsive insulator to permissively conductive plasma. The leader literally drills its way through the atmosphere toward the Earth's surface, one 200-foot (60-m) step at a time, at the leisurely – for lightning – rate of 100 miles per second (160 km/s). About 100 feet (30 m) above the Earth, the leader meets its missing piece: a streamer of positive charges nosing upward from the ground below. The result, as when flaming fuse meets waiting tube of dynamite, is spectacular. Lightning, in physical terms, is a massive and near-instantaneous electrical discharge. Negative charges by the trillion hit the ground, in a searing concussion known as a lightning strike, while positive charges rocket skyward from the streamer-leader contact point at the blinding speed of 75,000 miles per second (120,000 km/s). This upward-heading return stroke is what we see – for about 40 microseconds – as lightning.

What we actually *see* is fire. The plasma channel that

lightning bores through the air is narrow – less than an inch (2.5 cm) in diameter, about as big around as a pool cue. The rapid passage of the return stroke heats this channel to a phenomenal 50,000°F (28,000°C), five times hotter than the surface of the Sun. This blazing sliver of air is what our eyes detect as the lightning flash. Also air, when heated, expands; and the charbroiled air of the lightning channel, boosted in microseconds to immense temperatures, expands so rapidly that it literally explodes. The shock wave of the explosion – experienced, depending on how close you are to storm ground zero, as an ear-splitting crack or booming roar – is thunder, relayed belatedly post-flash because sound travels at a comparative putter, 900,000 times slower than the speed of light.

Perversely, thunder – Shakespeare's "deep and dreadful organ pipe" – is harmless, even though it often scares the daylights out of us and is likely to send the faint-hearted into the closet or under the bed. This condition, if taken to extemes, is referred to as brontophobia or tonitrophobia; dogs are particularly susceptible to it, though they can be cured, according to the Society for the Prevention of Cruelty to Animals, by playing a recording of firecrackers at feeding time. Mythological tradition suggests that people soon realized that thunder was more bark than bite: while lightning stories often involve admonitory bolts targeted directly at us, thunder tales usually involve somebody doing something loud directed somewhere else. Chinese legend, for example, held that the overhead peals and roars were the sound of ferociously battling dragons; Indians of the American Northwest blamed thunder on a pair of giant brothers hurling boulders at distant mountains.

No matter how vulnerable we feel when crouched beneath a thundercloud like ducks in a penny arcade, most lightning isn't aimed down. The bulk of lightning – about 80 per cent in each storm – stays in the sky; this, called cloud-to-cloud lightning, can wreak havoc with the electronic guidance systems in airplanes, but doesn't pose problems for people down below. A good deal of lightning, it turns out, also heads harmlessly straight up, bursting from the tops of clouds and soaring into the stratosphere. Such upper-story lightning was first discovered in 1990, when the late John Winckler and colleagues from the University of Minnesota serendipitously captured it on video camera. Winckler, a physicist and enthusiastic collector of weird lightning stories, had found some of nature's weirdest: fiery columns of salmon pink on pedestals of azure blue towering 30 miles (48 km) or more above the tops of thunderheads. He named the gaudy flashes sprites, after the flighty inhabitants of Prospero's island in Shakespeare's *The Tempest*.

Cloud-top lightning seems to be the result of electromagnetic whiplash, a reaction to the gargantuan discharge of conventional lightning striking the ground. Sprites are (just) visible to the naked eye, given a dark night, a flat viewing area, a major thunderstorm occurring 100 miles (160 km) or so away, and – since sprites only last about a millisecond – a resistance to blinking. They often appear as tentacled apparitions, compared, depending on the observer, to broccoli, carrots, angels, or jellyfish. Since their discovery, meteorologists have found a whole fantasy forest of related upper-atmospheric electrical phenomena, including blue jets, bright blue fountains shooting like sparklers from the cloud tops to heights of 25 miles (40 km); elves, immense spreading disks of light, 250 miles (400 km) or more across, that last a mere thousandth of a second; and trolls (for Transient Red Optical Luminous Lineaments), red spots with faint red tails, spewed up in the aftermath of particularly impressive sprites. In 2000, dazzled meteorologists added two more species to this celestial menagerie: gnomes, which are white spikes of light about half a mile (1 km) high, shooting from the tops of thunderheads; and pixies, flashing lights less than 300 feet (90 m) across.

It's still cloud-to-ground lightning, however – about 20 per cent of the output in every storm – that most directly concerns us. "Thunder is good, thunder is impressive," wrote Mark Twain, "but it is the lightning that does the work." Wind, rain, and hail also contribute to the storm's overall effect; thunderstorms are undeniably dangerous. At any given instant, there are 2000 thunderstorms battering away at the Earth's surface.

Some are accompanied by flash floods; a single thunderstorm can unload 125 million gallons (500 million liters) of water in less than an hour. All are accompanied by lightning, which strikes the earth 100 times a second, killing over 1000 people each year worldwide and injuring thousands more. Some people seem to positively attract it: the unenviable world recordholder, a former park ranger named Roy C. Sullivan, was struck seven times over a period of years, a feat puzzlingly listed in the *Guinness Book of World Records* under "Stunts and Miscellaneous Endeavors" rather than the preceding and more appropriate "Accidents and Disasters." And lightning starts fires. The catastrophic explosion of the *Hindenberg* over Lakehurst, New Jersey, in 1937, is thought to have been the fault of lightning igniting the flammable paint of the zeppelin's outer cover; and lightning is responsible for two-thirds of the world's forest fires.

The best protection from lightning is, of course, to avoid being struck at all. Pliny the Elder reported that sleeping in a sealskin tent would keep you safe from lightning strikes; the ancient Chinese favored jade amulets; and early Scandinavians protected their homes by setting acorns on the windowsills, since the oak – progenitor of acorns – was sacred to the thunder god, Thor. Peasants in the mountains of Japan protected their property by planting *kaminarikusu*, or "thundergrass;" and Japanese noblemen protected their persons by taking anti-lightning pills. In medieval Europe, it was believed that ringing church bells could keep lightning away, and many bells cast in the period were inscribed with a hopeful *Fulgura frango* – "I break lightning." The truth, unfortunately, was quite the reverse: bell towers, often the tallest thing around, were prime lightning targets. A treatise of 1784 titled "A Proof that the Ringing of Bells during Thunderstorms May be More Dangerous than Useful" definitively put a stop to the idea, listing 386 steeples struck and 103 bellringers killed in the line of duty.

Second best is to see that if lightning *does* strike, it at least does no harm, which is the theory behind the lightning rod. The lightning rod was invented by Benjamin Franklin, in a natural extension of his discovery of the electrical nature of lightning in the famous kite-flying experiment of 1752. Franklin's method of securing "Houses &c. from LIGHTNING" involved a series of pointed metal rods called points or air terminals arranged along the ridge of the roof, connected by brass wire and attached by cables to a pair of rods, or ground terminals, driven deeply into the earth at either end of the structure. The system effectively captured and channeled the "electrical fire," sending it safely into the ground.

Franklin's rods were not universally well received. Some clergymen rejected them as sinful, since they presumed to interfere with the delivery of what was obviously divine justice. When a Czech priest, a Father Divis, erected one anyway, superstitious villagers blamed it for a debilitating drought and tore it down. In the American colonies, an earthquake in 1755 was blamed on "the iron points invented by the sagacious Mr Franklin." (Tellingly, Boston, which had the most lightning rods, got the worst of the quake.) Acceptance of the rods was further complicated by political implications. Franklin, as ambassador from the American colonies to France in the sensitive days of the Revolutionary War, was linked in the popular imagination as both tamer of lightning and defender of freedom. A motto to that effect, coined by ex-finance minister Anne-Robert-Jacques Turbot – "He snatched the lightning from the sky and the scepter from tyrants" – was so often quoted that it reached the attention of Louis XVI. It annoyed the king to the point that he ordered the motto, with accompanying portrait of Franklin, to be printed on the bottom of a royal chamberpot.

In Britain, the inherent sinfulness of the rods was less of an issue than their mode of construction, since the Royal Society had recommended them as protection for the Royal Navy's sensitive powder magazines. The Franklin rod was pointed. The alternative model, promoted by rival scientist and portrait painter Benjamin Wilson, was "blunted" – that is, topped by a rounded knob. The pointed/blunted controversy soon became a bitter feud, described some years later by a Society historian as one of those "unhappy divisions which had so unfortunate an effect upon the Royal Society, and were so disgraceful to the cause of science and philosophy." In the wake of the Declaration of Independence, one's choice of lightning rod topper was a reliable index of political conviction. The Whigs, sympathetic to the American colonists' cause, opted for points; the Tories, who viewed the Declaration as the bleating of insubordinate rabble, called for blunts. The most vociferous defender of blunts was King George III, who demanded knobbed rods for the roof of Kew Palace.

Modern assessment of lightning rod effectiveness shows that the best solution lies somewhere in between the two: "moderately blunted," not knobbed, rods seem to be the most effective lightning receptors. Those on our roof ridge are a compromise of this ilk, shaped roughly like very dull bayonets. The summer they were disconnected, during a spot of roof repair, a lightning strike seared the chimney and torched the telephone, television, microwave oven, and computer. We were lucky. Each year lightning burns some 30,000 houses down.

Of the four elements, none is so unpredictable as fire. Fire deities are often irresponsible tricksters. The Norse fire god, Loki, was untrustworthy and relentlessly malicious; and the capricious Chinese fire deity Hui Lu, whenever the fit took him, sent bevies of flaming birds to set fire to the land. "Our planet is primed for ignition," writes Arizona State University fire historian Stephen Pyne, "stuffed with organic fuels, its atmosphere saturated with oxygen, its surface pummeled by lightning." A good deal of it, therefore, periodically goes up in a puff of smoke. Lightning and volcanoes are fire-starters; and in Australia, in the broiling days of summer, fires can burst into being through the action of the Sun's rays focused through bits of crystalline quartz. In the Middle East, flammable hydrocarbons seeping from seams beneath the ground may have touched off the firestorms that destroyed Sodom and Gomorrah and ignited Moses's mysterious burning bush.

A fire that burns out of control, whether ignited by nature or set by careless campers, is designated a wildfire; and such fires, which can gobble up millions of acres of land annually, are numbered among the planet's worst natural (or at least quasi-natural) disasters. Each year an average of 600,000 acres (240,000 hectares) of forested land burns in Alaska, variously started by lightning and people, including (once, in 1926) a group of children trying to scare a squirrel out of a tree. In the continental United States, the Yellowstone fires of 1988 incinerated 792,000 of the Park's 2.2 million acres of lodgepole pines (320,000 of 890,000 hectares); in 2000, a total of 123,000 fires accounted for 8.5 million acres (3.4 million hectares) of American forest. Fires in Indonesia in 1997 consumed 2400 square miles (7500 sq km) of forest and threw a health-threatening haze of smoke across all of Southeast Asia. Europe each year suffers some 45,000 forest fires, collectively scorching an area the size of Belgium.

A forest fire, writes Stephen Pyne, can release as much energy as an atomic bomb; and the sheer scariness of wildfire and the devastation left in its blazing wake precipitated, over much of the past century, a concerted campaign to prevent such things from happening. In 1937, Uncle Sam himself came out for fire prevention in the United States, appearing on a Forest Service poster in broad-brimmed hat and forest-ranger garb, pointing accusingly at a flaming tree. The message, printed across his boot, read "Our Forest – Our Fault." During World War II, fire prevention was touted as an act of patriotism ("Careless Matches Aid the Axis"); and in the 1950s, in a Forest Service publication titled "Forest and Flame in the Bible," it was promoted as a Christian duty. As of 1944, it acquired a popular spokesperson, still one of the most recognizable advertising icons in American history: Smokey Bear.

Smokey began life as a purely paper bear, the creation of animal artist Albert Staehle, who routinely depicted him on anti-fire posters in dungarees and ranger hat, clutching a bucket of water or a shovel. He was supplemented in 1950 by a real bear – a 4-month-old black bear cub, found clinging to a charred tree after a forest fire in New Mexico. Shortly thereafter, due to booming popularity, Smokey (both versions) became by Act of Congress the property of the US Department of Agriculture, under whose aegis a Smokey teddy bear was marketed (all profits to the government, for fire-prevention purposes), and Smokey was given his own zip code (for fan mail) and signature message: "Only YOU can prevent forest fires."

Though no one supports negligent match-tossing – much less the behavior of the distraught Terry Lynn Barton, who in 2002 stopped by the side of the road to burn a letter from her estranged husband, thus setting off the largest wildfire in Colorado history – wildfires, ecologists agree, are an essential component of the ecosystem. Too much fire prevention, in fact, is ultimately harmful. The forests of the American West thrive on fire; without one every five to ten years or so, immense amounts of brush and deadwood build up on the forest floor to the point where, when fire does inevitably strike, it's catastrophic. Periodic ground fires not only clear away the rubbish; they leave behind a soil-enriching layer of charcoal that promotes seed germination. Some plants literally need fire to survive – some pine cones, for example, such as those of the jack pine, lodgepole pine, and knobcone pine, will only pop open to release their seeds after exposure to a blaze – and many chaparral wildflowers blossom spectacularly in the wake of fire, among them orange fire poppies, lemon-yellow monkey flowers, scarlet larkspurs, and ground pinks.

Even animals can benefit from fire. The red-cockaded woodpecker, Kirkland's warbler, and black grouse do best in areas that experience regular burns, as does the black-backed woodpecker, whose charcoal-colored feathers make for perfect camouflage on charred tree trunks. The lure for the woodpecker is rich pickings: over forty species of insects, mostly beetles, are pyrophilic, which means that they chase fires with the same persistence that dogs chase automobiles. Beetles of the genus *Melanophila* – nicknamed "fire bugs" – lay their eggs in burned bark, which is the ideal food source for the developing larvae. The beetles can zero in on a fire from as much as

7 miles (11 km) away, tracking it to its source in part by smell and in part by means of sensitive infrared detectors, a pair of dimple-like organs tucked under their front legs.

And forget Bambi and his terrified forest friends, fleeing from fire as if it were a charging Godzilla. Though tragedies do happen, most small mammals can survive fires by ducking into the depths of their burrows; and larger animals can usually out-walk them. The average forest fire travels at the rate of about 2 miles (3 km/h) per hour, which leaves the average deer, moose, elk, and bear plenty of room to maneuver. At least such should have been the case in Bambi's fire, which seemed to be traversing level ground in a dampish deciduous forest.

The rate at which a fire travels depends on fuel, wind, and terrain. Fire spreads more rapidly if supplied with dry tinder – readily burnable stuff like grass, brush, and pine needles; and the more of it, the faster the fire. The speediest grass fires, which are almost entirely fueled by tinder, move at 10–12 miles per hour (16–20 km/h). Fires also, like airplanes and migrating ducks, travel faster with a tail wind; and – unlike people – they travel faster when heading uphill. The uphill advantage derives from the physics of heat, which rises, such that the fire preheats the leading upward slope, giving itself an extra edge. In fact, the steeper the slope, the faster the fire climbs up it, which makes it difficult for, say, mountain goats trying to get out of the way. On the other hand, fires have a hard time heading down; often they burn out altogether on hilltops.

Fire, though theoretically tamed, has always shown a disposition to run amok. Awful accounts of urban burnings punctuate history: Troy, Carthage, and Rome all burned, the last – on the night of July 18 in 64 CE – while the 26-year-old Emperor Nero reputedly admired the flames from the roof of his palace and played the lyre. Nero was accused of instigating the fire, being ambitious, wrote the historian Tacitus, "to found a new city to be called after himself," but this the emperor promptly and piously denied. Instead he blamed the conflagration on the city's Christians, many of whom he dramatically fed to the lions in Rome's lone remaining amphitheater in a clever political move to distract the angry public. The distraction was so effective that it kicked off a program of Christian persecution that was to last for the next three centuries. The fire can still be seen (in Technicolor) in the 1951 film *Quo Vadis?*, in which Peter Ustinov steals the show as Nero.

No one tried to fight the fire of Rome, wrote Tacitus, because "attempts to do so were prevented by menacing gangs." The gangs – who actually helped the fire along by flinging torches – may have been operating under the emperor's orders, or (Tacitus again, trying to be fair), "they may just have wanted to plunder unhampered." In any case, they effectively stymied the metropolitan fire brigades. These had operated in Rome since the days of Augustus Caesar, manned by *vigiles* whose job it was to patrol the streets, armed with ladders, ropes, axes, and leather buckets, keeping a weather eye out for smoke or flame. There were 7000 vigiles in total, divided into seven battalions, all under the authority of a prefect, who presided over the official hearing that followed each and every fire. Persons who were proved negligent were punished by anything from a whipping to death, depending on the magnitude of their neglect.

Fire and mushrooming population density don't mix; and fires were, and are, an ever-present urban danger. Rome burned not once, but many times, from 390 BCE when it was torched by the Gauls to 410 CE when it was incinerated by Alaric and his Visigoths. Similarly the history of London is a litany of fire. The young city burned to the ground in 60 CE and again in 125 CE; it also, according to Peter Ackroyd in his impressive *London: The Biography*, burned in 764, 798, 852, 893, 961, 982, 1077, 1087, 1093, 1132, 1136, 1203, 1212, 1220, and 1227. The city averaged 556 fires a year in the first half of the nineteenth century; and suffered 46,000 in 1993. All London's landmarks – London Bridge, the Houses of Parliament, the Royal Exchange – have burned down; and the famous "Great Fire" of 1666, though dreadful, was no greater than many of its predecessors. The

Great Fire was, however, notably better documented, since among the observers were John Evelyn and Samuel Pepys, two of the most prolific diarists of the times. Pepys, who watched most of the action from the shelter of an alehouse, described it as a "most horrid, malicious bloody flame, not like the fire flame of an ordinary fire." By the time it finally burned itself out, five-sixths of the city was in ashes; and molten lead from the roof of St Paul's Cathedral ran through the streets.

Almost every city has had a Great Fire. Chicago's Great Fire of 1871 was so impressive that it spawned an American national holiday, Fire Prevention Day, celebrated on October 9, the anniversary of the disaster. Actually the Great Fire began on the evening of the eighth, supposedly in the barn of Mrs Catherine O'Leary, whose cow kicked a flaming lantern into a pile of hay. Rapidly spread by a strong wind from the southwest, the fire soon reached the center of the city, where, now "a perfect ocean of blaze," it destroyed everything in its path, including the Courthouse Tower (the watchman, who had waited too long, escaped by sliding down the banister of a flaming staircase) and the offices of the Chicago *Tribune*, which had fondly been believed to be fireproof. The *Tribune* rallied manfully, managing to return to print two days later with all the dreadful details of "a conflagration which has no parallels in the annals of history." All told, over 17,000 buildings were burned and 250 people killed.

The frequency of great fires led to the early passage of multitudinous regulations intended to nip potentially disastrous outbreaks in the bud. Beginning in 872, a bell was rung in Oxford, signaling the time for all householders to "cover" or bank their fires for the night; from this regimented covering of fires – in Norman French, *couvre feu* – comes the modern term curfew. The Lord Mayor of London banned the building of wooden houses in 1189 ("No house should be built in the city but of stone and they must be covered with slate or tiled"); John Winthrop, governor of the Massachusetts Bay Colony, banned wooden chimneys and thatch roofs in 1631. Further ordinances banned the storing of flammables such as hemp, hay, straw, barley, or gunpowder within reach of any fire. By the eighteenth century, unauthorized bonfires were prohibited, as was smoking in the streets; by the nineteenth, firecrackers were deemed a hazard, and the citizens of Reading, Pennsylvania, not only were forbidden to use them on their own, but were invested *en masse* with the responsibility for confiscating and destroying any found in the possession of minor children.

No precaution ever proved adequate; and the fallback position, then as now, was the increasingly efficient professional fire company. *In extremis*, one could also turn to St Florian, the patron saint of fire. Florian was a young Roman soldier during the reign of the Emperor Diocletian, posted to Noricum – now Austria – under orders to persecute the resident Christians. This Florian refused to do, revealing to his appalled commanding officer that he himself was Christian. He stuck to it too, despite being offered a raise and a promotion if only he would recant. He was finally sentenced to death by fire; and seems to have gloried in the prospect, egging his erstwhile companions to light the pyre so that he could "climb to heaven on the flames." This rattled the commander so much that he had Florian drowned. Miracles followed his death. Prayers to Florian were found to protect those in danger from fire – presumably because he had been so fearless when faced with death by flames – and eventually the young soldier's spiritual reputation was such that he was canonized by Pope Lucius III in 1138. He became known as the patron of all those threatened by fire: firefighters, chimney sweeps, and soap boilers.

Fire has a primordial fascination. There's an irresistible and near-mesmerizing appeal to flickering flames, as anyone knows who has sat up late beside a flaring beach fire or whiled away a winter evening watching the last glow in the fireplace fade away. It's this universal allure, psychologists believe, that leads children to play with forbidden matches; and it's our ancient obsession with flame that lies at the heart of deliberate fire-setting behavior.

It can be a deadly obsession. Arson, historically, has been considered a crime on a par with murder; in the Middle Ages, perpetrators were burned alive. In fifteenth-century England, deliberate burning was deemed an act of high treason, though in the sixteenth, Bloody Mary – famed for her relentless burning of heretics – reduced it to a felony. In the nineteenth century, arson was subsumed under the "Malicious Damage Act" of 1861, with punishment meted out according to the nature of the structure burned. Setting fire to a church, chapel, dwelling house, manufactory, or farm building incurred a penalty of five years to life in penal servitude; burning crops, fourteen years; and burning peat or coal, life. Worst – after all, Britannia ruled the waves – was "wilfully and maliciously to set fire to any of His Majesty's ships or vessels of war," an act that condemned the offender to death.

According to the Texas Penal Code, explosion of a house by gunpowder or dynamite is also arson, which means that Timothy McVeigh, positioning his van full of homemade explosives next to Oklahoma City's Alfred P. Murrah Federal Building on April 19, 1995, was an arsonist. He shares the distinction with the only arsonist to be commemorated in a national holiday. Guy Fawkes, born in 1570 in York, England, son of a notary and minor landowner, was described by his contemporaries as a tall well-built man with thick reddish-brown hair, educated and even-tempered, "a man of great piety, of exemplary temperance, of mild and cheerful demeanour, an enemy of broils and disputes, a faithful friend, and remarkable for his punctual attendance upon religious observance." It was piety that ultimately drove Fawkes to fire. Angry at the persecution of English Catholics by Elizabeth I's successor, the Scottish-born Protestant King James I, Fawkes and a band of like-minded friends planned to blow up the House of Lords, killing in the process the king, the Prince of Wales, and a hundred or more peers of the realm. To this end, they hired a cellar beneath the Lords' chamber which they packed with thirty-six barrels of gunpowder. Fawkes, an ex-soldier with experience in munitions, was chosen to fire the lot; however, by the appointed date word of the conspiracy had been leaked to the authorities. Fawkes was apprehended and the arresting magistrate found upon his person a watch, "slow matches," and touchwood. The following morning – November 5, 1605 – Fawkes was brought before the king, where he sealed his fate by announcing untactfully that his intentions had been to blow all Scotsmen present back across the border. He was executed hideously the following January, by hanging, drawing, and quartering.

The celebration of the foiling of Fawkes's so-called Gunpowder Plot began the following year, eventually spreading to the American colonies, where it was known as "Pope's Day." More commonly called Guy Fawkes Night, or Bonfire Night, the occasion is still celebrated on the fifth of November, with bonfires, torches, fireworks, and the burning of Fawkes in effigy. Pope's Day in the United States is a thing of the past; it served as a rallying point for opposition to the Stamp Act during the eighteenth century and then was abandoned. By 1776, American revolutionaries presumably no longer regretted a spirited past attempt to blow up Parliament.

Fawkes chose arson for political purposes, as have, in recent centuries, the Ku Klux Klan, the IRA, anti-abortion religious zealots, and Islamic terrorists. Some arsonists burn for profit, a crime that dates at least to ancient Rome: Martial, for example, tells the story of a certain Tongilianus whose house – worth a modest 200 sesterces – burned in a "suspicious fire," only to be rebuilt at the greatly inflated cost of 10,000 sesterces, all collected by subscription from gullible friends. Other arsonists burn for revenge or for psychopathological reasons. Kids, according to the FBI, are responsible for approximately 40 per cent of intentionally set fires, including up to 75 per cent of the fires set in American public schools. Psychologists cite displaced anger and demand for attention as likely motivational candidates; and some propose that arson is directly related to sexual satisfaction. It's an intuitively logical connection: the language of sex is fraught with images of fire. In the 1960s, Jim Morrison and the Doors made rock-'n-roll fame singing "Come on, baby,

light my fire!" Arsonists perhaps simply take what all of us are
thinking one inflammatory step too far.

Fire, since ancient times, has been a vehicle of worship, supernatural appeasement, and spiritual communion. Offerings
were burned in the belief that their essence would rise to God
(or gods) with the rising smoke. Worshippers of Moloch in the
ancient Near East threw infants into a flaming furnace as a
sacrifice to their god (a practice roundly condemned by the
prophet Jeremiah). The ancient Chinese broiled wine, grain, and
silk in altar fires; and sacrificial fires burned in Greek temples,
and in the great Temple of Jerusalem. In Rome, fire was sacred
to Vesta, the Roman counterpart of the Greek Hestia, goddess
of the hearth. At her temple, a sacred fire was tended day and
night by six priestesses known as vestal virgins, chosen in
childhood and dedicated to the temple for thirty years. It was
a coveted position; these women were emancipated relative
to their male-controlled peers – they could own property and
sign legal contracts – and they were guaranteed front seats at
festival games. Provided, that is, that they remained virginal:
vestal virgins were allowed to marry once their term of office
was over, but if they anticipated the event, they were buried
alive.

Fire is sacred to the Zoroastrians – followers of the ancient
Persian prophet Zoroaster or Zarathustra – who worship in fire
temples where a holy flame burns eternally at the altar. A fire
ceremony is a feature of Hindu weddings; candles, a feature
of Jewish and Christian religious celebrations. Our custom of
lighting eternal flames on graves recalls ancient custom: a fire,
burning, still symbolizes remembrance. Linguistically, to keep
the fire burning is to keep faith in a cause alive.

Many cultures ceremonially incinerate their dead; and the
belief in fire as a conveyor belt to heaven was the impetus
behind the first funeral pyres. Among the Romans, an impressive pyre was a mark of utmost respect. (Those struck by

lightning, however, didn't get one; lightning victims, explains Pliny, had to be buried.) Vikings went gloriously to Valhalla in flames, consumed along with their burning ships; and Hindus for centuries have ritually burned the bodies of the deceased. Although in 1829, the occupying British ostensibly outlawed the practice, it was once the custom for the living to burn as well: Hindu widows immolated themselves on their husbands' funeral pyres in a rite known as "suttee" – an Anglicization of the Sanskrit *sati* meaning "faithful wife." (In a famous scene in Jules Verne's *Around the World in Eighty Days*, Phileas Fogg and his indomitable servant Passepartout rescue the beautiful Aouda from her rajah husband's pyre and flee by elephant into the night.)

Some early cultures used funerary fire in a protective sense, to fend off the evil spirits presumably responsible for or associated with death. Even the belongings of the dead were sometimes burned, to discourage malignant phantoms from lingering. In the traditional gypsy funeral, after the body is buried, the deceased's belongings are destroyed: dishes are smashed, household goods and caravan – even, in modern times, automobile and camper trailer – are burned. The lighting of candles at funerals today is thought to be a reflection of fire's ancient reputation as a bulwark between the living and the threatening spirits that surround the dead.

Fire is also the element of celebration: even in the sophisticated twenty-first century, fire still thrills. Every year on the Saturday night before Labor Day, a 50-foot-tall (15-m) wooden man is burned in the Nevada desert. The Burning Man celebration is described by its founders as "an annual experiment in temporary community, dedicated to radical self-expression and radical self-reliance." Thousands attend annually, bringing with them tents, food, water, and toilet paper, ready to join in a week-long art fest and to participate in one of dozens of interactive theme camps, such as the Moons of Mongo (bicycle laser tag and a stargazing dome), Barbie Death Camp and Wine Bistro (torch unwanted Barbie dolls while sipping cabernet), Philosophy Camp (intellectual discourse), Rainforest Refugee

Camp (get sprayed with mist and stamped with blessings), and Kidsville (a toned-down camp for families).

Older fire fests were far more serious: since ancient times throughout Europe, bonfires have been set on certain days of the year and people have danced around them, dashed through them, or leaped over them to ensure the fertility of crops, livestock, and each other. Traditionally such fires were lit in spring, midsummer, and fall. The Celts celebrated Beltane on the first of May and Samhain on 31 October – Halloween – with blazing bonfires, sometimes accompanied by human sacrifices. Julius Caesar, who conquered the Celts in Gaul and thus managed to observe the festivities at close range, described how criminals and war captives were imprisoned in immense images made of wickerwork and grass, then burned alive. The tradition of the wicker giants (sans criminals) persisted into modern times: Nevada's Burning Man continues to echo it, as do, just possibly, the monstrous human- and animal-shaped balloons of the annual Macy's Thanksgiving Day Parade.

According to Frazer's *Golden Bough*, folklorists propose two major theories of fire-festival origin. The first – the solar theory – argues that the fires were imitative sun charms, lit in summer to encourage lush, sun-drenched days for the fall harvest; and in midwinter to encourage the feeble sun of the cold months to regain its strength for the spring planting. The custom of the Christmas Yule log may have its roots in an ancient winter sun charm. The second – the purificatory theory, for which Frazer clearly has a soft spot – holds that the fires were meant to banish the forces of evil, fending off the witches, demons, and devils responsible for ill-fortune, disease, and death. The belief that fire can banish unseen dangers lurking in the night may be as old as man or older, reaching back to the time when prehistoric proto-humans, squatting around the fire pit, realized with astonished relief that dire wolves, cave bears, and saber-toothed cats were held at bay by flames.

And if wolves, why not witches? Belief in the purifying powers of fire fueled the witch hunts of medieval Europe, under the long and fearful arm of the Inquisition – from the

Latin *inquirere*, to investigate, though more specifically defined
by the *Catholic Encyclopedia* as "a special ecclesiastical institution for combating or suppressing heresy." "Are you a good
witch or a bad witch?" – the question put to the confused
Dorothy in L. Frank Baum's children's classic *The Wizard of
Oz* – was an oxymoron in the Middle Ages. In Inquisitional
terms, *all* witches were bad, as was anyone else who held
spiritual, ethical, political, or scientific views out of step with
those of the established Catholic Church. The burning of the
spiritually incorrect, sporadic through the Dark Ages, solidified into coherent policy beginning in the eleventh century. In
1022, King Robert of France ("the Pious") had thirteen suspect
citizens burned alive at Orleans. By the 1180s, Duke Philip of
Flanders, with the enthusiastic collaboration of William of the
White Hand, Archbishop of Reims, was burning "nobles and
commoners, clerics, knights, peasants, spinsters, widows, and
married women" and – a perquisite of piety – confiscating their
property. Heretics were immolated for what today seem like
reasonably negotiable doctrinal points: the Albigensians, for
example, equivocated over transubstantiation, the nature of
holy water, and the usefulness of confession; the Waldensians
and the Lollards were given to unsupervised preaching using
Gospels that had been translated into the vernacular; and the
Beguins ascribed too much importance to St Francis of Assisi.

Heretics, though depraved, stubborn, and misguided,
were not necessarily in league with Satan; witches, however,
an altogether more malevolent kettle of fish, were. When 19-
year-old Joan of Arc was burned in Rouen in 1431, her English
captors justified the act by accusing her of witchcraft and
attributing Joan's astonishing military victories to the devil.
While Joan's unquestionable piety may have given her judges
some pause, most suspects were barely given the benefit of a
doubt. Beginning in 1484, they were pursued mercilessly; in
that year, a papal bull of Innocent VIII, citing the appalling
prevalence of witches in northern Germany, sanctioned
aggressive witch-hunting. The bull was reprinted in 1486 as
the introduction to the standard guidebook for aspiring witch-

hunters – *Malleus Maleficarium*, "The Hammer of Witches" – a tome which described in chillingly matter-of-fact detail the processes of stripping and examination, interrogation, torture, and execution, with helpful hints for judges on how to address prisoners. During the heyday of witch-hunting in the late sixteenth and seventeenth centuries tens of thousands of victims – mostly women, though Innocent had even-handedly fingered "persons of both sexes" – went to the stake.

Church and state, suggests anthropologist Marvin Harris in *Cows, Pigs, Wars, and Witches*, may have deliberately promoted European witch mania in order to distract the peasant class from the realities of medieval politics: better that the people blame malevolent witches for their social and economic woes than turn their attention to corrupt prelates and incompetent princes. For whatever reason – fear, political gain, true belief – witches burned for over six centuries. The last in Europe – two "aged beldames" in Poland – were consigned to the flames in 1792. The Office of the Inquisition in the Vatican was officially closed in 1968, just eighteen months before men walked on the Moon.

Not only people, but their written works, were burned, since eliminating the heretical, the radical, or the intellectually disruptive did little good if the responsible party left disturbingly permanent books behind. Books, pointed out Voltaire in a satirical pamphlet titled "Concerning the Horrible Danger of Reading," can "dissipate ignorance, the custodian and safeguard of the well-policed state;" and for just this reason burning them has been a perennial and powerful tool of despots. In 231 BCE, the Qin Emperor Shih Huang-ti, reforming and consolidating the early Chinese empire, ordered a nationwide burning of the works of Confucius (except for one copy of each, to be preserved for historical purposes in the royal library). In 303 CE, the Roman Emperor Diocletian, in the interests of imperial solidarity, ordered the burning of all books by Christians; by the thirteenth century, the resilient Christians were burning the books of heretics, non-believers, and Jews – including, by order of Pope Gregory in 1239, all extant copies of the Talmud.

Fifteenth-century Italian religious reformer Girolamo Savon-arola, striving to transform the sinful city of Florence into a New Jerusalem, torched books, paintings, clothing, jewelry, dice, and carnival masks on towering "bonfires of the vanities;" ultimately his zealotry backfired, and he was excommunicated, tried (for false prophecy, religious error, and sedition), hanged, and then himself burned, along with copies of his sermons.

The most famous book-burning of modern times is almost certainly that sponsored by the German Nazi party in 1933, under the direction of Joseph Goebbels, Minister of Popular Enlightenment and Propaganda, during which impassioned university students – chanting "fire oaths" in support of national purity – tossed tens of thousands of "degenerate" books into the flames, among them volumes by Sigmund Freud, Albert Einstein, Karl Marx, Ernest Hemingway, Jack London, Helen Keller, and H. G. Wells. It should have been a compelling lesson; however, books continue to burn. During China's Cultural Revolution of the 1960s, the young Commu-nists of Mao Zedong's Red Guard burned all books suspected of upholding the repudiated "four olds:" old ideas, old culture, old customs, and old habits. Today fundamentalist Muslims – citing blasphemy – burn Salman Rushdie's controversial 1988 novel *The Satanic Verses*; and in the United States, funda-mentalist Christians – citing forbidden references to witchcraft – burn the books of J. K. Rowling's Harry Potter series.

Book-burning has always been a futile endeavor – "Every burned book enlightens the world," writes Ralph Waldo Emerson – but the image of a printed page in flames is a powerful and oppressive metaphor, an immediately recogniz-able symbol of the suppression of free thought and speech. Here, however, all-destroying fire fails to deliver. Despite all efforts, fire never vanquishes ideas; nor has it ever succeeded in purifying the diverse and contentious community of the mind.

In a biological sense, however, fire does purify. In the 1958 movie *The Vikings*, Eric, played by a young Tony Curtis wearing what appear to be leather hotpants, has his hand slashed off with a sword. Then, while the audience still gapes in shock, the bloody stump is seared with a torch. Eric howls. What he has just undergone is a form of cauterization, a primitive means of sealing and sterilizing exposed tissue, thus preventing infection. Cauterization, strictly speaking, is the application of a hot iron – not a flaming torch – to a wound, but the effect is the same. Until the late nineteenth century, cauterization was the treatment of choice for rabies. Bite victims were seared with red hot needles or gunpowder was sprinkled in the wounds and set alight. Louis Pasteur, who ultimately developed an effective vaccine for the disease, had a fearful childhood memory of a man bitten by a rabid wolf and treated, agonizingly, by cauterization.

The red hot poker, though not the most desirable method of wound sterilization, does kill microorganisms. Heat, even in the technological world of today, remains our best weapon against ubiquitous germs. Baking and boiling kill practically everything except the most resistant of bacterial endospores; and steam heat under pressure – the principle of the hospital autoclave or the stovetop pressure cooker – kills everything. The elimination of disease-causing microorganisms is a prime advantage of cooking; eradication of bacteria makes food safer and, by preventing spoilage, makes it last longer. Chances are, however, that early man adopted cooking because it made things taste good.

Cooking, wrote Charles Lamb in his tongue-in-cheek essay "A Dissertation upon Roast Pig," was discovered in ancient China. For seventy thousand years the Chinese had eaten their meat raw, until young Bo-bo, the son of a swineherd, played carelessly with fire and burned down the family cottage, along with the nine family pigs. The burned pig tasted so delicious that soon the neighbors were burning down houses right and left, eager for a share in the feast. Finally a sage discovered that it was possible to cook a pig without incinerating the house.

Thus the gridiron was invented, writes Lamb, and the roasting spit. However, he concludes, if ever there was a worthy pretext for setting houses on fire, it is – you can hear him licking his chops – "ROAST PIG." (The capitals are his.)

Most anthropologists agree that cooking was indeed discovered serendipitously, in a scenario that most likely involved some natural blaze baking nearby roots and tubers, rendering them appealingly edible. No one knows just when in human history this happened, but anthropological dogma generally places it between 250,000 and 500,000 years ago. Some, however – notably Richard Wrangham and NancyLou Conklin-Brittain of Harvard University – hypothesize that cooking was discovered a good deal earlier than previously suspected; and that fire, applied to prehistoric dinners, may have played a pivotal role in human evolution. The ability to roast a yam or broil a haunch of hyena may have been the crucial event that initiated the transition from ape-like australopithecine to humanoid *Homo erectus*. The earliest australopithecines appeared sometime between four and five million years ago. Lucy, the world's most famous fossil example, discovered by anthropologist Donald Johanson in the Afar Desert of Ethiopia in 1974, lived 3.2 million years ago: a slender adult female a little less than 4 feet (1.2 m) tall, who walked upright, but otherwise looked a good deal like a chimpanzee. Lucy and relatives had jutting jaws and large teeth, short legs, and long arms – probably still used for dangling from trees; and, like modern apes, they displayed marked sexual dimorphism. Females on average weighed just 60 pounds, less than half the size of the males. Both ate their food raw.

Their *H. erectus* descendants, much more akin to modern humans, boasted longer legs and shorter arms, flatter faces, smaller jaws and teeth, and bigger brains. Sexual dimorphism was substantially reduced. *H. erectus* females were just 15–20 per cent smaller than their male counterparts, about the same size difference seen between the human sexes today. All of this, argue Wrangham and colleagues, can be explained by the discovery of cooking and the concomitant increased availability of food.

Among the latest in California food fads is the Paleolithic diet – an "uncooked, unprocessed, unheated, and organic" plant-based meal plan in which everything is eaten raw. Dedicated raw foodists live solely on fruits, vegetables, and nuts, arguably a diet similar to that of our primitive forebears. (The original Paleolithic forager, on the other hand, was limited to what he/ she could scrounge from the landscape; modern Paleolithics have the option of raw restaurants with organic cotton table-cloths and wine lists, where the staff serves sunbaked flaxseed-meal pizzas, coconut noodles, and nut-milk ice cream.) In the long-term, however, the problem with the raw – both now and way back then – is nutritional stress. Caloric density in such a diet is low – a temporary plus for a pudgy *H. sapiens* hoping to drop a few pounds, but a detriment for continuing health. Raw foodists struggle with chronic energy deficits and vitamin deficiencies; and over 30 per cent of women on strict raw food diets cease to menstruate. Based on studies of modern dieters, Wrangham and Conklin-Brittain calculate that even to support an inactive couch-potato-like Western lifestyle, raw foodists would have to consume the equivalent of 9 per cent of their body weight daily – say, 14 pounds (6 kg) or so of berries and nuts.

Anthropologists estimate that the introduction of cooking more than doubled the daily prehistoric caloric intake, both by expanding the range of edible foods and by increasing the accessibility of nutrients. Cooked starch, for example, is ten times easier to digest, which means that one roasted yam, nutritionally, is worth several raw. And cooked food is a lot easier to chew. *H. erectus* shows a greater reduction in tooth size than occurs at any other point in human evolution, suggesting that every mouthful at dinner no longer needed to be effortfully gnawed. The acquisition of soft nutritious food – and more of it – could have led to the development of bigger and more fertile females as well as to crucial changes in social behavior. For example, cooking implies a place in which to do it. Home is where the hearth is: the first primitive dwellings may have grown up around a food cache and a cooking fire.

Cooking caught on, not because it was immediately recognized as nutritionally superior but because it made hitherto bland or tasteless foods mouth-wateringly flavorful. ("O father," cried Lamb's Bo-bo, mouth full of pork crackling, "the pig, the pig, do come and taste how nice the burnt pig eats.") The secret to the yummy burnt pig lies largely in a series of complex heat-induced chemical changes called browning reactions, responsible for the rich flavors of baked, fried, and broiled goods, among them coffee beans, dark chocolate, bread crusts, grilled steak, and Guinness stout. Chemists know these as Maillard reactions, after French biochemist Louis Camille Maillard, who identified them in 1912. Browning – a somewhat loose term, since Maillard reactions can generate any color from yellow to charcoal black – results from the interaction of small amounts of sugars and starches with proteins at high temperatures (300–400°F (270–370°C) and above). The process is still not wholly understood – entire scientific conferences are devoted to it – but the end result is an alluring cocktail of over six hundred different molecular products, among them colored polymers called melanoidins, a host of pyrazines, oxazoles, thiazoles, and organic acids, and dozens of fragrant volatiles. Flavor chemists today recognize sixteen principal flavor notes; of these fire-born Maillard reactions are essential for five.

Ever since prehistoric man gathered around a campfire to roast a likely slab of mastodon, fire has been the heart of homes, a gathering point for families. For much of human history, this was a necessity, since the central fire was the sole source of heat and light. "There is no place more delightful than one's own fireside," wrote the Roman orator Cicero in the first century BCE, and nearly 2000 years later, "fireside" remains a synonym for home. Poets and authors write affectionately of the friendly fireside, among them William Cowper, who gloried in "fire-side enjoyments, home-born happiness." In the mid-1800s, one group of authors noted for their family values, homespun verse, and uplifting sentiments, were known as the Fireside Poets: these included James Russell Lowell, Henry Wadsworth Longfellow, John Greenleaf Whittier, and Oliver Wendell Holmes. "Keep the Home Fires Burning" was a popular song of World War I, the lyrics written in 1915 by Lena Guilbert Ford, an American poet living in London. The theme appealed to homesick soldiers, for whom the home fire represented the familiar circle of family and friends they had left behind. In 1932, at the height of the Great Depression, when Franklin Delano Roosevelt began his presidency, he chose to communicate with the American people in "fireside chats:" informal discussions via radio that gave listeners the reassuring impression of a warm and intimate family talk around the home fire.

Keeping the home fires burning was never an easy trick. "The sheer difficulty of fire starting probably accounts for the reverence accorded perpetual flames in some cultures," write Margaret and Robert Hazen in *Keepers of the Flame*. The time-honored technique involves rubbing two sticks together, which is much trickier than it sounds: Tom Hanks, marooned on a desert island in the 2000 movie *Castaway*, struggles desperately before at last producing his first feeble flicker. Stick-rubbing, however, properly practised, can generate friction of sufficient heat to ignite tiny shreds of tinder, and the fire-starting devices of many primitive peoples – fire drills and fire bows, Polynesian fire plows, and Indian fire saws – are all basically improved sticks, designed to make more friction, faster. The fire plow, for example, consists of a slice of hibiscus wood laid flat on the ground while a narrow stick is pushed rapidly back and forth across it. Eventually the stick digs enough of a groove in the hibiscus to accumulate wood dust, which the heat of friction sets aflame.

H. erectus may have started his fires by rock banging which, while making stone hand axes, may have proceeded vigorously enough to generate sparks. By the Neolithic period, a variation on axe-banging may have been the fire-starter of choice: the 5000-year-old Iceman, discovered in 1991 by hikers in a glacial crevasse in the Alps, had a fire-starting kit in a leather belt pouch containing iron pyrites, flint, and tinder fungus. The technique seems to have involved grinding the fungus until it reached the fluffy fibrous consistency of fine cotton wool, piling it in a small container – mollusk shells used for this purpose have been found in some Stone Age fire kits – and finally striking sparks with flint and pyrite until the tinder ignited. The first feeble glow was probably boosted into flame by careful blowing; then the tiny blaze was fed more tinder – perhaps dried grass or moss, thistledown, or bullrush pith – until a steady fire developed. In historical times, steel was substituted for the iron pyrite, and this higher-tech combo – flint and steel – remained the fire-starter of choice throughout the seventeenth and eighteenth centuries. The colonial tinderbox packaged flint, steel, and tinder in one handy container; in theory, if the weather was right, an adept pioneer could go from spark to full-blown fire in under a minute.

A third possibility for early fire-starters utilized the Sun, focusing and concentrating its rays by means of concave polished metal mirrors. The burning mirror was reportedly a common household item in ancient Greece, though its most famous depiction dates to the siege of Syracuse in 214–212 BCE. Syracuse — the name comes from the Greek for "swamp" – is located on the southeast coast of Sicily, the island just off the toe of the aggressive Roman boot, noted for oranges, lemons, and the smoking cone of volcanic Mount Etna. During the Punic Wars – which raged between Rome and Carthage during the third century BCE – Syracuse threw in its lot with Carthage, thus bringing down upon itself the wrath of Rome, in the person of General Marcus Claudius Marcellus, backed by sixty quinqueremes and two legions of foot soldiers. That Syracuse managed to hold out as long as it did in the face of such military might and expertise is a tribute to Archimedes, whose contributions to the conflict reportedly included an ingenious array of burning mirrors that focused the Sun's beams on to distant enemy ships causing them to burst explosively into flame. The feasibility of Archimedes's mirror-generated death rays remains a matter of debate; and most historians cautiously refer to the story as "unsubstantiated." However, several experimenters have shown that it could have been done. In 1741 the renowned French naturalist the Comte de Buffon managed to set fire to a fir tree using an array of 128 Archimedean mirrors; with a set of forty-five, he melted a tin bottle. Even more effectively, in 1973 Greek scientist Ioannis Sakas, using a bank of seventy bronze-coated mirrors, set fire to a tar-coated rowboat 165 feet (50 m) away.

The ancient Chinese had bronze and copper burning mirrors, as did the Incas, Olmecs, and Aztecs of pre-Columbian South and Central America. According to a sixteenth-century description, the Incas lit their sacred fires by focusing the Sun with carefully aligned silver mirrors set in gem-encrusted gold frames. Ivan Watkins, professor of geosciences at St Cloud University in Minnesota, proposed that the Incas may even have used their mirrors to melt stone, thus creating the amazingly precise right-angled joints that characterize the stone walls of Inca architecture. These crisp geometric interfaces, suggested Watkins, could not have been made by hammers, but must have been seared out by a blast of concentrated heat, perhaps using a sort of mirror-driven primitive blowtorch. Unfortunately an experiment designed to test the hypothesis failed to melt rock, though Watkins did manage to ignite a wooden popsicle stick.

Since starting a fire was so prolonged and tedious a business, most people, once they got one going, were reluctant to let it go out again. The Iceman carried his fire with him: among his gear archaeologists found the remains of a stitched bark box, insulated with the scorched remains of damp leaves, a container for transporting embers. Carefully protected, an unfed fire can be preserved for a considerable period of time.

In early households, fires were routinely banked for the night, the hot coals buried protectively in ashes where they would continue to smolder for up to twelve hours. If the home fire died despite all best efforts, however, it was often easier – rather than starting from tinder and scratch – to borrow fire from a neighbor, a practice as common in the eighteenth and nineteenth centuries as the social borrowing of a cup of sugar was to become later.

The ability to start a fire – depending on where and when you lived – could be the difference between life and a frigid death. The classic tale of fire failure is Jack London's "To Build a Fire," first published in the *Youth's Companion* in 1902. The story is set in the Yukon. It's winter; it's seventy-five degrees below zero; and the nameless protagonist has just fallen through the ice and soaked himself to the knees. He has to start a fire or he'll freeze. Then one thing after another goes wrong: snow from an overhanging spruce tree smothers his first small fire; the second, its twigs scattered too far, flickers out; then his hands are too cold to hold his last matches. At the end, he succumbs to hypothermia, a conclusion to a far-north adventure that must have discouraged many a Youth.

Ideally in such a scenario, the matches would have made all the difference. The match was the technological triumph of its century. Textbooks of the 1800s compared it reverently to such life-changing inventions as the steamboat and the cotton gin; and Victorian philosopher Herbert Spencer called it "the greatest boon and blessing to come to mankind." It was, all in one skinny stick, a reliable heat- and light-provider, a fail-safe bulwark against the cold and dark, usable anytime, anywhere, by anybody. With it, man at last could feel himself in control of the essential but unreliable element, fire.

Chemically, matches depend on the element phosphorus, which name comes from the Greek for "light-bringing." Phosphorus was first isolated in 1669 by German alchemist Hennig Brandt, who was trying at the time to find the fabled philosopher's stone that could turn base metals into gold. The new element was unearthly stuff that glowed with an eerie greenish light, which outlandish feature made it much in demand for upper-class party tricks. (In 1677, King Charles II, himself a hobby alchemist, paid 1000 thalers – $30,000 or £20,000 in today's money – for a phosphorus demonstration at court.) The new element also spontaneously burst into flames. Such ultra-reactivity ensures that free phosphorus is never found in nature; instead it occurs primarily in the form of phosphate (PO_4^{-3}), in which each phosphorus atom is firmly bound to four oxygen atoms. The effect is the chemical version of tethering a maddened bull. Phosphorus, imprisoned by oxygen, is stable and benign; kick its oxygens loose, though, and it goes on a rampage.

Free phosphorus quickly clusters into tiny four-atom pyramids (P_4), a form known as white phosphorus, in which it is stable – just – provided it is kept cautiously underwater. Exposed to air, white phosphorus reacts explosively with oxygen, generating so much heat that it catches fire. On a small scale, this was the secret behind the early self-igniting match. "Self-igniting" is the key term here: matches had been around since ancient times, but all required a heat source to light. The Chinese and the Romans both had sulfur matches – thin strips of pine tipped with sulfur that, when touched to an ember, would burst into flame. Without an initiating ember, however, sulfur matches were just so much dead wood and, as such, not much use on an emberless cold morning. The self-igniting match, in contrast, generated fire out of thin air. The first of these – known as the "Phosphoric Candle" – was marketed in France in 1781. It consisted of a strip of paper impregnated with phosphorus and sealed in a glass tube; when the glass was smashed, the paper spontaneously ignited.

Alternatives were quickly developed. In 1786, French chemist Claude Berthollet produced *briquets oxygènes*, made from a dried paste of sugar and potassium chlorate that, upon addition of a tiny drop of sulfuric acid, exploded satisfyingly into flame. Variations on this theme included the Instantaneous Light Box, invented by Henry Berry of London in 1805, in which splints tipped with a potassium chlorate mix were dunked into a handy bottle of acid and pulled out ablaze; and the Vesuvian match, in which paper soaked in sulfur, sugar, and potassium chlorate was wrapped around a sealed glass tube of acid. When you needed a light, you crushed the tube with a pair of pliers.

All, though functional, were awkward; and acid-ignited potassium chlorate matches were not only messy but potentially dangerous, since concentrated sulfuric acid, if leaked, spilled, or carelessly splattered, could of itself cause severe burns. The solution was the friction match, serendipitously invented in 1827 by British pharmacist John Walker who, after preparing a mix of potassium chlorate and antimony sulfide, dropped a dollop on the floor and then stepped on it by mistake. The trodden-upon mix burst into flame, giving Walker – presumably after an initial yelp of dismay – the idea for a match ignited by the heat of friction. The new matches, lit by pulling the chemically coated match head through a folded scrap of sandpaper, were only a modest success; a possible drawback, explains John Emsley in *The 13th Element*, his absorbing history of phosphorus, was that, when ignited, they went off with a bang like a firecracker, tossing bits of blazing match head about the room. Another problem may have been their lumpish name; Walker had dubbed his product "sulphuretted peroxide strikables." In any case, Walker never patented the idea, which was a mistake. By 1830, the friction match had been co-opted by rival match-maker Samuel Jones who, in an inspired advertising ploy, coined a new name for the matches: lucifers. Lucifers made Jones and partners rich.

Despite the success of lucifers, friction matches only became truly reliable self-igniters with the addition of phosphorus to the match head, an idea attributed to at least four different chemists. The front runner, Charles Sauria of France, a student at the time, was too poor to patent his discovery, though at least two of his competitors (Joseph Kammerer of Germany and Stephan Romer of Austria) did; fifty years later, however, in 1884, by which time Sauria was in his seventies, he was officially recognized as the inventor of the phosphorus match and given a congratulatory medal by the French Academie Nationale Agricole.

In England, the new phosphorus-supplemented friction matches were initially called congreves, both to differentiate them from the phosphorus-less lucifers and in honor of William Congreve, inventor of the similarly fiery military rocket. Congreve's rocket was essentially a guided pipe bomb: a skinny 3-foot-long (meter-long) tube of cast iron packed with explosives. Congreves were used during the Napoleonic wars – the British used 200 of them in 1807 to set fire to Boulogne (a mistake; they were aiming for the French fleet) and 25,000 of them in 1808 to level Copenhagen (on purpose); and they were used to bombard Maryland's Fort McHenry during the American War of 1812, thus creating the rockets' red glare immortalized in excruciatingly high notes in Francis Scott Key's "The Star-Spangled Banner."

The word congreve, meaning match, was short-lived. Linguistic tradition is notoriously hard to budge (after all, we still measure automobile engine performance in horsepower, a term based on the hauling capacities of eighteenth-century Welsh mining ponies); and popular usage came down on the side of lucifer, which familiar name the public cheerfully transferred to the new phosphorus matches. The improved lucifers were much more efficient than their predecessors – in some cases too efficient, since phosphorus matches often ignited unwisely and too well. Rats, who rapidly discovered a fondness for the taste of phosphorus, were given to nibbling at unprotected match heads, thus starting numerous fires; and the matches also ignited if dropped, stepped upon, shaken together in the box, or even if left carelessly on a windowsill and exposed to the heat of the sun. Emsley quotes a story from the London *Times* of June 8, 1867 about the sad demise of the 19-year-old Archduchess Matilda, a young woman "endowed with rare gifts of person, mind and heart," who "died on Thursday last at 8 o'clock in the morning – of a lucifer match. She inadvertently trod on one which was lying at her feet, as she leant out at the window talking to one of her relatives. Her summer dress was in a blaze before she was aware of it and before anyone could run to her rescue she sank to the ground in an agony of pain, from which only death released her."

The frequency of such incendiary tragedies was sharply reduced with the substitution of the safer and stabler red phosphorus for the unreliable white. White and red phosphorus are allotropes – that is, different structural forms of the same element, displaying widely different physical and chemical properties. (Prime examples are the carbon allotropes, which include diamond, graphite, and the cage-like buckminster-fullerenes.) In white phosphorus, individual atoms are bound together in groups of four to form pyramidal molecules; in red, formed when white phosphorus is heated for several days in a closed container, the atoms interlock in an extended random network. ("Red" phosphorus can actually be anything from orange to purple, depending on minor variations in network structure.) The network configuration has a stabilizing effect on ordinarily hyperactive phosphorus: red phosphorus, a far stodgier and safer construct, neither glows in the dark nor bursts spontaneously into flame when exposed to air.

In the match industry, the transition from white to red phosphorus was not fully implemented until the twentieth century, having been resisted by manufacturers on the grounds that red phosphorus, being much more difficult to produce than white, was markedly more expensive. Eventually, however, even the most adamant of match barons was forced to give way, as by the 1890s work with white phosphorus was officially recognized as a "dangerous trade." Unofficial recognition had come far sooner, since white phosphorus is an unsubtle killer. A lethal dose for a human adult is about 100 mg – that is, about the amount found in the average box of nineteenth-century matches, which is why depressed Victorians sometimes committed suicide by eating match heads. Death resulted from liver destruction. Even exposure to sub-lethal doses of phosphorus – as experienced by nineteenth-century match-makers, most of them women and children – was ultimately disabling, inducing a hideous condition known as phosphorus necrosis or "phossy jaw." The condition began with gum abscesses, followed by loosened and decaying teeth, and finally a painful destruction of the jawbone, accompanied by a persistent foul-smelling ooze of pus. The disease was so awful that over half those suffering from it, according to a French study in 1858, killed themselves. Hans Christian Andersen's pitiful Little Match Girl must certainly have been at risk of this; by comparison, her heart-wrenching death of cold (still clutching a handful of burned-out lucifers) looks merciful. In 1906, with all this in mind, an international convention in Berne, Switzerland, formally outlawed the manufacture of white phosphorus matches. It was signed by all attendees except the United States, which pleaded constitutional conflicts, though, to be fair, the Americans promptly dealt with the white phosphorus problem at home by taxing its users out of existence.

The safety match was invented in 1855 by Johan Lundstrom of Jönköping, Sweden – which city still styles itself the "Match Capital of the World" and boasts the world's only museum devoted exclusively to matches. Lundstrom's match was safe because it was a chemical amputee. Rather than incorporating all necessary flammable components in the match head, Lundstrom removed the red phosphorus, mixed it with powdered glass, and glued it to a panel on the side of the matchbox. Safety match heads contain a mix of sulfur or antimony trisulfide and potassium chlorate. When scraped across the phosphorus-impregnated striking surface, heat of friction converts the red phosphorus to white, which ignites spontaneously upon contact with air. The heat of the flaming phosphorus in turn ignites the sulfur or antimony trisulfide of the match head, while liberating oxygen from the potassium chlorate to further fuel the blaze.

Today about 500 billion matches are struck annually, most of them from Sweden, which remains the world's leading match producer. ("Close cover before striking" is said to be the most printed phrase in English.) The match's major competitor today is the pocket lighter, fueled with butane, and ignited by a spark. The spark, in a neat historical example of what goes around comes around, is generated by flint scraping against steel.

Modern humans, born in the age of matches, tend to take fire for granted. In fact, the closest most of us get to the experience of Stone Age man, frustratedly striking sparks into his mollusk shell of shredded tinder, is the seasonal experience of lighting the backyard barbecue grill. The average backyard barbecue is fueled with biscuit-shaped charcoal briquettes, ordinarily ignited with a spritz of lighter fluid and a match. After the initial whoosh, the hopeful barbecuer then waits until the coal-black briquettes turn ashy gray, signaling the establishment of a heat-radiating bed of coals suitable for broiling hamburgers, hotdogs, and sauce-slathered pork ribs. Provided, of course, that all goes as planned, which it often doesn't, charcoal briquettes being legendarily reluctant to burn. Innovative approaches to this problem are the challenge of the "Great Engineer BBQ Contest," originated in 1994 by George Goble, a senior systems engineer at Indiana's Purdue University. The contest grew out of an annual Indiana summer picnic – the sort with red-checked tablecloths, potato salad, beer, and badminton – in this case primarily attended by engineers, who soon devised a number of techniques to accelerate the lighting of charcoal for grilling the picnic hamburgers. The ultimate solution, of which any number of fire deities would have been proud, was a bucket of liquid oxygen, the stuff of rocket fuel, which, when dumped on to 60 pounds (27 kg) of charcoal and topped by a single smoldering cigarette, exploded into a monstrous fireball, reaching a temperature of 10,000°F (5500°C) and vaporizing the barbecue grill. Videos of this event used to appear on the world wide web; when I last checked, however, they had been removed, presumably by regulatory personnel who felt it set a poor example for the barbecuing general public. It was, however, a thrilling example of the chemical process of combustion.

Of all the four elements, fire, chemically and physically, has proved the most enigmatic; and scientists struggled for centuries to determine just what it was. In many ways, fire had aspects of living creatures – it was born, required food to grow, and, if not fed, died – and some ascribed its behavior to tiny swarming "igneous Beings" or "igneous animalculae." Jean-Baptiste Robinet, author of *De la Nature* (1766), even claimed to have examined some of these under a microscope, where he discovered fire to consist of "little shiny worms." More commonly, however, eighteenth-century scientists explained fire chemically rather than biologically. Most subscribed to phlogiston theory, first proposed around the turn of the seventeenth century by Georg Ernst Stahl, a professor of medicine and court physician to the Frederick William I, King of Prussia.

Stahl is described in J. R. Partington's *A Short History of Chemistry* as "morose in disposition," which may have been a result of his four marriages. In his portrait, for which he wore an enormous curled wig and an elaborately knotted cravat, he has a tight-lipped and suspicious expression; it's easy to see him as a man not readily amused. His scientific writings were kept in an obscure mix of German, Latin, and Greek, in which last language the word phlogiston first appears. Phlogiston – the term comes from the Greek meaning "to burn" – was hypothesized by Stahl to be an inherent constituent of anything flammable. When something burned, it lost phlogiston to the air; when it stopped burning, either it had exhausted its store of phlogiston, or the air had become so phlogiston-saturated that it was incapable of accepting any more. Smelting was understandable in terms of phlogiston: when an ore was heated with charcoal, phlogiston was transferred from combustible charcoal to inert ore, thus generating phlogiston-rich metal. Calcination, on the other hand, in which metal, heated in air, was reduced to a powdery rust or calx, was due to the loss of phlogiston. Limestone (calcium carbonate), when heated, formed quicklime (calcium oxide); the transformation resulted from a gain in phlogiston, picked up by limestone from the fire.

Phlogiston theory made reasonable sense, and it perceptively linked a number of phenomena – combustion, smelting,
and rusting or calcination – previously believed to be wholly
unrelated. As a logical construct, it held together well enough
to dominate the science of chemistry for nearly a century. In
practice, however, phlogiston had issues. For one thing, the
elusive substance was annoyingly inconstant. In calcination,
for example, the powdery calx that resulted from burning
metal was heavier than the parent metal – not lighter, as would
be expected from a material losing phlogiston. Charcoal, in
contrast, which was presumably practically *all* phlogiston,
since nothing was left of it after burning but a smattering
of ash, lost weight as expected. Perhaps there were different
phlogistons, researchers suggested, with markedly different
behaviors, or even – to explain the peculiar weight gains in the
course of calcination – a phlogiston of negative weight, such
that losing it would effectively make a substance heavier.

The process that we know now as combustion was eluci-
dated by French chemist Antoine Lavoisier, a brilliant theoreti-
cian and obsessive fussbudget in the matter of measurements,
and not a man to be lightly taken in by such concepts as phlogistons of negative weight. By 1777, after performing a number
of carefully calibrated experiments, he had scuttled phlogiston
theory altogether, conclusively demonstrating that burning
and rusting were chemical peas in a pod. Both were the result
of substances combining with oxygen. Burned substances lost
weight not because they lost phlogiston, but because oxygen
combined with components in the fuel, converting them to
volatile carbon dioxide and water vapor. Rusted metals gained
weight because oxygen combined with surface molecules,
forming a crusty coat of metal oxide.

Most fire, chemically, is the reaction of a hydrocarbon fuel
– for example, heaps of charcoal briquettes – with oxygen,
producing carbon dioxide and water, and emitting light and
heat. Though straightforward in essence, burning can still be
a complex process. Even the burning of a simple five-atom
molecule can involve over a hundred interim chemical reactions

before the substrate is at last reduced to carbon dioxide and water vapor; and the incineration of a more complex organic – say, a Yule log or a Cuban cigar – may involve literally thousands of chemical interactions. The process has fascinated many, although perhaps none more than nineteenth-century British scientist Michael Faraday.

Faraday was a serious scientist – among other things, he discovered electromagnetic induction and invented the battery, the transformer, and the electric generator. He was also the nineteenth century's equivalent of such children's television celebrities as Mr Wizard and Bill Nye the Science Guy, famed for a popular science lecture series punctuated with exciting hands-on experiments that were performed as treats for the young at Christmastime. (The lectures, still equally appealing, continue to this day.) Best known of Faraday's presentations was "The Chemical History of a Candle," an enthralling explanation of combustion theory, first presented in 1860. The "Chemical History" consisted of six lectures variously covering candle structure, fuel and flame, combustion products, the nature of the atmosphere, and "Respiration and Its Analogy to a Candle." The performance included illustrations, anecdotes, and dramatic demonstrations, among them a sizzling game of snapdragon with raisins and flaming brandy, the launch of a hot-air balloon, and the explosion of a sample of gunpowder. Children loved it, as did many adults, among them Prince Albert and Charles Dickens.

Faraday – using, for demonstration purposes, tallow candles, beeswax candles, paraffin candles, a mauve candle made with the new coal-tar-based chemical dyes, and even a candle dredged up from the wreck of the *Royal George*, a ship sunk at Spithead in 1782 – explained that candles burn because the heat of the lit wick melts the candle wax. The liquefied wax is then sucked up the wick by capillary action. (Faraday illustrated capillary action with a story of dribbling shrimp that makes for an intriguing glimpse into the nineteenth-century ducal dinner party: "The late Duke of Sussex," he writes in a footnote, "was, we believe, the first to show that a prawn might be washed upon this principle. If the tail, after pulling off the fan part, be placed in a tumbler of water, and the head allowed to hang over the outside, the water will be sucked up the tail by capillary attraction, and will continue to run out through the head until the water in the glass has sunk so low that the tail ceases to dip into it.")

When the long hydrocarbon polymers that make up the wax reach the heat of the candle flame, they are vaporized. Then, in gaseous form, they diffuse outward and combine with oxygen in the air. This is the true combustion process: fuel and oxygen are converted to carbon dioxide and water vapor, with accompanying release of energy in the form of heat and light. You can prove that it's the vaporized fuel that burns, Faraday told his awestruck audience, by blowing out the candle and then holding a lit taper to the air two or three inches above the wick. The flame runs greedily through the hydrocarbon-saturated air to re-ignite the wick. (Try it; it works.)

Vaporized wax forms immense numbers of different reaction products, among them ball-shaped carbonaceous buckminsterfullerenes – named for architect Buckminster Fuller and colloquially known as buckyballs – and ring-shaped molecules called polycyclic aromatic hydrocarbons which clump to form particles of soot. Soot is responsible for the soft illumination of romantic candlelit dinners. Heated carbon becomes incandescent; the yellow light in the tongue of the candle flame is the glow of broiling soot. Soot also burns up altogether, which generates heat; and, when the candle is extinguished, soot spirals upward as thin black smoke – when a lovely flame dies, it's polycyclic aromatic hydrocarbons that get in your eyes. The blue base of the candle flame, on the other hand, results from chemiluminescence, light energy emitted by the rapid vibrations of overheated carbon–carbon and carbon–hydrogen bonds in wax polymers en route to vaporization and combustion. The blue is the hottest part of the flame and the worst place to stick your finger – around 1670°F (1400°C). The yellow, by contrast, runs around 1200°C; the reddish center of the flame, about 700°C.

A candle flame is a creature of gravity. Hot gases rise, leaving an empty low-pressure zone beneath them; fresh air rushes into the gap, fueling the flame with a fresh supply of oxygen. The steady upward flow creates the elongated El Greco-like shape of the flame. If there are a lot of flames, as in a full-fledged forest fire, the rising hot air and resultant pressure differential can create roaring winds of up to 100 miles per hour (160 km/h). In the microgravity of space, where none of this applies, candles burn with faint blue spherical flames like miniature will-o-the-wisps, barely visible unless viewed in the dark.

Fire, our sole source of artificial light prior to Thomas Edison's light bulb, is always an uneasy companion; and the same updraft that shapes the romantic dining-table candle flame may be the crucial factor in spontaneous human combustion (SHC). Spontaneous human combustion, in which unlucky victims suddenly and inexplicably burst into flames – a favorite topic of tabloid newspapers today – has appealed to our prurient interests for centuries. The first known collection of SHC stories dates to 1763; titled *De Incendiis Corporis Humani Spontaneis*, it was the work of Frenchman Jonas Dupont, who was reportedly inspired by a court case in which the accused, a M. Millet, was acquitted of the murder of his wife, having managed to convince the jury that she had died of spontaneous combustion. The very possibility of such a hideous end tickled the sensational tastes of the nineteenth century: it was a favored demise in numerous penny dreadfuls – the Victorian equivalents of the modern comic book; and in the 1832 novel *Jacob Faithful* by Captain Marryat, the lead character's mother died of it (reduced, wrote Marryat gruesomely, to "a sort of unctuous pitchey cinder"). Even Charles Dickens used it, to polish off Krook, the drunken rag-and-bone dealer in *Bleak House*.

Interest in the phenomenon resurged in the 1950s with the

case of Mary Reeser, a 67-year-old widow of St Petersburg, Florida, whose remains were discovered by her landlady on the night of July 1, 1951. Nothing was left of the unfortunate Ms Reeser but her left foot, still clad in its bedroom slipper, and the charred springs of her easy chair. In the substantial literature of SHC, she is often referred to as the "cinder woman." The dreadfulness of Ms Reeser's end and a rash of subsequent highly popularized equivalent events led to a rash of hypotheses: ball lightning was implicated, as was unsuspected build-up of flammable methane in the intestinal tract. Larry Arnold, author of *Ablaze!: The Mysterious Fires of Spontaneous Human Combustion* (1995), variously blames imaginary subatomic particles dubbed "pyrotrons;" extreme emotional stress; or, perhaps most worrisomely, "preternatural combustibility," by which he means that sometimes the body's cells, for no apparent reason, become as explosively flammable as celluloid. The Victorians, who leaned toward the latter view, felt that SHC occurred in heavy drinkers, brought on by a saturation of bodily tissues with highly combustible alcohol.

The most likely explanation for such grisly events, most forensic scientists guess, is the so-called "candle" or "wick" effect. In SHC, usually initiated by dropped cigarettes, smoldering embers, or carelessly handled matches, the human body functions as a sort of inside-out candle. The slow blaze dissolves body fat which soaks into clothing and furniture upholstery, both of which can function as wicks, continually feeding the flames. Over a matter of hours, the body is reduced to greasy ash. "Even a lean body contains a significant amount of fat," reports Joe Nickell ominously in "Not-So-Spontaneous Human Combustion," a report on the phenomenon by *Skeptical Inquirer* magazine in 1996, implying that none of us is safe, especially if we're suffering from stress or are full of brandy.

On the other hand, some of us seem to be fireproof. In the early 1800s, Frenchman Ivan Chabert – otherwise known as the "Human Salamander" or the "Fire King" – impressed audiences worldwide with his ability to stride into a blazing oven holding a pair of raw steaks in his bare hands; he would emerge, unharmed, with the sizzling steaks cooked to perfection. (In England, in lieu of the steaks, he used a rump roast and a leg of mutton; these were subsequently eaten for dinner and pronounced delicious.) The Fire King could also bathe his feet in boiling lead, grasp red-hot iron bars, and eat torches "with as much gusto as other of his countrymen devour frogs." Chabert was one of many: other flame-resistant performers included Signora Josephine Girardelli, the "Fire-proof Lady," who cooked eggs by holding boiling oil in her hands; Signore Lionetto, the "Incombustible Spaniard;" and J. A. B. Chylinski, the "Polish Salamander," noted for sticking his head in a blazing charcoal fire.

The secret to such startling fire resistance, according to modern scientists, is physics. The juggling of hot coals, grabbing of hot iron bars, dunking of feet in molten lead and the like are made possible by the Leidenfrost effect, the same physical quirk that allows you to lick your finger and test the surface of a hot flatiron without getting burned. The explanation behind the effect – named for scientist Johann Gottlob Leidenfrost, who published a detailed paper about it in 1756 – is creative insulation. When the wet finger, hand, or naked foot contacts a hot object such as a hot iron or a bucket of liquid lead, the water is quickly vaporized, generating a protective layer of steam. The steam, a poor conductor of heat, acts as a buffer between skin and heat source – provided, of course, that contact is brief.

Those that survived the medieval ordeal of trial by fire, in which defendants proved their innocence by hoisting red hot bars of iron or walking over hot coals, also owed their success at least in part to the Leidenfrost effect. The most famous example in history is probably that of the eleventh-century English Queen Emma, wife in succession to Aethelred the Unready and Canute the Great, mother to the saintly Edward the Confessor. Emma, who seems to have been a hot ticket (the Archbishop of Canterbury called her "a wild thing"), was accused of unseemly dalliance with Aelfwine, Archbishop of Winchester, and threatened with imprisonment and loss of property. She defended her virtue by walking barefoot over nine red-hot plowshares without receiving so much as a singe; her shamed son apologized for doubting her, and she got all her manors back.

The hot coals effect is somewhat different: the unscathed soles of firewalkers appear to be a matter of thermodynamics. Firewalking, in which the faithful, fanatical, or foolhardy pace barefoot through a bed of hot coals, has been practised worldwide for centuries; the earliest known reference to it comes from an Indian manuscript of 1400 BCE. The fakirs of India, the !Kung tribesmen of Africa's Kalahari Desert, and the Yamabushi mystics of Japan still participate in ritual firewalks; and today in the United States and Europe, firewalking has caught the fancy of New Agers, who view it as an enhancer of spirituality, a cure for phobias, and an exercise in mental control. It's also an impressive demonstration of what you can get away with in the presence of lousy conductivity.

Heat is transmitted in one of three ways: by convection, which is how water is brought to a boil on a stovetop burner; by radiation via electromagnetic wave, which is how heat travels to the Earth from the Sun; and by conduction or direct transfer, which happens when a hot object touches something cold. In conduction, the rapidly vibrating molecules of the hot object transfer energy to the sluggish molecules of the cold, thereby making it hotter. How well this transfer works depends on the thermal conductivity of the objects. Air, for example, is a poor thermal conductor, which is why you can poke your (relatively cold) hand into a hot oven without getting fried; metal, on the other hand, is an excellent conductor, which is why – unprotected – you don't want to grab the sizzling meatloaf pan. Wood is on the low end of the thermal conductivity scale, 500 times less conductive than iron, 1000 times less than aluminum, 3000 times less than copper. Wood undeniably gets hot – the smoldering embers of the average firewalk average 500°C (960°F) – but the poor conductivity of wood, charcoal, and ash ensures that this heat is not transferred effectively to the firewalkers' feet. Again, there are limits: firewalkers need to move along at a reasonable clip, such that the bare foot is never in prolonged contact with the hot coals. Prolonged contact is anything over a second, which is the time it takes for sufficient heat transfer to burn shoeless flesh. At average walking pace, however, the average step takes half a second, which is well short of the cut-off point. Provided the fire is properly prepared and the walker keeps moving, the laws of physics say the feet should be fine.

Fire is so all-consuming that we traditionally view anyone or anything resistant to it with awe. "When I was about five years of age," wrote Italian goldsmith and sculptor Benvenuto Cellini in his autobiography (1558–66), "my father, happening to be in a little room in which they had been washing, and where there was a good fire of oak burning, looked into the flames and saw a little animal resembling a lizard, which could live in the hottest part of the element. Instantly perceiving what it was, he called for my sister and me, and after he had shown us the creature, he gave me a box on the ear. I fell a-crying, while he, soothing me with caresses, spoke these words: 'My dear child, I do not give you that blow for any fault you have committed, but that you may recollect that the little creature you see in the fire is a salamander, such a one as never was beheld before to my knowledge.' So saying he embraced me and gave me some money."

The legendary fire-dwelling salamander puts scampering human firewalkers – even the phenomenal M. Chabert, with his hand-held rump roasts – to shame. The name "salamander" comes from the Greek for "fire lizard;" ancient tradition – supported by such authorities as Aristotle and the irrepressible Pliny the Elder – insisted that fire was its natural element. The superstition apparently arose because salamanders were often seen, as in the Cellini household, crawling out of the flames. This occurs because salamanders, who actually prefer life dark, cool, and damp, often hide in the crevices beneath fallen logs; when a salamander-sheltering log is thrown on a fire, the outraged salamander scuttles for safety, giving the erroneous impression that its home is in the heart of the fire.

The salamander and its supposed fiery habitat were a persistent medieval legend: a salamander in flames, variously a symbol of protection, constancy, or fortitude in the face of suffering, appeared on assorted coats of arms, most prominently that of Francis I of France (1494–1547), who decorated his castles with it in bas-relief. In Christian legend, the salamander represents virtue, righteously triumphing over the fires of passion; and according to the *OED*, salamander, as of 1711, was a synonym for a woman who "lives chastely in the midst of temptation."

Sometime in the late twelfth century, a letter began to circulate among the kingdoms of Europe purporting to be from a fabulous monarch of the East named Prester John. The John legend flourished for centuries: his elusive kingdom was said to be in India, or possibly Ethiopia. The king himself was said to be a descendant of one of the Three Wise Men; as a devout Christian, he was continually said to be sending armies to aid the Crusaders, but these were always being sidetracked or delayed. Marco Polo, who inquired after him in the course of his thirteenth-century journey to Cathay, records that he ruled a vast tract of land in Mongolia and had been killed in battle by Genghis Khan, who wanted to marry his daughter. On one point, however, all stories agreed: Prester John was filthy rich. He lived in a palace of gold, with crystal windows and a roof of ebony, where he was waited upon by seven kings; his throne was studded with emeralds and pearls; his bed was made of sapphires. He had giant ants that mined gold, a spring whose waters conferred eternal youth, and a sea of sand and gravel that none the less contained fish. He also had garments made of salamander skins. "Our realm yields the worm known as the salamander," one of the John letters read. "Salamanders live in fire and make cocoons, which our court ladies spin and use to weave cloth and garments. To wash and clean these fabrics, they throw them in the flames."

The marvelous flame-resistant fabrics were real. Marco Polo reveals, in tones of astonishment, in his *Travels* that the supposed salamander wool was really "stuff dug out of the mountain, which when crumbled and made into cloths, emerges unscathed, indeed, cleansed, by fire." The stuff was, in fact, asbestos, a class of fibrous silicate minerals known since ancient times, the most common of which is white asbestos or chrysotile. The name "asbestos" comes from the ancient Greek for "unquenchable," in recognition of its fire-retardant properties – nicknames for it included "mountain leather," "rock floss," and "stone wool." Charlemagne is said to have had a tablecloth made out of it; and Benjamin Franklin, as a thrifty teenager, made himself an asbestos purse, ostensibly to keep money from burning a hole in his pocket. (It could certainly have done so: asbestos has a melting point of 800–850°C (1470–1560°F).) In the 1939 movie version of L. Frank Baum's *The Wizard of Oz*, Margaret Hamilton, the Wicked Witch of the West, flew across the silver screen on an asbestos broom.

Asbestos usage peaked in the mid-twentieth century, when the versatile substance was used for an immense range of products and purposes, among them insulation, building materials, fireproof suits, brake linings, theater curtains, toothpaste, and the lumpy mattresses used in sleeper cars on passenger trains. At the New York World's Fair of 1939 ("Building the World of Tomorrow"), Johns-Manville – then the world's largest manufacturer of asbestos products – sponsored an impressive pavilion featuring a gigantic Asbestos Man, proudly touting the value and versatility of the company product.

Asbestos has since plummeted in popularity, taking with it Asbestos Man. By 1963, fallen on hard times, he appears in a Marvel Comics issue as a costumed villain, an analytical chemist gone bad, who – in an attempt to win the respect of the criminal underworld – challenges the upright Marvel superhero the Human Torch to a duel. It's a suspenseful battle. The Torch can throw fireballs; Asbestos Man, however, by virtue of his midnight blue asbestos suit, shield, and silly-looking pig-snouted helmet, is impervious to flame. In fact, with his net of "skin-covered nitrogen strands," he can hurl the fireballs right back again, with explosive force. At one point, he even extinguishes the Torch by dropping him evilly through a trapdoor into a moat. Ultimately, however, the Torch wins and Asbestos Man ends up behind bars. As the cell doors clang shut, the Torch is unable to resist a few taunting last words: "Remember, professor, you can never play with fire without ... getting burned!"

Asbestos Man was arraigned for bank robbery; asbestos itself was ultimately condemned as a life-threatening health hazard. As early as the first century CE, Pliny the Elder pointed out that slaves laboring in the asbestos mines died young of lung disease; and modern medical studies have shown that inhalation of tiny glassy fibers of asbestos leads to lung damage (asbestosis), lung cancer, and mesothelioma. Claims against asbestos manufacturers over recent decades have driven most out of business.

True salamanders are damp, thin-skinned creatures, denizens of shady woodland areas with nearby streams and ponds. The brightest of the bunch is the so-called "fire salamander," *Salamandra salamandra*, an 8 to 10-inch-long (20–25-cm-long) animal whose black body is patterned with gaudy spots and splotches of orange or yellow. Despite its flame-friendly moniker, it, like any other sensible salamander, reacts to surrounding fire with horror. In this sense, salamanders are not unlike people. We too find it hard to keep our cool when our abode, be it hut, cabin, village, fort, or presumably impregnable castle, bursts into flames. This is why, of all the elements, fire over the centuries has been a terrifyingly effective weapon.

According to J. R. Partington's *History of Greek Fire and Gunpowder*, all "savage races" who know how to use a bow have invented an incendiary arrow. "Savage," by this criterion, encompasses practically everyone: the Greeks used incendiary arrows against each other in the Peloponnesian War; the

Vandals and Visigoths used them against the Romans; the Arabs used them against the Crusaders; and in North America, the native tribespeople used them against the European settlers. If flame-tipped arrows could be destructive, whole flame-filled pots could be more so: Assyrian carvings dating to the ninth century BCE show containers of burning pitch being flung from the walls of besieged towns on to the attackers below. Aineias in a treatise *On the Defense of Fortified Positions*, written around 360 BCE, describes flaming pots of pitch, sulfur, tow, pine chips, and granulated frankincense which were hurled on to the wooden decks of enemy ships; a particularly nasty aspect of such fires, he adds, is that they could only be extinguished with vinegar. The Arabs favored incendiary tubs of naphtha, flung at distant objectives with catapults. Surviving Crusaders told horror stories about the unquenchable flaming oil with which the Saracens immolated their siege towers.

The fuel for such primitive fire-bombs and incendiary grenades came from natural seeps of petroleum. Petroleum (from the Latin for "rock oil"), tar, and natural gas all ooze unaided from the ground at points where oil-bearing rocks – certain shales, sandstones, and limestones – are exposed to the surface by erosion. Not surprisingly, most of the seeps used by ancient peoples dotted what today is prime oil-producing territory. Ancient historians, among them Herodotus, Strabo, Dioscorides, Pliny, and Plutarch, describe the oil pits and pitch springs of the Middle East, a foreshadowing of OPEC, power politics, and the Gulf Wars.

The natives of the New World were also familiar with pitch. The three largest pitch lakes in the world are all in the Americas; number one, first described by Sir Walter Raleigh, who encountered it in 1593, is on the southwest coast of the island of Trinidad. The lake is a sulfurous and sticky pool 100 acres (40 hectares) in area, surrounded by cashew and breadfruit trees, and containing up to ten million tons of asphalt – at present rate of consumption, a good 400-year supply. According to the local Chaima Indian tribe, the lake came into existence as a punishment from the gods, after a chief at a victory feast unwisely ate a sacred hummingbird. The insulted gods, in retribution, buried him and his village in simmering tar. Next in size after Trinidad's lake (today cruelly referred to as the "ugliest attraction in the Caribbean") are California's La Brea Tar Pits, first described by eighteenth-century Spanish explorers, who spoke gingerly of exploding bubbles, and Lake Bermudez in Venezuela.

Pitch, also known as bitumen or asphalt, is a crude oil solid, usually brownish-black or black in color – hence "pitch-black" or "black as pitch." Chemically, pitch is a mixed bag of hydrocarbon molecules, a complex conglomeration of high-molecular-weight asphaltenes, polycyclic aromatics, waxy saturates, resins, and volatile gases. The chemical fingerprints of various pitches are so distinctive that archaeologists are able to trace samples from artifacts to their original geological sources, thus documenting trade routes throughout the ancient Middle East. Pitch is initially exuded as a thin liquid, which, as its volatile components evaporate, gradually solidifies to a slurpy jelly, a viscous goo, and eventually a rocky substance tough enough in places to support animals, people, and even, in Trinidad, the laboring bulldozers of Trinidad & Tobago Lake Asphalt, Ltd. Appearances, however, can be deceiving: tourists straying from the designated paths in Trinidad have been known to plunge through the thin crust and sink up to their waists; and the La Brea Tar Pits are a fossilized record of disastrous miscalculations, to date generating the remains of over sixty different ancient species of mammals, including saber-toothed cats, dire wolves, camels, bison, and mastodons.

Pitch was used by the Sumerians, Assyrians, and Babylonians to waterproof boats and baskets, as mortar for their clay-brick walls and palaces, and as an adhesive for mosaics and inlaid jewelry. Leonard Woolley's excavation of the royal tombs of Ur – the ancient Mesopotamian city once located on the Euphrates River southeast of Baghdad in Iraq – brought to light exquisite artifacts made with bitumen, among them a cosmetics box, a game board set with silver lion's heads, and a lyre inlaid with lapis lazuli, silver, and mother-of-pearl. According to the Bible, pitch was used to seal Noah's Ark and Moses's floating basket of bulrushes; and it served as mortar for the ill-fated Tower of Babel and the terraces of the Hanging Gardens of Babylon. The ancient Egyptians used immense amounts of it for gluing together the multiple linen layers of mummy wrappings; the Greeks and Phoenecians caulked their ships with it; and the Romans used it to seal their reservoirs, aqueducts, and baths.

In the ancient world, prime sources of pitch were the tarry Fountains of Is near the modern city of Hit in Iraq, and Israel's Dead Sea. The Dead Sea – known to the Romans as Lacus Asphaltites, or Lake Asphalt – generated floating lumps of jellied oil that were harvested, rolled in sand, packed in leather bags, and loaded on to camels for export, primarily to Egypt. The chief inhabitants of the region were the Nabateans, an Arabic people who occupied the Negev Desert, the shores of the Dead Sea, and the bulk of modern-day Jordan from the sixth century BCE through the first century CE, at which point they were absorbed by the Roman Empire. During most of this period, the Nabateans dominated the eastern Mediterranean bitumen trade, using their oil profits to build the fabulous city of Petra, carved from the red rocks of the Jordanian desert. (Viewers of Steven Spielberg's 1989 movie *Indiana Jones and the Last Crusade* have seen it; it was there that Harrison Ford, Sean Connery, and a band of pursuing Nazis went in search of the Holy Grail.)

Pitch was flammable, but naphtha – from the Arabic *naft*, meaning "crude oil" – was more so. This lighter and more volatile oil, bubbling up through fissures in the ground, is thought to be the source of ancient eternal flames. The city of Baku on the western shore of the Caspian Sea, a rich area for natural petroleum springs, was famed for thousands of years for its perpetual fires, most likely first ignited by lightning. (Baku was lucky; some archaeologists suggest that the Biblical tale of sinful Sodom and Gomorrah reflects an ancient incineration brought on by a malignant combination of oil springs and lightning.) The temple at Baku, with its oil-fueled central altar, was sacred to the fire-worshipping Zoroastrians, whose religion dominated Persia from at least 600 BCE. Marco Polo, who visited Baku in the late thirteenth century, mentioned that the holy flame was visible for miles across the desert; he also commented that "Near the Georgian border, there is a spring from which gushes a stream of oil, in such abundance that a hundred ships may load there at once." The oil, he adds, is not good to eat, but is excellent for burning and efficacious for treating camels afflicted with mange. Today Baku is located in Azerbaijan, which name comes from the ancient Persian meaning "garden of fire."

The ancient Greeks knew naphtha as "oil of Medea." In Greek mythology, Medea was a princess of Kolchis, the oil-rich region between the Black and Caspian Seas. She was also a witch, and an expert in the uses of herbs and potions, which she put to good effect when the handsome Jason and his Argonauts arrived to steal her father's prized Golden Fleece. With help from Medea (some of it morally suspect; in one version, she murders and dismembers her brother), Jason manages to pinch the Fleece, returns safely home, and becomes King of Corinth. He and Medea marry and, in any respectable fairytale, would then proceed to live happily ever after. They don't: Jason's attention wanders and he decides to divorce the devoted Medea, now mother of his two children, in favor of the younger and prettier Glauke, daughter of King Creon of Thebes. Medea, furious but quiet about it, prepares a coronet and a beautifully embroidered wedding gown for Glauke, and sends it to her rival via her two young sons. When the delighted Glauke puts the gown on, it clings poisonously to her flesh and burns her to death.

The lethal wedding gown may have been impregnated with naphtha, writes Partington; and Adrienne Mayor, author of *Greek Fire, Poison Arrows, and Scorpion Bombs*, suggests that the fatal recipe was a mix of naphtha, sulfur, and lime. This last has also been proposed as the much-disputed recipe for Greek fire, the horrific combustible supposedly invented by

Kallinikos, a refugee from Arab-occupied Syria who ended up, sometime in the late seventh century, in Constantinople. (An alternative story circumvents Kallinikos and holds that the recipe was given by an angel to Constantine the Great.)

Greek fire was certainly some form of petroleum – everyone agrees that it was an incendiary liquid or semi-liquid that could burn on water, and was only extinguishable by vinegar or urine, presuming either of these was available in sufficient quantity. It could be efficiently delivered from fire ships by "siphon" – a pump-operated spray device that cascaded fire on to enemy decks. For a period of several hundred years, Greek fire rendered the Byzantines nearly undefeatable. Using it, they repeatedly repulsed attacks; Igor of Kiev, explaining his ignominious defeat in 941, wrote, "The Greeks possess something like the lightning in the heavens, and they released it and burned us. For this reason we did not conquer them."

One can see why Igor wanted to go home. The closest modern equivalent to Greek fire is napalm, invented by Harvard University scientists in 1942.

Fire – our "good servant, but bad master" – shows its latter side most plainly in war. Conquest, since ancient times, has gone hand in hand with flame. The Trojan War ended in conflagration. Scorched-earth policies date at least to the time of Romans who – feeling vengeful toward Carthage – not only burned the place down, but sowed the earth with salt. William the Conqueror torched the fields of rebellious Saxons; Sherman, marching to the sea, burned Georgia; Josef Stalin burned the Ukraine. The term firepower today, however, means artillery rather than torches; and trial by fire, rather than red-hot plowshares, means guns.

Gunpowder, most historians agree, was invented by the Chinese. The crucial explosive mix of sulfur, charcoal, and saltpeter (potassium nitrate) seems to have been discovered sometime around 800 CE, and somewhat anti-serendipitously, since the intention had been to devise an elixir of immortality. It was obvious almost immediately that the new invention meant trouble. By 850 CE a Taoist alchemical text, *The Classified Essentials of the Mysterious Tao of the True Origin of Things*, was warning against incautious experiments: some investigators, heating the touchy ingredients with honey, had burned their houses down. None the less, over the next centuries the Chinese persevered, developing fireworks, rockets, bombs, and, arguably, the first guns.

Chinese gunpowder seems to have been first used for firecrackers. Pre-powder firecrackers were simply lengths of green bamboo; when tossed in the fire, rapid expansion of steam blew the woody bamboo stems apart with a loud crack, a process analogous to the popping of popcorn kernels. The noise was believed to scare off demons and devils. Marco Polo, whom it evidently scared out of his skin, reported that the bamboo crackers "burn with such a dreadful noise that it can be heard 10 miles [16 km] at night, and anyone who was not used to it could easily go into a swoon or die." Polo, writing with an eye to his audience, may have exaggerated; alternatively his "dreadful noise" may have been truly dreadful, the explosions exacerbated by packing the hollow innards of the bamboo tubes with gunpowder. By the twelfth century, gunpowder-laden firecrackers were common in China, often assembled into strings and connected by a single fuse, much like the interconnected strings of explosive red-wax-coated paper tubes set off today to celebrate the Chinese New Year. In recognition of their organic origins, the Chinese still refer to firecrackers as *pao chuk* – that is, "burst bamboo."

Just how the gunpowder recipe reached Europe remains a matter of debate. Most historians hold that it arrived in the West via the Arabs, who got it from the Mongols, who got it from the Chinese. Alternatively it may have been invented independently in several different locales. In England, Roger Bacon, a Franciscan monk noted for his early forays into experimental science, recorded a possibly original recipe for gunpowder around 1268 in a treatise titled *On the Marvelous*

Power of Art and of Nature and Concerning the Nullity of Magic.
Bacon was cagey with his instructions – the proper proportions for the ingredients were couched in the form of a Latin
anagram of such obtuseness that it was only unscrambled by
a British artillery colonel in 1904 – which suggests that, even if
Bacon did invent a form of gunpowder, the public may have
learned about it from alternative sources. Or Bacon may have
been one of many to have acquired the recipe from someone
else: he is said to have spoken fluent Arabic and thus could
have picked it up from a knowledgeable contact from the
Middle East.

Another gunpowder claimant is the German alchemist
Berthold Schwartz or Black Berthold, sometimes nicknamed
the "Powder Monk," who is said to have invented both
gunpowder and the cannon sometime in the thirteenth or
possibly fourteenth century; in one version of the story, he
came up with the latter while attempting to make gold paint.
Most modern accounts, however, dismiss Berthold as purely
legendary; and J. R. Partington suggests tartly that he was
invented by boastful Teutonic historians for the sole purpose
of providing Germany with an artillery first.

The origin of the gun, for all its overwhelming impact on
human history, is also muddily uncertain. A bas-relief in the
temple of Ta-tsu in China's Szechuan province, dated to 1128,
depicts what some archaeologists argue is an early – if not the
earliest – gun: the carving shows a squatting horned demon
clutching a pot-shaped weapon about the size of a guitar, its
flared barrel spewing fire. The first mentions of cannon in
Europe date to the early 1300s, where early models seem to
have been similarly pot-shaped, since they were often referred
to as "firepots." An illustration of such a piece appears in
Walter de Willamete's "On the Majesty, Wisdom, and Duties
of Kings" in 1326: it shows a rather squeamish-looking gunner
in chain mail igniting a gun with a piece of hot iron. The gun,
shaped like an immense vase or garden urn, lies on its side on
a trestle table, stuffed with gunpowder and loaded with an
oversized arrow.

The squat pots were soon exchanged for bigger and more accurate tube-shaped artillery pieces. By 1453, when Constantinople fell to the 21-year-old Ottoman Sultan Mehmet II, the Turkish victory was a triumph of the latest wave of artillery, among them a 26-foot-long (8-meter-long) behemoth capable of firing half-ton (500-kg) stone balls. Despite the defenders' desperate attempts to cushion the blows by hanging bales of straw and leather from the battlements, Constantinople's walls, after fifty-four days of steady gunpowder-based bombardment, fell down; and along with them the thousand-year-old Byzantine Empire was blasted to bits, a victim of the gun. In Europe, the debut of the gun put an end to a military era. With its acquisition, the legendarily romantic age of chivalry was over; the armored knight, the sword, the shield, and the walled and moated castle all were now definitively obsolete. In 1520, Niccolò Machiavelli wrote in his now-classic *Art of War*, "No wall exists, however thick, that artillery cannot destroy in a few days."

The Chinese apparently never pursued the gun with the enthusiasm of the West, though they did devise gunpowder-powered "fire-lances," bombs, incendiary rockets, and a creative weapon known as a "ground-rat," originally designed as an entertaining firework, that seems to have crossed the line from the civilian arena to the military. The rat was a bamboo tube filled with gunpowder and perforated at one end such that, when lit, it would race insanely around on the ground. These were popular by at least the thirteenth century, since records of a fireworks display in 1264 describe how an unfortunately aimed ground-rat chased the emperor's mother; luckily for all concerned, the dowager empress was apparently both fast on her feet and a good sport. Such incidents suggested to discerning military observers that the rats might be equally effective at routing enemy horses, and they soon became a staple of Chinese cavalry battles.

The Chinese brought fireworks to a high art. Sixteenth-century Jesuit visitors described gunpowder that could be made to explode in dazzling colors; and there are mentions of fire trees, flame flowers, glittering peach blossoms, and sparkling wheels, all early forms of pyrotechnical displays. These were the progenitors of the fireworks featured in so many national holidays today: Victoria Day in Canada, Guy Fawkes Night in England, Bastille Day in France, and the American Fourth of July are all fireworks festivals. Queen Elizabeth I adored fireworks, as did Louis XIV, Queen Victoria, and Peter the Great. Fireworks, historically, were a feature of royal birthdays, coronations, weddings, anniversaries, and jubilees, and a celebratory signal of military triumphs. The army of Charles V, Holy Roman Emperor from 1519 to 1556, included pyrotechnical "fireworkers" whose sole purpose was to implement victory firework displays.

Among the most renowned of such displays were those designed to celebrate the end of the War of Austrian Succession (1744–8). The English, to commemorate the event, ordered the construction of an immense wooden machine, 410 feet (125 m) long and 114 feet (35 m) high, from which were to be fired 11,000 pinwheels, flares, and rockets, all designed for the occasion by the famous Italian fireworks maker Gaetano Ruggieri. Sharing responsibility for the display with Ruggieri were several English pyrotechnicians, among them Captain Thomas Desaguliers, the royal firemaster. A confusion of conflicting orders ensued, and a portion of the machine exploded, causing its outraged builder, Cavalieri Servadoni, to attack one of the offending pyrotechnicians with his sword. Despite all this excitement, the show, which had attracted an audience of 12,000 and created a three-hour traffic jam on London Bridge, was deemed a fiasco. Most of Ruggieri's fireworks were never ignited; the Duke of Richmond used all the unexploded leftovers three weeks later at a garden party. None the less, this pyrotechnic fizzle has an enduring claim to fame because it was immortalized in music: George Frideric Handel's *Music for the Royal Fireworks* was composed for the occasion.

The sister celebration in France was far worse: again a dispute arose among rival pyrotechnicians, "who, quarreling for precedence in lighting the fires, both lighted at once and blew up the whole," killing forty spectators and injuring three hundred. It may have been this incident that gave Madame de Pompadour, Louis XV's politically influential mistress, her lifelong distaste for fireworks; her attitude was certainly exacerbated by a display in which a stray spark from a Ruggieri concoction set fire to her hat.

The Chinese held that the explosive nature of black powder came from a particularly effective union of yin and yang: that is, female (yin) saltpeter and male (yang) sulfur, mated, made fire. If so, saltpeter – like the gigantic threatening females of Thurber cartoons – is by far the dominant sex; gunpowder ideally contains about 75 per cent saltpeter, 12 per cent sulfur, and 13 per cent charcoal. Saltpeter, to the ancients, was Chinese snow; to modern chemists, it's potassium nitrate or KNO_3. The nitrogen-bound oxygens of potassium nitrate are the secret to gunpowder: these provide the oxygen that supports the explosive combustion of the two fuels, sulfur and charcoal.

Historically saltpeter was by far the most difficult of the crucial trio to obtain. Charcoal could be had for the asking, by the controlled burning of wood. Sulfur, found in elemental form in nature, was mined from volcanic beds – most, in the ancient world, came from immense deposits in Sicily – or could be produced by roasting iron pyrites ("fool's gold" or iron sulfide). Saltpeter, however, had to be extracted from organic material of the sort that early recipes delicately refer to as "manurial soils" and painstakingly purified. It was thus the limiting ingredient in gunpowder manufacture and access to it, from the fifteenth through the mid-nineteenth centuries, was a matter of strategic importance, akin to the possession of oil today. The British victory over France in the Seven Years' War (1756–63) – known in America as the French and Indian War – was assured by a clever maneuver in the saltpeter trade; in 1761, the British ousted the French from India, then the world's prime saltpeter source. The Treaty of Paris – which, among other things, ceded Canada to Great Britain – was signed at least in part because the French were running dangerously low on gunpowder.

Saltpeter, sulfur, and charcoal, loosely mixed, create at best a relatively feeble gunpowder. The secret to more effective firepower was grinding, which interspersed the particles of the three components such that fuel and oxygen source were more intimately connected. Gunpowder ground dry, it was soon discovered, had a lamentable tendency to blow up in the grinder's face. If liquid were added to the mix, however, the powder could be safely ground as a wet paste, then dried. The process was known as "corning," since the outcome was a granular powder: "corn" is an archaic term for grain. (Corned beef, for example, is prepared by preserving the meat in coarse grains, or corns, of salt.) Corned powder was powerful stuff – made more so, fifteenth-century dogma held, if the ingredients were ground in brandy; almost as good was the urine of a heavy wine drinker and/or a clergyman.

The finer the powder was ground, the faster the burn and the more explosive the reaction; an eighteenth-century French recipe, for example, called for a twelve-hour grinding process until the final mix is dauntingly potent and as "as fine as women's face powder." Such batches were too much for the average fifteenth-century gun; the French face powder could blow a primitive cannon to smithereens. Twenty-nine-year-old James II of Scotland – colloquially known as "James of the Fiery Face" for a prominent red birthmark – was killed by an exploding cannon at the seige of Roxburgh Castle in 1460. It was the first time the king had used cannon in battle and, proud of his new artillery but ignorant of black powder's potential, he was standing too close to the loaded gun.

The lethal explosion of James's cannon was the result of an exothermic reaction: that is, a chemical reaction that rapidly releases large amounts of energy in the form of heat. The components of gunpowder, ignited, produce nitrogen gas; heated, the gas expands at a fearsome rate, generating pressures up to 6000 atmospheres in mere thousandths of a second, which is the force that propels the arrow, cannonball, or bullet out of a gunpowder-loaded gun. About 40 per cent of the products of a black powder explosion are gaseous; the remaining 60 per cent – in the form of potassium carbonate and sulfide – are particulate solids, dispersed as clouds of white smoke.

Before the invention of smokeless powders, the recently fired gun smoked. The "smoking gun" of figurative speech is a synonym for incontrovertible evidence, the equivalent of being caught red-handed – that is, found obviously guilty, fresh from the crime, hands still stained with the victim's blood. "The chaplain stood with a smoking pistol in his hand," wrote Arthur Conan Doyle damningly in 1894; and the long-running television western *Gunsmoke* – set in Dodge City, Kansas, in 1873 and starring James Arness, Milburn Stone, and assorted criminal gunslingers – should have been, for authenticity's sake, smoky. Field artillery belched massive clouds of smoke, blackening soldiers' faces and covering eighteenth- and nineteenth-century battlefields in impenetrable murk.

The smokeless powders used in firearms today had their beginnings in a kitchen in Basel, Switzerland, in 1845, where chemist Christian Friedrich Schonbein, who had been strictly forbidden by his wife to conduct experiments in the house, was tinkering with mixtures of nitric and sulfuric acids. In the process of his investigations, he spilled some of his samples. Fearing detection – Frau Schonbein must have been an intimidating personality – he hastily snatched up his wife's cotton apron, mopped up the evidence, and hung the wet cloth over the stove to dry. Shortly thereafter, with a flash and a bang, the apron exploded. The astonished Schonbein had discovered nitrocellulose, an explosive molecule commonly called *schiessbaumwolle*, or guncotton.

Cellulose, a long-chain sugar polymer, is a major component of plant cell walls. People can't digest it, which is why we are unable to survive on a diet of leaves and grass; termites and cows can, but only with the help of gut bacteria who produce the enzymes necessary to cleave the recalcitrant inter-sugar bonds. Cotton, a plant fiber, is 90 per cent cellulose. When cotton is dunked in nitric acid, nitro (NO_2) groups bind to numerous sites on the cellulose molecule. The more nitration, the more explosive the nitrocellulose, which meant that before chemists learned to control the outcome of the reaction, guncotton was a horrifyingly unpredictable proposition. Schonbein never profited from his discovery – too many of his factories blew up – but in the 1880s Frederick Abel and James Dewar successfully used it to make cordite, the first practical smokeless powder. The Spanish–American War of 1898 was the last to be fought with black powder; conflicts thereafter were settled with variations of Abel and Dewar's fiery mix of nitrocellulose and nitroglycerine.

Though black powder awed the Middle Ages, it hardly began to plumb the awesome power of fire. In modern terms, black powder is a "low" explosive, capable of propelling a speeding bullet, but not much use for blasting things to bits. Firepower with real explosive oomph is the province of the so-called "high" explosives, such as nitroglycerine, dynamite, and TNT. Nitroglycerine, a nervy melding of glycerine (now known as glycol) with nitric and sulfuric acids, is so unstable that it explodes when struck with a hammer (or even, in some cases, when poked with a feather). Where black powder generates 6000 atmospheres of pressure in thousandths of a second, nitroglycerine generates 270,000 atmospheres of pressure in millionths of a second. Ascanio Sobrero, a chemistry professor at the University of Turin, who discovered it in 1847, was horrified by it and refused to commercialize it.

Alfred Nobel, however, like the accommodating Barkis, was willing. Nobel – a photograph shows him with a jutting jaw, receding hairline, and rather wild-eyed stare – was born in Stockholm in 1831, but raised and educated in St Petersburg, Russia. By the age of seventeen, he was fluent in five languages – Swedish, Russian, French, English, and German – and had developed intense interests in poetry and chemistry. His father, appalled by the poetry, sent him abroad to study chemical engineering, in much the same spirit that young women of that era who formed unsuitable romantic attachments were packed off on prolonged sea voyages. The ploy was successful: Alfred met Sobrero, became interested in nitroglycerine, and sidelined his Muse.

The Nobel family fortunes began with nitroglycerine factories, an industry persistently fraught with danger. Explosions abounded, to the point where concerned officials banned Nobel from the city of Stockholm, forcing him to set up his laboratory on a barge anchored in nearby Lake Malaren. Nobel's research was aimed at taming the ferocious nitroglycerine monster, rendering it controllable by packaging it in a form that would explode when wanted rather than at unexpected and undesirable points in between. The solution was dynamite. Nitroglycerine, when mixed with an inert solid – Nobel's favorite was *kieselguhr*, or diatomaceous earth – formed a stable putty-like stuff that could be shaped into cylindrical sticks and wrapped in protective paraffin-coated paper. Dynamite – the name came from the Greek *dynamis*, meaning power – was soon in such demand that by 1868 Nobel had established nitroglycerine plants worldwide. Most have long-since exploded – while dynamite is stable, nitroglycerine manufacture remained a risky business – though one surviving factory in Norway has been preserved as a working museum.

Fire is relative: where gunpowder burns, dynamite detonates. The awful effectiveness of a high explosive is the result of a monumental shockwave. The shockwave, in turn, is the result of the violent expansion of gases; and the gases expand at such a phenomenal rate because explosions are hot. In a chemical reaction, heat is the result of an energy difference between reactants and products: that is, when high-energy reactants re-combine to form low-energy products, the leftover energy is released as heat. In nitro-compound explosives, such as dynamite's nitroglycerine, the prime product of the reaction is nitrogen gas (N_2), an extraordinarily stable molecule of such sluggishly low energy that chemists refer to it as "inert." Since nitrogen is so stable – a Rock of Gibraltar in the chemical world

– huge amounts of energy are released during its formation. The more energy, the more heat; the more heat, the more rapid the expansion of the generated gas. In a nitroglycerine explosion, the gas-driven shockwave can travel at the rate of 20,000 feet per second (6000 m/sec), shattering bedrock and blasting buildings out of existence.

Alfred Nobel was a pacifist. His explosives, he fondly hoped, would be used for peaceful purposes; and dynamite indeed was the force behind such socially hopeful projects as the Canadian Pacific Railway, the American Transcontinental Railroad, the Hoover Dam, and the Panama Canal. Unfortunately, Nobel flatly rejected the one undeniably peaceful use of nitroglycerine: as medicine.

The physiological effects of nitroglycerine were first noted by its discoverer, Ascanio Sobrero, who tasted a bit and promptly developed a gargantuan headache. Nitroglycerine, ingested, breaks down, releasing nitric oxide, a small molecule with numerous metabolic effects, among them the dilation of blood vessels. This property, the root of Sobrero's migraine, proved a boon for angina sufferers, relieving the chest pains brought on by constricted blood vessels supplying the muscles of the heart. (More recently the blood-vessel-swelling activity of nitric oxide was exploited for the development of the anti-impotence drug Viagra.) Alfred Nobel, a victim of angina, refused to sample nitroglycerine, insisting that it couldn't possibly work; his reluctance may also have stemmed from repeated experience with agonizing nitroglycerine-induced headaches. He died of a cerebral hemorrhage on December 10, 1896, at the age of sixty-three, leaving his enormous fortune for the establishment of annual prizes in the fields of chemistry, physics, medicine, literature, and peace.

Peace, sadly, remained an impossible dream; and Nobel's belief that the development of truly horrific explosives would act as a permanent deterrent to war proved wholly unfounded. Rather than making wars impossible, bigger and better bombs simply made them worse. The explosive of choice during World War I was another nitro compound – trinitrotoluene, otherwise known as TNT; the Germans got it first, which gave them a considerable military edge until the Allies managed to catch up. By World War II, scientists had translated Albert Einstein's esoteric theory of relativity into hardware; in 1945 America dropped the first atomic bombs on Hiroshima and Nagasaki.

The Earth itself periodically explodes.

In the fall of 1979, my husband – not long out of graduate school – passed through the Cascade Mountains of Washington on a camping trip. He stopped along the way at a turn-off beside the road to view in the distance the symmetrical snow-topped peak of the 9677-foot (2950-m) mountain known as the "Fujiyama of America." Beside him was an informative sign, confidently erected by the National Park Service. It read: "Mount St Helens. Extinct Volcano." A few months later, on May 18, 1980, jiggled into wakefulness by an earthquake that measured 5.1 on the Richter scale, Mount St Helens blew up. The eruption blasted 1300 feet (400 m) off the top of the mountain – 23 cubic miles (96 cubic km) of earth and rock; threw up an 80,000-foot-high (26,000-meter-high) ash cloud, and destroyed 230 square miles (600 sq km) of surrounding forest. It also killed fifty-seven people and millions of animals, including an estimated 7000 deer, bear, and elk, and 12 million Chinook and Coho salmon.

In Shusaku Endo's 1980 novel *Volcano*, the volcano is used as a metaphor for human life: "In youth it gives rein to passions, and burns with fire. It spurts out lava. But when it has grown old, it assumes the burden of past evil deeds, and it turns quiet as a grave." Sometimes indeed it does; a volcano which has no historical record of eruption and is seen to be deeply weathered and eroded is considered to be safely extinct. Sometimes, however, there's unexpected blood in the old man yet. My husband regrets that sign. He wishes he'd stolen it.

About 80 per cent of the world's active volcanoes occur

along subduction zones – that is, regions where one crustal plate dives beneath another, carrying solid rock into the molten depths of the Earth. Mount St Helens is one of these, a charter member of the volcano-rich Pacific Ring of Fire that extends from New Zealand up the eastern coast of Asia, through Alaska's Aleutian Islands, and south along the western coasts of North and South America. Some of the world's most famous volcanoes dot the Pacific Rim: Mount Pinatubo in the Philippines; Japan's Mount Fujiyama, the most photographed mountain in the world; Mexico's Popocatepetl, whose name comes from the Aztec for "Smoking Mountain," and Paricutin, the 1345-foot-tall (410-meter-tall) cinder cone that sprang up out of Dionisio Pulido's newly planted cornfield in 1943. A second band of fiery activity stretches from the Mediterranean through Iran, and then – after a short jump – through Indonesia and into the western Pacific. At the Indonesian end of the line, where the Indo-Australian plate slides beneath Eurasia, smolder the phenomenal Tambora, whose explosion in 1815 is considered the largest in recorded history, and Krakatoa, whose blast in 1883 was heard nearly 3000 miles (4800 km) away, on Rodriguez Island in the Indian Ocean.

Their opposite numbers in the Mediterranean, where the African plate meets Europe, include Vesuvius, nemesis of Pompeii, and Thera, on the Aegean island of Santorini, whose violent eruption in the sixteenth or seventeenth century BCE is thought to have led to the myth of the tragic lost continent of Atlantis. In the sunny Caribbean, the island of Martinique – birthplace of Napoleon's Empress Josephine – is balanced on the tectonic border where a slice of the North Atlantic plate is being subducted beneath the Caribbean plate. There Mont Pelée erupted in May of 1902, obliterating the city of St Pierre along with all 30,000 of its inhabitants. The sole survivor, Auguste Ciparis, had been in jail at the time of the eruption, but was protected from the scorching effects of the blast by the thick insulating layers of volcanic ash that covered the one tiny window of his cell. He was rescued four days later, was promptly pardoned, and spent the rest of his life as a featured attraction with Barnum & Bailey's Circus. Another escapee was an Italian ship, which headed out to sea leaving half its cargo still unloaded on the day the volcano began to belch ash and roar. The captain, a native of Naples, had been raised in the shadow of Mount Vesuvius and knew what was coming.

Another 15 per cent of the world's active volcanoes are located along the global rifts, primarily in Iceland and east Africa. The remainder are hot-spot volcanoes, popping up peculiarly in the middle of crustal plates over some deep and stationary heat source, most likely a thermal plume from the mantle. Hot spots were first identified in the early 1960s by a geophysicist from a cold country: J. Tuzo Wilson (1908–93) of Canada's University of Toronto. It had been known since the early nineteenth century that the islands of the Hawaiian island chain – the balmy archipelago that Mark Twain described as "the loveliest fleet of islands that lies anchored in any ocean" – grow progressively younger moving from northwest to southeast. Kauai, at the northwest end of the line, is nearly five million years older than the "Big Island" of Hawaii on the southeast, with Oahu, Molokai, Lanai, and Maui falling in chronological sequence in between. Island mythology echoes this geological age range: ancient Hawaiian legend explains that when Pele, the beautiful and tempestuous goddess of fire, came to Hawaii, her canoe first landed on Kauai. Before she managed to establish herself, however, she was driven away by her angry older sister Namakaokahai (back home in Polynesia, Pele had seduced her sister's husband). Pursued by the furious Namakaokahai, Pele moved southward from island to island, finally ending up on Hawaii where, with her magic digging stick, she gouged out the fiery crater of Kilauea and settled down.

The Hawaiian islands are the leading edge of a much longer island chain, dotted in a straight line across the Pacific from Kauai to Necker (10 million years old), Gardner (12 million), Pearl (21 million), and Midway (28 million). The line then makes an abrupt turn and dunks underwater, forming a chain of seamounts – the Emperor Ridge – that extends north to the Kamchatka peninsula. Such islands form, explained Wilson, because, while tectonic plates move, hot spots stay put. As the Pacific plate moves slowly over a buried hot spot, a rising plume of fiery magma hot enough to melt the thin rock of the crust from underneath. Eventually enough rock melts that the hot spot punches through, exuding lava in such quantities that a volcanic mountain forms, slowly rising out of the sea. In half a million years, the sub-Pacific hot spot has built Mauna Loa on the island of Hawaii, the world's largest volcano, 33,132 feet (10,099 m) tall from its peak to its base on the ocean floor, nearly a mile (over a kilometer) taller than Mount Everest. Then the plate moves on and the hot spot goes to work again. Forming now, some 25 miles (40 km) south of Hawaii, is a new volcano, Loihi, which – given another 10,000–30,000 years of continual eruption – will emerge from the depths of the sea.

Pele, when enraged, is said to cause Hawaii's spectacular volcanic eruptions; and her worshippers placated her for centuries by flinging breadfruit, pigs, and fish – and later tobacco and bottles of brandy – into her mountaintop fire pits. A tantrum isn't a bad analogy; just as build-up of emotional pressure can prod even the mildest adult into blowing his/her stack, increasing gaseous pressure can pop the top off a mountain. A volcanic eruption begins when magma, lighter and more buoyant than the surrounding rock, begins to rise toward the Earth's surface, like globs of wax blebbing mesmerically upward in a lava lamp. As the magma rises, dissolved gases begin to separate out of the mix, collecting at the top of the melted mass and building pressure, just as carbon dioxide gas collects at the top of an explosive bottle of New Year's champagne. Finally the pressure becomes too much for the confining vessel. The cork pops, the gas explodes outward, expanding as it goes, and fountaining shattered rock, ash, and blazing lava into the air.

Magma, once it leaves its volcano, becomes lava, and – generally speaking – the more viscous the composition of the lava, the more awful the accompanying eruption. Basaltic lava, a dark syrupy stuff with the consistency of melted tar, is the most liquid of lavas; this tends to produce rather understated eruptions, gooey extrusions of molten rock that dribble down the mountain sides. Such lavas are common in the Hawaiian islands where, laid down in layers over time, they create immense dome-shaped shield volcanoes. Monstrous Mauna Loa – the name in Hawaiian means "Long Mountain" – is a classic example. One geologist compares such volcanoes to giant turtle shells, which is just what they look like. Silicic or trachytic lava, however, which contains over 70 per cent silica, is less tractable. This, the consistency of soft cement, is the sludgiest of lavas, and is capable of generating violent explosions, sometimes ripping the volcano apart and hurling out tens of cubic kilometers of rock.

Lava, however, is not necessarily the worst output of an erupting volcano. Volcanoes also vomit pyroclastic debris. If this roars down the slopes of the mountain, clinging to the ground, it is called a pyroclastic flow. If airborne, it's called tephra, from the Greek for ash, though tephra can also contain chunks of rock the size of washing machines. Ash is the finest of volcanic particles, and often the most insidious: ash falling from Vesuvius in 79 CE buried the streets of Pompeii ten feet (3 m) deep and asphyxiated much of the unlucky populace, while ghoulishly preserving them where they fell. Clouds of it can remain in the upper atmosphere for months, even years. The eruption of Tambora on the Indonesian island of Sumbawa in 1815 tossed enough ash into the air to cause a two-year global cooling. Winters were unusually cold and characterized by a particularly dreary brownish snow, colored by volcanic dust in the atmosphere; and 1816, in which killer frosts struck in July, was known as the "Year Without a Summer." Lord Byron, Percy Bysshe Shelley, and his wife Mary all fled to Lake Geneva to escape Tambora-generated chill and gloom, only to find that the miserable weather was universal. Byron amused himself by writing a dismal poem called "Darkness;" and Mary used her time to write the Gothic horror novel *Frankenstein*.

Tephra – variously in the form of smothering ash; cinders, which can measure up to half an inch (1 cm) across; lapilli, volcanic lumps the size of walnuts; blocks, which are brick-like chunks of volcano-tossed rock; and bombs, which are molten masses of lava that solidify into cannonball-like missiles in midair – is collectively fearsome, but pyroclastic flows are worse. These are fiery avalanches of superheated gases and blazing fragments that thunder along the ground at speeds of up to 300 miles per hour (500 km/h), incinerating everything in their path. Mont Pelée destroyed St Pierre by means of a pyroclastic flow, the first to be recognized and described. Appalled French geologists named the phenomenon a *nuée ardente*, or glowing cloud. "The side of the volcano was ripped out, and there was hurled straight toward us a solid wall of flame," wrote Charles Thompson, who watched the eruption from the deck of the steamship *Roraima* in the harbor. "It sounded like thousands of cannons. The wave of fire was on us and over us like a lightning flash . . ."

For all its flaming output and toll of human life, the eruption of Mont Pelée, according to the official Volcanic Explosivity Index, only rates a 4. The Index is the volcano's version of the Richter Scale [see Earth], in which eruptions are given points for performance, based on volume of ejecta and height of explosion cloud. The scale runs from 0 to 8, in which 0 and 1 are essentially a few rumbles and gentle ooze. The 1980 eruption of Mount St Helens rates a 5; Mount Vesuvius (79 CE), Thera (*c.* 1620 BCE), and Kratatoa (1883), all rate 6; and the phenomenal eruption of Tambora in 1815 – an event referred to by geologists as "colossal" – a 7. Toba, however, for which geologists reserve the word "humongous," rates an 8.

The world itself will not end in volcanic fire, but there's a possibility that we might. In fact, we've already come uncomfortably close to volcanic extinction. Toba erupted 74,000 years ago in northeastern Sumatra, a catastrophic blast 10,000 times more powerful than the eruption of Mount St Helens. It left behind a caldera, now filled by Sumatra's largest lake – Lake Toba, 55 miles (85 km) long and 1740 feet (530 m) deep – and an immense quantity of ash. Foot-deep (30-cm-deep) layers of Toba ash have been found on the floors of the South China Sea and the Indian Ocean, 1400 miles (2400 km) away. In the aftermath of the eruption, immense clouds of ash and aerosols blocked incoming sunlight, dropped global average tempera-tures by 21°F (10°C), and created a multi-year volcanic winter. And it nearly did the human species in. Sometime between 70,000 and 80,000 years ago, the human population of the planet appears to have plummeted to just 5000–10,000 individuals, almost to the numerical point of no return. This population estimate is based on studies of mitochondrial DNA – unique molecules that each of us inherits solely from our mothers, found in the mitochondria of each body cell. The known mutation rate of mitochondrial DNA, paired with an analysis of distribution of mutations, allows geneticists to estimate the size of past populations. By such criteria, the range of differ-ences in human mitochondrial DNA is nowhere as large as it should be. Something disastrous happened in our distant past, creating a population bottleneck. Many geologists are betting that that something was Toba.

Toba was not just a volcano, but a supervolcano. The run-of-the-mill volcano forms when a column of magma rises from the Earth's mantle and erupts on to the surface, its lava hardening in layers to form a mountain. Supervolcanoes, in contrast, form superholes. In the supervolcano, magma, rather than rising in a constricted column, accumulates in a vast subterranean reservoir, gradually building up tremen-dous pressures over periods of thousands – even hundreds of thousands – of years. When it finally erupts, emptying its reserve in one catastrophic blast, the result is horrific. Think of Mount St Helens as a water pistol and Toba as Niagara Falls.

No supervolcano has erupted on the planet in history; and very few such exist – luckily – on Earth. The United States, however, has one, and three million tourists blithely visit it every year, where they picnic on top of it and take pictures of bears. Beneath their sneakered feet, the hot spot that fuels Yellowstone National Park has built up a seething chamber of magma almost the size of the park itself, 42 miles (70 km) long and 18 miles (30 km) wide. In the past, this lethal reservoir has erupted approximately once every 600,000 years, the last – the Lava Creek Eruption – 620,000 years ago. Some geologists insist hopefully that Yellowstone's activity is waning, but others claim that it's alive, healthy, impatient, and overdue. If Yellowstone blows, it could bury half the United States in several feet of ash, instigate a volcanic winter, and push us all to the brink of extinction. Subterranean fire may just be the death of us all.

This is knowledge, I realize, that I'd rather have done without.

In some cultural traditions, the afterlife of the sinful is dismal and cold; Christians, however, envision eternal punishment of the wicked by fire. In the Middle Ages, the entrance to Hell was said to be through a volcano – specifically through Hekla, the most active and presumably hottest volcano in Iceland, a rocky 4911-foot (1497-m) peak rising out of the bleak highlands east of Reykjavik. Flocks of ravens hung around it, which was thought to be significant; some said these were the souls of the dead, en route to their final destination. Into the volcanic vent dead souls were said to be dragged each day – it occurred to me that daily raven tallies should have been kept, but no one seems to have done so – and one believer wrote that "whenever great battles are fought or there is bloody carnage somewhere on the globe, then there can be heard in the mountain fearful howlings, weeping, and gnashing of teeth."

The Christian Hell, gruesomely described in the Book of Revelations, is blazing hot – a lake that burns with fire and brimstone – though the name itself comes from Hel, Norse goddess of death, whose underworld realm was cold. Daughter of Loki, the tricky god of fire, Hel was a repulsive hag, her face half black and half blue, her body that of a decaying corpse. Her kingdom was populated with those hapless Vikings who did not die in battle and thus didn't qualify for the glorious gleaming halls of Valhalla.

Hot, cold, or indifferent, most conceptions agree that the

afterworld – especially the negative aspects of it – is located somewhere *down*. Dante's Inferno is a pit, with its center directly beneath the city of Jerusalem; Hades is a subterranean abyss; and in Chinese mythology, the land of the dead is a series of dank underground caverns through which ghosts gloomily wander, awaiting their turn to be reborn. When John Milton's Lucifer fell from Paradise, he plummeted for nine days into the depths.

Strictly speaking, the temperature of the infernal nether regions should depend on just how deep down they are. Caves, for example, all heavily insulated with overlying layers of rock and earth, have remarkably stable ambient temperatures, generally equal to the average yearly surface temperature of the area of the world in which the cave is found. In most cases, this is coolish, hovering around 55–60°F (13–16°C). The world's deepest caves, however – Voronya Cave in Abkhazia, Georgia, and the Lamprechtsoften-Vogelshacht cave system near Salzburg, Austria – are only a bit over a mile (1.5–2 km) deep at most, impressive from a spelunker's standpoint but in geological terms still practically on top. Deeper yet is hotter.

The Earth's mantle – extending 30–1700 miles (50–2700 km) beneath the surface – has temperatures ranging from 1600°F (870°C) at the top to 4000°F (2200°C) or more at the bottom; the outer core, below it, reaches 9000°F (5000°C), and the inner core, over 13,000°F (7200°C). According to the calculations of Henry T. Wensel, chief of the US National Bureau of Standards Pyrometry Section from 1917 to 1946, this lands the biblical Hell someplace in mid-mantle. Wensel based his numbers on the telling lake of brimstone. Brimstone, or sulfur, boils at 831°F (444°C), which means that at or above that point the seething sulfurous lake could not exist, having been converted en masse into green gas. On the other hand, below 235°F (113°C), the melting point of sulfur, the burning lake would freeze solid, such that, at the very coolest, the realm of the damned must still be hotter than the boiling point of water. Critics soon pointed out that Wensel had not taken pressure into proper account: at Hell's presumed great depth, intense pressures would elevate the temperature at which sulfur stays liquid, boosting it to as much as 1900°F (1040°C), making Hell even hotter. Wensel, using a descriptive passage from Isaiah and the Stefan-Boltzmann equation, then proceeded to calculate the temperature of Heaven, which he found, embarrassingly, to be 977°F (525°C), some 150°F (80°C) hotter than Hell. What all this indicates to the discerning taxpayer is that Henry T. Wensel didn't have enough to do, though he did try to cover up by publishing his results in 1972 in the technical journal *Applied Optics*.

The hellish heat at the Earth's core has traditionally been attributed to the leftover fires of creation. The inner core, a 750-mile (1250-km)-thick ball of solid iron and nickel, bleeds primordial heat into the outer core, a 1300-mile (2200-km)-thick molten wash of liquid iron and nickel, whose ceaseless swirl powers the Earth's geomagnetic field and maintains the convection currents in the mantle that keep the crustal plates in motion. In this scenario, the inner core operates rather like the hot potatoes that nineteenth-century children were once given to keep their cold hands warm on the long trek to school: a hot conductive lump, transmitting thermal energy outward, keeping us safely cozy. In time, however, our global hot potato will slowly cool. In four billion years or so, geologists predict, the core will have cooled to such an extent that our geomagnetic field will no longer function. Without it, we're goners: the geomagnetic field is all that stands between us and the deadly high-energy particles that bombard us ceaselessly from the Sun.

An alternative theory gives us a shorter lifespan but a more exciting mid-section. Geophysicist J. Marvin Herndon of San Diego, California, suggests that the heat beneath our feet comes from a central five-mile-wide (8-km-wide) ball of uranium, a vast natural nuclear fission reactor. There's precedent for such spontaneously initiated reactors. In 1972, French scientists discovered the remains of an immense natural nuclear reactor in a vein of uranium in west Africa's Republic of Gabon; and to date, fifteen such fossil reactors have been identified in Gabon's Oklo uranium mines. All have left evidence behind in the form of isotopes – that is, atoms of a given chemical element that vary in molecular weight. Most of Earth's uranium consists of two radioactive isotopes: U-238 and U-235. In any given sample of uranium, about 99.3 per cent of the total consists of U-238, 0.7 per cent of U-235. Today U-235 is the most in demand, since it is best capable of sustaining a chain reaction; a kilogram of U-235 can release 20 trillion joules of energy, the equivalent of 1500 metric tons of coal. The French, mining at Oklo to supply the needs of their burgeoning nuclear program, were startled to discover that their ore was sharply depleted in the desirable U-235. Oklo's uranium, it turned out, had already been used, processed some two billion years ago in a massive naturally instigated chain reaction that had continued for over 200 million years.

Herndon's theory of a natural nuclear reactor at the Earth's core is appealing because it explains many previously inexplicable phenomena. The giant planets – Jupiter, Saturn, and Neptune – are brighter than they ought to be, indicating that they generate more energy than they absorb from the Sun. The possible explanation: internal nuclear fission reactors, supplying vast amounts of energy from the inside out. The Earth's unpredictable magnetic field, which waxes, wanes, and occasionally does a startling global somersault and reverses itself altogether, may reflect the state of its buried reactor. Herndon's postulated uranium core – a mix of U-238 and U-235, constituting some 63 per cent of the world's uranium – may function as a breeder reactor, not only generating energy from splitting uranium atoms, but simultaneously creating its own new fuel, in the form of fissionable plutonium and other smashable actinides. Such a subterranean chain reaction is not perennially constant, however; as it proceeds, it inevitably begins to stew in its own juice. As byproducts accumulate, they begin to absorb the atom-splitting neutrons, yanking them out of circulation and gradually slowing the process to a crawl. Simultaneously the world's magnetic field wavers, grows feeble, and remains so until the reactor-choking byproducts – all lighter than heavy uranium – gradually diffuse outward and upward, away from the site of the action. The chain reaction then cranks up again, a distant internal furnace rumbling back to life.

The bad news, according to Herndon, is that our presumptive internal uranium is rapidly being used up. Over the 4.5 billion years of the Earth's existence, about 75 per cent of it has been consumed; at the present rate of power output – about four terawatts – we've got at best another 2 billion years to go, and maybe a good deal less. (The iron–nickel core proponents, whose model is more optimistic, predict another 4 billion years.) In either case, the prognosis for life is dreary. The planet itself, however, will sturdily outlast us; geology ultimately has greater staying power than biology. Ultimately, however, the Earth will vanish too, torched by the fires of the Sun.

In his poem "The Hollow Men," written in 1925, T. S. Eliot posits that the world, like a dud firecracker, will go out with a disappointing fizzle, "not with a bang but a whimper." Robert Frost, in "Fire and Ice," gives us a bit more credit, teetering between a world ending in either fire or ice. Eventually he opts for fire, which puts him in the camp of many global mythologies. For the most part, we don't see ourselves dwindling gently into that good night. We see ourselves going out in a blaze of glory.

The end of the world, according to the ancient Norse legend, begins in ice, but ends spectacularly in conflagration. First the Fimbulwinter, the winter of winters, will freeze the land, and the wolf Skoll will swallow the Sun. Then will follow a supernatural free-for-all, in which the frost giants aboard the ship Naglfar, a vessel made from dead men's fingernails, meet a coalition of the Aesir, led by Odin astride his eight-legged horse, the Midgard Serpent, the Fenris Wolf, and the fire demons of Muspellheim. In the course of the battle, almost all the gods, the giants, the Serpent, and the Wolf are slain; finally, in an apocalyptic coup-de-grâce, the fire demons destroy the world by setting it ablaze. The Norse referred to this collective disaster as Ragnarok, the "Doom of the Gods;" in German, the term is Gotterdammerung. In Hindu mythology, the present age of the world will end when the god Vishnu, sweeping the world clean of wickedness and vice, will drink up all the water on Earth, then ignite the parched remains in a tornado of flame. In the Christian Apocalypse, the seas become blood and the Earth is scorched by fire, accompanied by thunder, lightning, earthquakes, and hail.

While the ultimate end of the human species – whether by war, pestilence, famine, or sheer environmental idiocy – is unpredictable, astronomers are convinced that a fiery finish will be the inevitable fate of our planet. Unless a lethally aimed asteroid gets here first, we're doomed to be swallowed by the Sun. This seems ungrateful behavior on the Sun's part, since from time immemorial awed human beings have viewed our nearest star as an object of veneration. Cultures worldwide have given the Sun a central place in their mythologies, and many depict it as an all-powerful person. To the Australian aborigines, for example, the fiery Sun was a glorious goddess painted with red ocher carrying a bark torch across the sky. The paint gave the sunrise and sunset their brilliant colors; and the torch, carried underground at night as the goddess returned to her camp in the east, warmed the earth for the growing of plants. The Aztec Sun god Huitzilopochtli, depicted as a blue-skinned warrior decked in gold bracelets, gold bells, and a gold headdress shaped like a hummingbird beak, continually battled the dark forces of night to ensure the survival of his people; each sunrise was seen as a victory for the contentious god, who required the blood of human sacrifices to keep up his stamina. The Egyptian's falcon-headed Sun god Ra was the creator of all things. Pliny the Elder, who claimed to be quoting Homer, called the Sun the ruling principle of the Universe, splendid and supreme, all-seeing and all-hearing, who not only lights the stars and governs the change of seasons, but "dispels the gloomy aspect of heaven and lightens the clouds over men's minds."

Many ancient scholars firmly believed that the Sun and stars were living beings; and, in a dismal reflection on human progress, a 1989 survey of Americans found that 70 per cent of respondents still thought so. From a scientist's point of view, though the Sun is not alive in any biological sense of the word, it is definitely up and running. Every second the Sun converts five million tons of matter into energy, via nuclear fusion reactions in which hydrogen nuclei are welded together to create helium. The result is the energetic equivalent of the detonation of 100 billion one-megaton nuclear bombs. Earth gets only a minuscule fraction of this fabulous energy output – about two-billionths of the total – but it's still enough to douse us with 130 watts of power per square foot (1400 watts sq m) of planetary surface.

For all that, in stellar terms, the Sun is a small star. Its relatively lean mass determines its destiny: five billion years from now, when the Sun has consumed all it has in the way of hydrogen and helium fuel, it will expand into an immense red giant, sequentially engulfing Mercury, Venus, Earth, and Mars, and eventually extending as far as the orbit of Jupiter. By this time the cooling Sun will be energetically unstable, gradually shedding its gaseous outer layers – among them the disembodied atoms of planet Earth – until nothing is left but the solar inner core. It will then have become a white dwarf, the stellar equivalent of doddering old age. White dwarves are small – about the size of the present planet Earth – but amazingly dense. Surface gravity on a white dwarf is 100,000 times greater than that on Earth, and density over a million times greater: a single teaspoon of a white dwarf would weigh upwards of 5.5 tons.

By now, its fiery inner furnace dead, the Sun will have nowhere to go but out. The white dwarf will slowly cool, radiating heat into the chill of outer space, until finally – billions of years later – it will flicker definitively out, becoming a black dwarf, a cold dead lump of cosmic carbon.

Fire, like water, is the element of life. There's no life without fire.

Part V

Earth

*Being thus arrived in a good harbor, and brought safe to
land, they fell upon their knees and blessed the God of
Heaven who had brought them over the vast and furious
ocean, and delivered them from all the perils and miseries
thereof, again to set their feet on the firm and stable earth,
their proper element.*

William Bradford
Of Plymouth Plantation

Earth, that stodgiest of elements, is perhaps the one closest
to our hearts. After all, we're land animals. Our personal
subset of the globe is terrestrial; when deprived of it, we
miss the ground. In Kevin Costner's 1995 film *Waterworld*,
the worst of global warming has come to pass: the entire
planet is flooded; there's nothing but water from pole to
pole; and Costner, as the mutant Mariner, skims the waves in
a jerry-rigged catamaran, desperately seeking that ultimate
of desirables, dry land. Dry land is where we belong. Storm-
tossed travelers who finally attain the beach fall down on
their knees and kiss the ground – "their proper element,"
wrote William Bradford in 1620 of the relieved Pilgrims, at

last safe ashore in Massachusetts after a tempestuous three-month crossing of the Atlantic.

Earth beneath our feet makes us feel secure. Some argue that it does even more: "All the parts of the earth's surface on which we tread will fulfill a particular service of life for the health of the body," wrote James Bain in 1914 in *The Barefoot League: The Manifold Nutriments Obtained for the Body through Barefoot Walking*. "Thus if we walk on the young and living grass we shall receive of its fresh and living, yet soothing, virtue. If we walk on the mountain turf, hot in the sun's rays, we shall receive of the very strength of the mountain ..." To walk barefoot, argue the proponents of the present-day Society for Barefoot Living, is to break down the leathery laced-up barriers between ourselves and nature, and to meld with the earth through the soles of our feet. They may have a point. Despite a sobering accident at the age of six involving bare feet, a pile of leaves, and a broken Mason jar, I'm not fond of shoes. Going barefoot is a simultaneously liberating and subversive experience; after all, bureaucrats don't do it; and stockbrokers always keep their wingtips on. Maybe our delight in shoelessness is primal memory, a distant recollection of how our hominid ancestors felt, padding barefoot along the muddy creeks of Africa. Or perhaps Bain and the Barefoot League are right; through naked flesh, we draw sustenance from earth. Perhaps in a sense we're all lesser versions of Antaeus, the giant of Greek mythology, who derived his power from contact with the ground. Earth – the original steroid hormone – turned Antaeus into an undefeatable wrestler. He was finally beaten by Hercules, who held him up in the air until he weakened, then strangled him.

In many mythologies, earth is not only our proper, but our seminal element. The Micmacs of eastern Canada claim that the original human beings were made from sand; the Hopis, that we were made from colored mud; and the Dusun people of Borneo, that we were molded from the dirt of a termite mound. According to the Old Testament's Book of Genesis, the first man was made "from the dust of the ground;" and the first woman, made from the first man's rib, was ground once removed.

One current scientific theory, originally popularized by British chemist Graham Cairns-Smith in the early 1980s, claims that life, quite literally, came from earth. In Cairns-Smith's model, clay served as the supportive template upon which the first biologically active molecules were synthesized. By virtue of its microcrystalline physical structure, clay was able to act as a sort of organic flypaper, inducing precursor organic molecules to stick to it, thus concentrating lots of them all in one place and encouraging intermolecular binding to create increasingly larger and more complex biopolymers. James Ferris of New York's Rensselaer Polytechnic Institute hypothesizes that the crucial clay may have been montmorillonite, a hydrated aluminosilicate shown in laboratory experiments to anchor individual nucleotides – the building blocks of DNA and RNA – and to promote the synthesis of linked chains up to fifty nucleotides long. Such proto-RNA molecules, Ferris suggests, may eventually have taken the form of self-replicating enzymes, first able to reproduce themselves, then to direct the synthesis of proteins, and finally – some 3.5 billion years ago – to orchestrate the sudden and startling proliferation of microorganisms known to evolutionary biologists as the Cambrian explosion. The name of the biblical first man – Adam – translates as "man of earth." Life's first molecules, similarly, may have been born of elemental earth, strung together on a supportive bed of clay.

Water, wind, and fire, usually by scaring us silly, earned our fearful respect, but earth appeared to early humans in a somewhat gentler mode. The earth, traditionally, is a mother, a nurturer, provider of food and shelter, patroness of babies and harvests. The term "earth mother" these days conjures a vision of caftans and organic tomatoes, but the original mother was an awesomely powerful parent who expected her numerous offspring to toe the line. In ancient carvings, she's a dead ringer for the middle-aged Queen Victoria: autocratic, prolific, short, fat, and unamused.

The first earth mothers – at least we think they're earth mothers – date to the late Stone Age. The best-known of these, a stubby four-inch statuette called the Venus of Willendorf, is a fat little figure in limestone, roughly shaped like a potato, depicting a faceless and footless female with a bulging belly, enormous behind, and fancy braids. Dozens of these stone women have been found at Neolithic sites throughout Europe – the remnants, archaeologists guess, of a prehistoric cult of fertility and sex. The Stone Age fat ladies gradually evolved into a range of svelter and more seductive earth goddesses, among them the Sumerian Ninhursag, the Babylonian Ishtar, the Egyptian Isis, the Roman Tellus – whose name survives in element 52, tellurium, and the Germanic Ertha or Erda, whose lumpy Teutonic moniker is the source of our modern word "earth."

Earth mothers (or, occasionally, fathers), like participants in some cosmic witness protection program, tend to have mysterious origins. Most simply appear willy-nilly out of chaos, infinite darkness, or primal fog. Once here, they often suffer from conflicting and shifting identities; the Greek Earth Mother, Gaia, for example, is sometimes portrayed as a person, sometimes as a place, and occasionally as an awkward mixture of the two. An alchemical illustration of 1618 shows her as a planetary chimera: an ungainly woman whose torso is the globe of the Earth, with head growing out of the North Pole, legs out of the South, and breasts positioned roughly over Siberia and Hudson Bay.

The Greeks were not alone in equating the Earth proper with a human body. In Chinese mythology, the creator god Pan Ku spent 18,000 years designing the Universe; upon his death (from exhaustion), his head became Earth's mountains, his blood the rivers, his flesh the soil, his bones the rocks, and his expiring breath the wind and clouds. In Scandinavian mythology the Earth and heavens are the remains of the slaughtered frost giant Ymir; and in Tahitian legend, the god Ta'aroa – who created the world in a temper tantrum – eventually tore himself apart in fury, leaving his flesh to become the earth; his backbone, the mountains; and his internal organs, the clouds, except for his intestines, which were transmogrified into lobsters, shrimps, and eels.

Gaia, on the other hand, managed to do it all and still remain alive and breathing. For this reason, in the 1970s, British chemist James Lovelock and American microbiologist Lynn Margulis commemorated her in the Gaia hypothesis, which argues that the Earth, rather than an inanimate lump, is itself a living organism. Not all planets qualify as organisms – Mercury, Venus, and Mars, for example, are nothing more than rocks in space, as dead as dodos. Earth, however, is biologically self-actualized. It possesses a biosphere – a resident population of biological life forms – which cooperatively interact to create a global self-regulating system. All of us, from weasels to banana trees, unconsciously participate in the life of the Earth, collectively maintaining air, water, and soil in quantities and compositions to suit ourselves. Earth is a living planet, in other words, because living things are on it, collaborating to produce a favorable climate in the same manner that lungs, liver, heart, and brain cooperate to maintain a functional human being.

There may be a few less admirable parallels, too. "It is far from easy to determine," wrote Pliny the Elder, "whether she [Nature] has proved to man a kind parent or a merciless stepmother." Gaia, according to her ancient biographers, was no maternal role model: she committed incest and adultery, bore enormous numbers of hideous children, and had her husband, Uranus, castrated. This is Gaia's dark side: nature, red in tooth and claw. The biosphere cooperates globally, but on the individual level its members go to unpleasant lengths in order to survive. Lions eat zebras; polar bears eat seals; snakes swallow mice; and sharks – though very rarely – even eat us, at least if given half a chance. Chimpanzees commit infanticide; incrowd chickens peck outsiders to death; female spiders and praying mantises, after a bout of consensual sex, eat their erstwhile mates. To Gaia, it's all one. Mother Earth happily dispenses flowers and fruit; then with cheerful equanimity, she dices us up for dinner.

Of all the elements, earth intrudes itself most insistently into our daily lives. Water, air, and fire – the sea within our body's cells, the 300-mile-high (480-km-high) column of gas above our heads, the fire at the core of the planet – are easy to take for granted; we can pursue our days' occupations without giving them so much as an appreciative thought. Earth, on the other hand, is hard to ignore. Earth is the element of solidity and substance; we deal with it every time we stand up, sit down, jump over a fence, or drop a coffee cup. We are constrained and confined by weight; and, for the most part, we resent it. We speak morosely of the weight of the world resting on our exhausted shoulders; sorrow makes our hearts heavy; and Shakespeare's schoolboys, trudging toward school with heavy looks, looked miserable. Our perennial earth-enforced heaviness, hypothesizes Italo Calvino in *Six Memos for the Next Millennium*, explains our persistent fantasies of lightness, our dreams of flying carpets, winged horses, magic broomsticks, and phials of dew that, evaporating, lift us to the Moon. Joy makes our spirits light; and to be uplifted is to be intellectually, morally, or spiritually exalted. To be dragged down, on the other hand, is to descend, kicking and screaming, into failure and destitution.

Earth-dwellers, however, need to be dragged. The Earth, simply by being there, affects everything upon or around it. Its physical bulk keeps the Moon in orbit, holds the atmosphere in place, and, down here on the planet's surface, ensures that whatever goes up inevitably falls right back down again. The force with which the massive Earth yanks us firmly toward its center is gravity – from the Latin *gravis*, meaning heavy – though the concept of gravity was initially postulated by Aristotle, a Greek. Aristotle envisioned an Earth composed of interlocking concentric spheres, at the center of which was a heavy ball of elemental earth, encased by increasingly lighter spheres of water, air, and fire. Within these layers, objects, like depth markers, were thought to seek their own level: hence stones, being rich in earth, plummeted downward through air and water; rain (water) fell downward from air; and smoke (air) and sparks (fire) rose upward from earth. Downward movement, in Aristotle's scheme, was governed by the force of gravity.

The mathematical description of the force that holds us all so firmly down was formulated in 1667 by Isaac Newton, then an awkward and ill-tempered twenty-three, home from Trinity College, Cambridge, in order to avoid an epidemic of plague. During his year of enforced exile on his home estate in Lincoln-shire – a period now known to scientists as the *annus mirabilis*, or year of miracles – he proved the binomial theorem, invented calculus, and developed color theory, as well as elucidating the mathematics of gravity. The nature of this last is said to have coalesced while Newton, mooching about the family orchard in a "contemplative mood," observed the fall of an apple. The end result of that providential fall was the Universal Law of Gravitation.

Gravity, as revealed by Newton, is reciprocal, a two-way tug-of-war, operating between any two bodies over distance: between the Earth and the Moon, between two billiard balls upon a table, between you and your next-door neighbor, between you and the Earth. In some cases – the unfair contest between you and the Earth, for example – the two-way tug is grossly unequal. Gravity, ultimately, is proportional to the size of the bodies concerned and to the distance (strictly speaking, the inverse square of the distance) between them. Thus the bigger the body and the closer the distance, the greater the gravitational pull. The Earth, with a mass of 6×10^{24} – that is, 6 septillion or 6,000,000,000,000,000,000,000,000 – kilograms, is simply colossal compared to the mass of the average human being, which means that the Earth pulls on us with consider-ably more force than we pull on it. The Earth pulls so hard, in fact, that we're hard put to get away from it; a rocket has to reach a speed of 40,000 km/h (24,000 mph) just to escape Earth's gravitational field.

Gravity can be depressing after periods of high-caloric

merrymaking, since the weight of our often too too solid flesh is a matter of gravity. Mass, as in the total amount of matter contained in object, is an independent quality, remaining stubbornly constant regardless of position in the universe. Weight, on the other hand, is interactive; and ours, as Earth-dwellers, is distinctly the fault of our planet. Multiply your mass by the Earth's gravitational pull and you've got what registers each morning on the bathroom scale. The closer to the center of the Earth you are, the greater this grabby effect. Your weight at sea level is therefore greater than your weight on a lofty mountaintop, since by climbing you've put a mountain's-worth of extra distance between you and the center of the Earth. Similarly your weight at the equator will be slightly less than your weight at the North Pole, since the Earth bulges a bit around the middle. If you plan to binge on chocolate eclairs, best to do so in such forgiving gravitational regions as Ecuador, Indonesia, or Zaïre.

Gravity is an annoyance for the unpleasingly plump; it's also a force to be reckoned with when toting barges, lifting bales, or pedaling a bicycle up a steep hill. And it can be fatal. One G, the gravity that most of us live with day by day, is equal to an acceleration of 9.8 m (32 feet)/sec^2, which is the rate at which anything, dropped, hurtles toward the ground. The dangers inherent in falling contribute to its negative image: we fall out of favor, from grace, and into bad habits, for example; and even falling in love, though usually delightful, implies a certain emotional helplessness. The remorseless aspect of gravity impels us to defy it. High-wire acts, high dives, bungee jumps, rock climbs, and trips over Niagara Falls in a barrel are all attempts to tweak gravity and see just how much we can get away with. The answer is not much: according to the US National Safety Council, about 16,000 Americans each year die of falls, variously off ladders, out of buildings, into holes, down stairs, or just by tripping over the sidewalk curb. Falls, in fact, are a leading cause of unintentional injury deaths, second only to motor vehicle accidents.

A force of one G is about all we're equipped to handle, even when standing on solid ground. Much more than that and we're in trouble, as can be seen by the Dali-esque distortions on the faces of astronauts aboard accelerating rocketships. By 2 G, we're feeling squashed; by 3 G, we're losing vision; and by 5 G, we're unconscious. On the other hand, troublesome as gravity can be, we have a hard time living without it. Deprived of gravity, as in the near-weightless environment of space (micro-gravity, or 10^{-6} G), the human body begins to deteriorate: bones and muscles atrophy and blood volume decreases. Astronauts on long missions risk losing half their bone mass or more; and their blood volume can drop by as much as 22 per cent, with attendant risk of cardiac muscle atrophy and permanent heart damage. When poet Robert Frost wrote musingly in "Swinger of Birches" that he'd like to get away from Earth awhile, he failed to consider the potential difficulties in getting back. Earth's persistent pull helps us survive.

Gravity – or rather lack of gravity – has been shown to affect a wide range of biological systems. Cells cultured in space show developmental abnormalities. Quails hatched in space are perpetually disoriented, unable to fly or perch properly; infant rats raised in space have truncated limb muscles which lead to difficulties in walking or swimming. Plants – which normally depend on gravity for directional growth cues (leaves up; roots down) – have poor growth rates in space. Life as we know it seems to be neatly calibrated to heft of this particular planet, to the ceaseless tug of Earth.

When the ancients spoke of the Earth, they as yet had no grasp of how very much there was of it, or how its lethargic peregrinations influence how and where we live. The ground – as in the 30-mile (48 km)-thick layer of solid stuff upon which we all creep about from day to day – contains some 220,981,000 square miles (572,490,000 sq km) of rocks. Collectively these make up the Earth's lithosphere – from the Greek *lithos*, which means rock, and as such figures in such rock-laden terms as megalith, monolith, lithography, and Paleolithic Age. If you think of the Earth as an immense egg, the lithosphere is its rigid shell. It's not a bad analogy: comparatively, the lithosphere is eggshell-thin, a mere 0.8 per cent of the Earth's radius and 0.6 per cent of its total mass. If the Earth were the size of a beachball, the lithosphere would be just five-hundredths of an inch (1 mm) thick – thinner than a coat of paint; and Mount Everest, the most impressive terrestrial feature upon it, would be an infinitessimal pimple, eight-thousandths of an inch (0.02 mm) tall. In the planetary scheme of things, terrestrial topography averages out to smooth.

The lithospheric eggshell is not only thin, but – like the fallen Humpty Dumpty – broken. Earth's crust is fragmented into a dozen or more ragged chunks known as plates, whose leisurely but relentless movements are responsible for the topography and distribution of the continents and for such geological excitements as earthquakes, mountains, and volcanic explosions. The first scientist to propose that our seemingly stationary land masses wandered erratically about the surface of the planet was the nineteenth-century German astronomer/ polymath Alfred J. Wegener, a minister's son, born in Berlin in 1880. Wegener obtained his doctorate in astronomy from the University of Berlin in 1904 – he is still noted for some early experiments on the origin of lunar craters which involved dropping lumps of plaster on to trays of powdered cement – but his interests soon rapidly expanded to encompass meteorology, climatology, geology, topology, hot-air ballooning, and Arctic exploration. A favorite photograph shows him in wool jacket and furry hat, a meerschaum clenched in his teeth, eyes fixed on the chilly distance. The Arctic was to be his downfall: he died in a blizzard on his fourth expedition to Greenland in 1930, at the age of fifty.

Wegener's eclectic mix of scientific disciplines combined in his controversial *Origin of Continents and Oceans*, the first description of the theory of continental drift, published in 1915. In it Wegener postulated that the continents were analogous to the floating ice floes in the northern ocean, formed annually from the break-up of sea ice. The suggestively interlocking shapes of Africa and South America were reminiscent of the matching chunks of a newly broken ice sheet. Such recip-rocal break patterns, along with the similarities of fossils and geologic strata on opposing sides of the Atlantic – the rocks of North America's Appalachian Mountains, for example, look an awful lot like those of the Scottish Highlands – led Wegener to conclude that global land masses move around. Our present continents, he suggested, were once clumped together in a single massive lump, dubbed Pangaea, from the Greek for "all-earth." Over time, propelled by tidal forces, they separated and gradually drifted apart. Wegener envisioned them plowing indomitably through the rock of the ocean floor like ponderous shark fins carving a path through water, sepa-rating from each other at a rate of approximately 8 feet (2.5 m) per year.

Actually continents don't move; plates do. The Earth's lithospheric plates are all restlessly, if imperceptibly, in motion, variously bashing into each other, pulling away from each other, or sidling past each other in uneasy parallel like so many outsize bumper cars. These movements are collectively known as plate tectonics – after Tekton, the master builder in the *Iliad*; the word shares a root with such construction-related terms as architect and architecture. The linguistic rationale here is that interactions among shifting plates are responsible for the shaping or building of the Earth's surface.

Continents, in effect, are the frosting on the crustal plates, the lightweight tips of rocky icebergs. All (Europe, Asia, Africa, Australia, Antarctica, and the Americas) are composed primarily of low-density granite – the grainy igneous rock that figures so prominently in monuments, mausoleums, and the presidential faces on Mount Rushmore. In the early days of the Earth's formation, as the primordial molten planetary blob slowly began to cool, its constituents separated from one another by weight, the lighter elements rising toward the planet's surface, the heavier sinking toward the center. The granitic continents thus rest on a much heavier base of basalt, the solid dark igneous rock that makes up the ocean floors. Plates are distinctly bottom-heavy. The ocean floors are on average 2.4 miles (4 km) below sea level and the continents on average 400 feet (125 m) above because basalt is weighty stuff. While continental granite is about three times denser than water, basalt is five times denser – which means that the mighty Himalayas are positively frothy compared to the leaden bottom of the sea. We are where we are by virtue of the relative density of granite; if it weren't light enough to ride above the waves on its pedestal of basalt, ours would be a water world and dolphins would doubtless be the masters of the Earth.

Plates move because they're sitting on slippery stuff. The plates of the Earth's crust float like rafts on a vast gooey layer of semi-molten rock known as the mantle – the upper, and particularly squishy, portion of which is called the astheno-sphere, from the Greek *asthenes*, meaning weak. What it lacks in strength, it makes up in quantity: the mantle, which extends from about 30 miles to 1700 miles (50–2700 km) beneath the earth's surface, comprises the bulk of the planet – 82 per cent of the Earth's volume and 68 per cent of its mass. Deeper yet is the Earth's core, which reaches from the nether edge of the mantle all the way to the center of the Earth, some 4000 miles (6400 km) down. The core has an outer layer of liquid iron, whose sluggish sloshings are responsible for the Earth's magnetic field; and a solid inner layer of iron and nickel. Structurally simplified, our planet is an immense ball bearing, surrounded by hot mud.

The innards of the Earth are hot, and the deeper you go, the hotter it gets. Temperature increases, on average, about one degree C for every 15 feet (4.5 m), reaching, at the center of the planet, a sizzling peak of 15,700°C (28,800°F). This fire far beneath our toes keeps the Earth from freezing solid; and at the same time provides the impetus for the leisurely realign-ments of the global surface. The movement of the plates is driven by heat – more precisely by thermal convection, the same mechanism by which water comes to a boil on top of the stove. Set a water-filled pot on a hot burner and the liquid at the bottom, closest to the heat source, will warm first. As it does, its density will decrease and, as it lightens up, it rises to the top of the container. There, in contact with the cooler liquid at the top of the pot and the still cooler air outside, the heated blip cools, its density increases, and it sinks – back toward the heat-generating burner to start the whole process over again. This looping pathway (heat–rise/cool–fall) is called a convection current. Such currents are set up in any fluid (water, soup, molten subterranean sludge) caught between a heat source and a cold source. In the case of the planet, the heat source is most likely a combination of radioactive decay and primordial heat from the Earth's core – the broiling leftovers of Creation; and the cold source is the outer rock of the Earth's crust, the atmosphere, and ultimately the frigid reaches of space. Heat, roiling to and from the surface from the Earth's core and mantle, determines how much and how fast the plates, with their lofty cargo of continents, will move. During the Archaean Era, 2.5 billion years ago, when the Earth was much hotter, geologists hypothesize that surface plates were both more mobile and more numerous – perhaps a hundred or more, as opposed to the mere dozen that characterize the planet's sedate middle age.

Today calculations based on measurements from satellite global positioning systems (GPS) indicate that the plates move an arthritic 2–4 inches (5–10 cm) per year – about as fast as our fingernails grow. Two inches doesn't sound like much – the average snail, without unduly exerting itself, can cover it in half a minute – but the implications, given sufficient time and persistence, are substantial. Inch by subtle inch, plate tectonics has torn India away from Madagascar, thrown up the Himalayas, closed off the isthmus of Panama, popped open the Strait of Gibraltar, and ripped apart the shores of the Red Sea. Geology is the quintessential tortoise, living proof that slow and steady, if not precisely winning the race, can certainly wreak havoc with the topography of the finish line.

Earth is the ultimate purveyor of real estate. Just where your particular bit of continent happens to be positioned on the globe determines your personal environment, decrees whether your wardrobe is heavy in shorts or long underwear, dictates your diet, prescribes your social and cultural life, and channels your career opportunities. Nowadays, of course, people have potential for considerable mobility – Lapp reindeer herders, should it strike their fancy, can relocate to Jamaica – but in the long term, it's Earth that calls the tune. I own snow shovels and boiled-wool socks because of tectonic drift: some 200 million years ago, during the Mesozoic Era, the North American plate slithered northwest.

Earth also sculpts the local scenery. Mountain chains, for example, are the products of aggressive plate tectonics, brutal crustal collisions known to geologists as orogenies (mountain-building events, from the Greek *oros*, mountain). The effect is analogous to pushing on a tablecloth to form overlapping folds, crumples, and wrinkles. In fact, such folds in compacted rock are often called *nappes*, from the French for tablecloths. The peculiarly folded structure of the Earth's mountains caused considerable geological heartache prior to the advent of plate tectonic theory: there seemed no logical reason for mountains – whole ranges of them – to simply poke up out of the ground where they did, like so many magnified molehills. The best guess, according to John Van Dyke's *The Mountain*, originally published in 1916, was planetary shrinkage: that is, as the interior of the Earth cooled and contracted, its surface wrinkled like the withered skin of a dried apple, forming wizened rows of mountains. This hypothesis, unfortunately, was unable to explain why, if the entire planet were shrinking, there weren't *more* mountains, and left geologists at a loss to account for the existence of plains.

The Alps, Appalachians, and Himalayas are all folded mountains: the Alps, the result of a broadside collision of Europe and western Asia with the African plate; the

Appalachians, the remnants of an encounter between North America and Eurasia; and the Himalayas, the spectacular product of a 40-million-year-old smash in which India, riding northward on the Indo-Australian plate, encountered the continent of Asia, poking out of the Eurasian plate, and sitting stolidly in its way. Mountains ranges in general are the aftermath of what amounts to a prolonged geologic gulp and swallow. As a crustal plate nudges aggressively against its neighbor, the dense leading edge – the heavy basalt of the ocean floor – sinks, diving under the opposing plate to be swallowed into the interior of the Earth, in a process known as subduction. The Andes of South America were formed in this fashion, as subduction of the encroaching Nazca plate crumpled the continent from Colombia to Tierra del Fuego. When the interacting plates both carry a cargo of continental crust, subduction is complicated by head-on collision, as high-riding continents – too lightweight or too thick to sink – crunch forcibly together. The result is accretion, a welding together of two separate land masses, now zippered down the middle with a massive range of mountains. The Urals of Russia are the result of such an impact, a weighty Paleozoic crunch that pasted Europe inextricably to Asia, leaving a lumpy 3000-mile (4800-km)-long line of peaks to mark the border in between.

Not all mountains are the progeny of proper mountain-building processes. Japan's Mount Fuji, Africa's Mount Kilimanjaro, Mexico's Mount Pinatubo, and Washington's Mount St Helens are all volcanoes and thus – though they distinctly *look* like mountains – are in a geologic class by themselves. The mountains of the Moon – though named after such respectable earthly ranges as the Alps, the Carpathians, the Apennines, and the Pyrenees – are peaks thrown up in the wake of massive meteoric impacts; the Moon, solid all the way through, has no tectonic activity. The largest mountain in the Solar System – the 15-mile (24-km)-high Olympus Mons on Mars – is a volcano; the Martian crust, like that of the Moon, is immobile; the planet is as solid as a bowling ball. Only Earth boasts mountains generated by crustal collisions; our mountains – like the brave, deluded, or insanely optimistic souls who attempt to climb them – are unique.

The uses of mountains, according to Elie Bertrand of Switzerland, who wrote an essay on that very topic in 1754, are many and varied. They serve as foundations for the Earth; they enforce boundaries between nations; they are beautiful and inspirational to look at; they produce winds and springs; and they contain providential caves for the shelter of persecuted Christians. Historically, such a positive spin was a switch. Through the early eighteenth century, mountains were popularly held to be horrid, ugly, misshapen, and unnecessary, lumpy monstrosities dumped upon the Earth as manifestations of divine displeasure. One theory held that the Earth, prior to the Flood, had been smooth as a bowling ball, and still would be if early human beings had only managed to behave themselves. King George III found mountains unsightly and Samuel Johnson called them sterile outcasts of nature.

Our majestic mountains, in global perspective, may be pretty small potatoes, but from the average human standpoint – King George aside – they're awe-inspiring. They're also surprisingly rare. Geologists demonstrate this using a mathematical construct called a hypsometric curve – a graceful graph in which percentage of the Earth's surface area is plotted as a function of altitude. The results show that about 70 per cent of the Earth's surface has – from a land animal's point of view – essentially no altitude at all, being underwater. Another 20 per cent or so hovers right around sea level. Less than 9 per cent of Earth is 2000 feet (600 m) or more above sea level, and a mere 3 per cent measures 10,000 feet (3000 m) and up.

Immensity and rarity are a telling combination. Earth, piled into lofty peaks, pinnacles, crags, and cliffs, inspires worship and wonder. "I will lift up mine eyes unto the hills, from whence cometh my help," sings the psalmist of the Old Testament. In many mythologies, the gods were believed to live on the inaccessible peaks of the mountains. The ancient Hindus claimed that Siva lived in the upper reaches of the Himalayas, wrapped in gold and purple clouds and sitting on a bed of diamonds. The Greeks believed that Zeus hurled his lightning bolts from the top of Mount Olympus; and the Plains Indians that the Great Spirit dwelt on the summit of the Rockies, where he was given to sitting in meditation in the evenings, smoking a pipe, the white feathers of his war bonnet streaming behind him in the wind.

Worldwide cultural traditions are filled with intrepid souls climbing mountains in search of enlightenment. Moses climbed Egypt's Mount Sinai to receive the Ten Commandments; Mohammed climbed Saudi Arabia's Mount Hira and there experienced the vision that led to the birth of Islam; and Jesus delivered Christianity's most famous sermon on a mount. The Buddhists climb mountains to attain nirvana: half a million pilgrims each year scale the 12,389 feet (3776 m) of Japan's Mount Fuji. For over 3000 years, Chinese emperors made ceremonial pilgrimages to Shandong Province's Mount Taishan, foremost of China's Five Holy Mountains. A successful ascension of Taishan is said to guarantee the climber a lifetime of a hundred years; despite this incentive, only five of the imperial visitors actually made it to the top, among them the notably athletic Qing emperor Qianlong, who scaled it eleven times. (He died in 1799, aged 88.)

Climbing is a common metaphor for overcoming intellectual or spiritual trouble: Christian of John Bunyan's *Pilgrim's Progress* has to hike up the Hill of Difficulty and clamber up Mount Zion before he finally attains the Celestial City. Sometimes, however, we climb simply to test our physical selves. Mountain climbing for sport became popular in the first half of the nineteenth century, during which European climbers flocked to the Alps in such numbers that a flourishing tourist industry developed, complete with hotels, restaurants, tour guides, and even, for the less physical, portered chairs. Guidebooks proliferated, with a wealth of helpful hints for novices, including dietary cautions (beware of toasted cheese), preventives for snowblindness (green crepe), and remedies for sore feet (soak in neat brandy). Clothing became increasingly imaginative, though never satisfyingly dry or warm. While the

5300-year-old Iceman set out on his Paleolithic climb dressed in goatskin leggings and a grass cape, nineteenth-century alpinists wore tweed, with gaiters, quilted waistcoats, knitted mittens, and hobnailed leather boots.

The enthusiasm for altitude suffered a sharp setback in 1865, after the four companions of Edward Whymper, the conqueror of the Matterhorn, were killed during the descent when a rope snapped, plummeting them into a glacial crevasse. News of the tragedy, with awful artistic interpretations of the accident, was broadcast around the world, arousing shock and dismay, and leading the horrified Queen Victoria to propose that mountain climbing be outlawed altogether. It was too late, however, to stem the upward-bound tide. Whymper himself returned to the Alps to climb the Matterhorn twice more; in 1871 he published a book about his exploits – *Scrambles Amongst the Alps* – which became a classic of mountain-climbing literature and remains in print to this day. And other climbers were already turning their attention farther afield, to the mountains of the Americas and to the ultimate challenge, the towering Himalayas, home of the tallest peaks on Earth.

The world's ten highest mountains are all Himalayan, and the highest of all, the awesome Everest, towers 29,035 feet (8850 m) – five and a half miles (nearly 9 km) – into the sky. Its name in Tibetan – Chomolungma, "Mother Goddess of the World" – suits it better than the prosaic Everest; the English name was bestowed in 1865, in honor of Sir George Everest, British surveyor-general of India in the 1830s and 1840s, who may never actually have seen it. If he did, he knew it, even more prosaically, as it was marked on the original British surveyors' maps, as Peak XV. To the avid mountaineer, Everest's great height, steep slope, frigid temperatures, incessant winds, and – towards the top – dangerously low atmospheric pressure and lack of breathable oxygen, were so many challenges to be faced and overcome. It took until 1953, however, for Sir Edmund Hillary and Tenzing Norgay to finally reach the peak, a feat commemorated by Hillary with the triumphant words, "Well, we knocked the bastard off!"

By the time George Leigh Mallory set out on his final fatal attempt to climb Everest in 1924, he was clearly sick of being asked "Why climb?" His answer to a *New York Times* reporter some months earlier had been the terse and much-quoted "Because it is there." In 1922, however, he gave a fuller response to the question, which conveys a real sense of what we find on Earth's mountaintops. Climbing, said Mallory bluntly, "is no use. There is not the slightest prospect of any gain whatsoever. Oh, we may learn a little about the behavior of the human body at high altitudes, and possibly medical men may turn our observation to some account for the purposes of aviation. But otherwise nothing will come of it. We shall not bring back a single bit of gold or silver, not a gem, nor any coal or iron. We shall not find a single foot of earth that can be planted with crops to raise food. It's no use. So, if you cannot understand that there is something in man which responds to the challenge of this mountain and goes out to meet it, that the struggle is the struggle of life itself upward and forever upward, then you won't see why we want to go. What we get from this adventure is just sheer joy. And joy is, after all, the end of life."

It's a sentiment that even far lesser climbers share. Our mountains here in Vermont, compared to the lordly Himalayas, are dwarves; the tallest of the lot, Mount Mansfield in the Green Mountain range, reaches to just 4393 feet (1340 m) at its highest point, familiarly known as the Chin, since it looks a bit like one in profile. From the Chin on a clear day, you can see a 360-degree panorama of mountains, valleys, and hills: the Adirondacks, the Taconics, the White Mountains of New Hampshire, the distant blue glimmer of Lake Champlain, and even, far to the north, the silvery skyline of Montreal. It's a 9-mile (14-km) trail from foot to summit. Still, standing there, realizing that you've done something awful to your calf muscles, you're suffused with accomplishment and a Mallory-esque rush of joy. You've triumphed over difficulty and the mountaintop is your reward. Perhaps it's the closest we can come these days to understand how it once felt to worship the still-magnificent Earth.

The Himalayas, at 40 million years old, are mere tots as mountains go; California's Sierra Nevadas are twice as old, and the Appalachian range over six times as old, its weathered peaks dating back 250 million years. The youthful Everest is still on its way up, growing at a rate of 3–5 millimeters each year – which means, in effect, that every climber who manages to struggle to the top sets a new record. Older ranges, in contrast, beset by weathering and erosion, are clearly on their way down. The combined effects of wind and water – a deadly duo – can level a mountain range in a few tens of millions of years, busily scouring it down to size, from mountain to hill to featureless plain. This process provides the central theme for Christopher Monger's 1995 movie *The Englishman Who Went Up a Hill But Came Down a Mountain*, in which a pair of English surveyors – played by Ian McNeice and Hugh Grant – arrive in a small Welsh town in the early days of World War I to measure the height of the much-vaunted local mountain. Measured, the mountain proves to be just 984 feet (300 m) tall – it's (horrors) a hill, 16 feet (5 m) short of official mountainhood – and the outraged townspeople, armed with rocks, dirt, and wheelbarrows, set out to make up the difference, converting their hill to a mountain once again.

It's a chore that the Earth's crust ordinarily performs for us. Historically, to geologists, the fact that massive mountains wear down is less amazing than that they survive as long as they do. The 250-million-year-old Appalachians, for example, should by all rights have vanished 200 million years ago – and would have if it weren't for the lithosphere, providing a helpful boost from underneath. The unexpected longevity of mountains is due to the delicate balance between floating and sinking, a phenomenon known as isostatic equilibrium. Land, in effect, behaves like a raft. If heavily loaded from the top – as with a cargo of multi-mile-high mountains – it sinks deeper into the slush of the underlying mantle. Conversely, if its mountainous load is lightened, the crust bobs up, irrepressible as the unsinkable Molly Brown, riding proportionately higher on the mantle's molten sea. Erosion fights a losing battle; the more rock is shaved away from the top, the greater the corresponding lift from below.

Isostasy works because there's a lot more to the average mountain than meets the eye. The bulk of every peak is buried. A mountain floats on Earth's mantle much as an iceberg floats in the polar oceans: one-tenth of its mass soaring skyward, nine-tenths sunk deep in the Earth's crust. As the upper reaches of a mountain weather away, hidden roots rise up to replace the loss. Eventually, sadly, erosion triumphs; the crust can rise no more, and even the loftiest purple mountain's majesty is inexorably reduced to fruited plain. Cratons, the stable central nuclei of the continents, are the ground-down remains of ancient mountain roots: examples include the vast (and flat) Canadian, Amazonian, and Australian Shields. Earth is a work in progress – or perhaps more accurately a work in continual turmoil, incessantly building up and tearing down, erecting and reducing to rubble.

This concept – known to schoolchildren today as the rock cycle – was first proposed in 1788 by Scottish geologist James Hutton, who later expanded his theory into a massive multi-volume illustrated text titled *Theory of the Earth* (1795). The Earth, in Hutton's view, was an endlessly cycling "world machine." The cycle begins as the land is worn away by erosion. The eroded fragments are then deposited as sediment on the ocean bottoms, where they are slowly pressed into solid rock again, compacted and consolidated by heat and pressure. Finally, the new rock layers are lifted up to form new continents, upon which the whole process begins again. None of this happened fast; Hutton's world machine, like the mills of the gods, ground slowly. His theory, generally known as uniformitarianism, holds that the present-day topography of the planet is the result of excruciatingly slow but constant processes, inexorably reshaping the Earth's crust over immense periods of time. Its opposite number is catastrophism, the belief that changes in the Earth's surface occur rapidly, through periodic

Blitzkrieg-like disasters. Nineteenth-century catastrophists proposed anywhere from four to twenty-seven such disasters over the course of Earth's history, the most recent being Noah's Flood, described in destructive detail in the Book of Genesis. At the time, Hutton's work garnered less attention than it deserved. It was left to another Scottish geologist, Charles Lyell, to popularize it a generation later, in his landmark *Principles of Geology*, published between 1830 and 1833. The book won Lyell a medal from the Royal Society and eventually, somewhat unfairly, earned him the title "Father of Geology."

The rocky skin of our planet is largely hidden. Much of it lies at the bottom of the sea; and, prior to the mid-twentieth century, the curious had no efficient methods for exploring it. It was only in the years following World War II that geologists finally obtained tools that enabled them to study the ocean floors – notably echo-sounders and magnetic anomaly detectors, devices used during the war to troll for lurking enemy submarines. The foremost of the early ocean-investigating geologists was Harry Hammond Hess – in the 1940s, Captain Harry Hess of the assault transport *Cape Johnson*, assigned to duty in the Pacific. Between battles – the *Cape Johnson* participated in fighting in the Marianas, the Leyte Gulf, and Iwo Jima – Hess and crew used their state-of-the-art equipment to study the ocean bottom. This was humans' first glimpse of the Earth beneath the sea: a startlingly complex world of plains, plunging chasms, and steep underwater mountains, which last Hess, perhaps dreaming of peaceful academic days past, named guyots, after Arnold Guyot, the first geology professor at Princeton, his home university.

The Earth's longest mountain chain is underwater. In the 1950s, cruising the billowing Atlantic, researchers from Columbia University's Lamont Geological Observatory (now the Lamont-Doherty Earth Observatory) mapped the Mid-Atlantic Ridge, the longest mountain range in the world, stretching some 9000 miles (14,400 km) from the north of Greenland to a point south of continental Africa. Similar ridges trail from Mexico south through the east Pacific (the East Pacific Rise) and meander southwest beneath the

Indian Ocean (the Carlsburg Ridge) – a total of 37,200 miles (59,500 km) of undersea mountain ranges worldwide. Down the center of each ridge, distinct as the white stripe down a skunk's back, runs a deep valley, some 6000 feet (1800 m) deep and 8–30 miles (13–48 km) wide. This rift represents what geologists call a divergent boundary, a long torturous crack in the Earth's surface where the planet is slowly tearing itself apart. The Arctic and Mid-Atlantic Ridges are separating at the rate of about an inch (2.5 cm) per year; the East Pacific Rise – which brushes past Easter Island, where the natives used its volcanic rock to carve enormous heads – at 6 inches (15 cm) per year. Separation, however, is a deceptive term; a more apt label is "sea-floor spreading." Ridges don't simply pull apart; they simultaneously fill in the hole they leave behind. At the site of each mid-oceanic gap, magma wells up from the Earth's mantle, solidifying upon contact with cold seawater and creating new stretches of ocean floor. This process can be observed on dry land in Iceland, which uneasily straddles the Mid-Atlantic Ridge. While California is slowly crumbling into the sea, Iceland, in a continual state of rift and volcanic eruption, annually expands its territory, steadily growing outward from the middle.

Hess's concept of sea-floor spreading, originally published in 1962 as "The History of Ocean Basins," was so revolutionary that the author himself described it nervously as "an essay in geopoetry." The hypothesis was subsequently confirmed, however, by a pair of young British geophysicists, Drummond Matthews and Fred Vine of Cambridge University, who were studying magnetism. Matthews and Vine knew that the Earth's magnetic poles periodically reverse themselves, a startling global flip in which the North Pole suddenly becomes the South. On the Sun, such flips are common – solar magnetic poles reverse themselves every eleven years, in synchrony with the sunspot cycle – but terrestrial reversals are much less predictable, occurring at intervals of anything from 5000 to 50 million years. The last took place about 740,000 years ago, in the depths of the Pleistocene.

Magnetic reversals leave evidence of themselves behind, clear as a line of muddy footprints across a kitchen floor, embedded in igneous rock. Molecules of iron in newly formed igneous basalts align themselves in the direction of the Earth's current magnetic field, taking on the imprint of the polarity under which they were born. As lava slowly oozes from mid-ocean rifts, the result is a rocky videotape of Earth's magnetic history: a line of parallel magnetic stripes in an elegant zebra pattern of sequential magnetic flip-flops extending outward in opposite directions from the undersea ridges, each side the mirror image of the other. Submarine rock solidifying at ridges today points directly toward the current magnetic North Pole – a gently peripatetic entity, presently in the region of Ellef Ringnes Island in the Canadian Arctic. Stripes increasingly distant from the ridge are concomitantly increasingly older, their molecules still resolutely pointing in the direction of magnetic poles long past. The magnetic evidence proved that Harry Hess's hypothesis was geofact: the sea-floor begins at the ridges and crawls outward.

Left to themselves, the global ridges – perpetually fed by subterranean magma pools – could in theory slowly increase the surface area of the Earth, swelling the planet like a balloon. In practice, however, they don't: what comes up eventually goes back down. As new crust is created, old crust sinks back into the planet's interior to be destroyed. This occurs at convergent boundaries or subduction zones, where one plate boundary dives down beneath another, dragging its rocky constituents into the mantle to be melted. The Pacific's Mariana Trench is a particularly impressive example – a 35,820-foot-deep (10,918 m) hole where the Pacific Plate, shoved by the East Pacific Rise, dives beneath the Philippines. Hutton's world machine is a conveyor belt, spewing new rock out at the ridges, cranking it across the ocean floor, and disposing of it at the trenches, where – processed into sludge – it oozes its way ridgeward once again.

The whole process, from creation at the ridge to destruction at the edge, takes at most about 200 million years, which means that the bulk of the Earth's crust is, in geologic terms, freshly minted stuff. The continents, in contrast, floating aloofly above the sea like so many aged rubber ducks, are exempt from the global recycling process. Dry land is far older than the bottom of the sea: the rocky floor of the Grand Canyon is 570 million years old; the rocks of South Dakota's Black Hills, 1.46 billion years old; and the rocks of western Greenland, 3.8 billion years old. The oldest mineral matter so far identified on Earth – a scattering of tiny zircon crystals – come from the hills of western Australia and are 4.4 billion years old, almost as old as the non-tectonic Moon. "Old as the hills" – an expression which, according to *Bartlett's Familiar Quotations*, comes from Sir Walter Scott – is, in terrestrial terms, as old as anything on Earth gets.

Nothing, not even the wind that blows, is so unstable as the crust of this Earth.

Charles Darwin

The Greek elements all have their dark sides. Water is floods and tsunamis; air is hurricanes; fire, incinerating forests, is wildfire. Beside such elemental furies, earth seems solid, stodgy, and reliable, even a bit dull. To the ancients, earth was bedrock, fixed and immobile, the stable foundation of matter. People with a psychological affinity for earth are said to be dependable but inflexible, stubborn, bull-headed, and impossible to budge in argument. They're also patient, methodical, and fond of long-term relationships; and they're all literal-minded pragmatists, the sort of party-poopers who refuse to clap for Tinker Bell.

As we've seen, however, the Earth isn't as immobile as all that. It continually moves under our feet; it just does so slowly. Uniformitarianism is leisurely to the point of imperceptibility. Such behavior, from a human perspective, is ploddingly monotonous: a lot of barely detectable nudging, rubbing, and dripping that, given a few million years or so, may eventually get us somewhere. Just don't hold your breath.

On the other hand, sometimes catastrophism takes a hand. Earthquakes, according to the ancient Japanese, were caused by the thrashing tail of a gigantic catfish; and according to the Mongolians, by the awkward stumbles of an enormous world-supporting frog. The American Indians blamed them on angry gods; and the ancient Greeks specifically attributed them to Poseidon in a tantrum, pounding on the sea floor with his trident. The result, in all cases, is the same: a violent shaking of the land itself, with attendant collapse of anything human beings have been foolish enough to erect upon it. The cost in human lives can be horrific: 200,000 in Yokohama, Japan, in 1923; 50,000 in Quetta, India, in 1935; 70,000 in northern Peru in 1970; 242,000 in Tangshan, China, in 1976; 41,000 in Iran in 2003. An earthquake toppled Pompeii in 62 CE; its harassed citizens were still repairing the damage seventeen years later when Vesuvius erupted, eradicating the city once and for all.

Over 100,000 earthquakes shake the Earth each year and almost all are surprise attacks. Despite decades of research, our ability to predict earthquakes remains about as reliable as gypsy fortune-telling. Our ability to evaluate an earthquake while in the throes of one, on the other hand, has become increasingly precise. The first known quake –measuring device – a primitive seismograph called an "earthquake weathercock" – was invented in China in the second century CE. It was an elaborate urn-like container made of bronze, topped by eight dragon's heads, each holding a metal ball in its jaws. When jarred by earth tremors, the dragons would drop their balls into the open mouths of a circle of bronze toads positioned below. Later instruments utilized pendulums, their builders having noticed the sensitivity of these to geological jiggling: church bells, for example, often rang during earthquakes, and pendulum clocks, jostled out of synchrony, wobbled to a halt. The first successful Western seismograph, invented in Italy in 1873, used two pendulums: one to detect north–south movements, and one to monitor east–west shifts, plus a weighted spring to measure movements up and down. Modern instruments are still often pendulum-based, though more advanced models feature electronic sensors implanted in bedrock.

Early earthquake measurements were complicated by difficulties in comparing one quake to another. Assessments of survivors reflected a dreadful similarity: all quakes inevitably were devastating, calamitous, and the worst. "… The most awful spectacle I ever beheld," wrote Charles Darwin of the 1835 quake in Concepción, Chile. "Not in history has a modern imperial city been so completely destroyed," wrote Jack London of the 1906 San Francisco earthquake. In 1883, Italian seismologist Guiseppe Mercalli, in an attempt to quantify the hyperbole of disaster, devised the Mercalli Intensity Scale, a rating of earthquakes based on specific and common observational criteria, such as what shook, broke, slid, or fell down. The scale runs from I to XII in Roman numerals, with the first real intimation of trouble appearing at IV, when windows rattle in their frames. By VI, furniture falls over and plaster cracks; at VII, chimneys fall down; by IX, rifts appear in the ground and general panic prevails; and by XII, destruction is total, complete with objects thrown into the air and large rock masses displaced. Thus the earthquake that struck Pompeii in 62 CE in which statues cracked and people were "deprived of their wits" sounds like Mercalli IX, while the "extraordinary and horrible" Santa Barbara, California, earthquake of 1812, in which adobe walls fell over, may have been a mere Mercalli VII.

Earthquake magnitudes nowadays are commonly measured on the Richter Scale, originally devised by Charles Richter and Beno Gutenberg, seismologists from the geologically unstable California Institute of Technology, in 1935. The Richter Scale, based on seismograph data, runs from 0 to 9 in logarithmic jumps, which means that each Richter level is ten times greater than the last. Thus a Richter magnitude of 7.0 is ten times more powerful than a magnitude of 6.0, and a million times more powerful than a magnitude of 1.0. A quake with a Richter magnitude of 7.0 is definitely a big one: the quake that crumpled the freeways of San Francisco in 1989 registered 6.9 on the Richter Scale. On the other hand, some are bigger yet: the quake that leveled Duzce, Turkey, in 1999, hit 7.2; the Mexico City quake of 1985, 8.1; the Good Friday Earthquake in Alaska's Prince William Sound in 1964, 9.2; and the Great Chilean Earthquake of 1960, an appalling 9.5.

Most earthquakes are perpetrated by the Earth itself, but a few we've set off ourselves. The underground nuclear explosions so popular during the Cold War weapons race occasionally generated moderate-sized earthquakes; and in 1969 David Stone, a geophysicist with a sense of humor, warned that the Chinese might do the same with nothing more than foot power. If the entire population of China were to jump simultaneously from a six-foot (2-m) stand, Stone wrote, the impact should release an amount of energy equivalent to an earthquake of magnitude 4.5 on the Richter Scale; and if such jumping were to take place in close proximity to the edge of the Pacific Plate, it might be capable of setting off an earthquake halfway around the world, on the vulnerable coast of California. Though synchronized jumping was never adopted as a Chinese military strategy, in the fall of 2001 a million British schoolchildren gave it a try, participating in a coordinated one-minute giant jump in schoolyards across the country. Calculations indicated that the jump should generate a collective two billion joules of energy, the equivalent of a magnitude 3.0 earthquake – a minor but detectable tremor. A few concerned citizens worried that concerted jumping by Scottish students over the Great Glen fault might instigate more than that even, potentially, a disturbance of the putative monster in the depths of Loch Ness, beneath which the fault runs. Seismograph records, however, indicated that the bouncing kids barely caused a planetary twitch.

For all our unquenchable hubris, human-generated quakes aren't a patch on what the Earth delivers. Over 90 per cent of earthquakes are tectonic quakes, occurring at the aggressive edges of plates, where chunks of crust – attempting to move over, under, or past each other – suddenly stick, locked together by friction along the ragged borders of faults. There they sit, straining against each other like rock-bound sumo wrestlers, until finally something pops loose with a jolt that shakes a good portion of the planet. The biggest earthquake in United States history – XI or XII on the Mercalli Intensity Scale – took place in 1811 in New Madrid, Missouri, and was felt as far away as Pittsburgh, Boston, and Washington, DC. The Prince William Sound earthquake of 1964 rocked 3.4 million square miles (8.8 million sq km) of territory stretching from the Arctic Ocean to southern California; and the Sumatra–Andaman Islands quake of December 26, 2004 – cause of what may be the world's most destructive tsunami to date – was detected in Oklahoma, on the opposite side of the globe.

Such catastrophic wriggles in the planet's crust release phenomenal amounts of energy. A quake of 5.0 or so on the Richter Scale generates the same amount of energy as 32,000 tons of TNT; a quake of 8.0, the energy equivalent of 1000 megatons of TNT or 100 simultaneously exploding hydrogen bombs. The 2004 tsunami-triggering quake in the Indian Ocean registered 9.0 on the Richter Scale and generated an estimated 2×10^{18} joules of energy – that is, the explosive punch of 32,000,000,000 tons of TNT – enough energy to boil 40 gallons (150 liters) of water for every person on the planet. The quake was so massive that it caused the Earth to wobble slightly on its axis, boosted the rate of the Earth's rotation (the length of the day, post-quake, was shortened by 2.68 microseconds), and shifted the northern tip of the island of Sumatra 118 feet (36 m) to the southwest.

Earthquake-generated energy travels in waves. The first and fastest of these are known as P, or primary waves, which rumble through the depths of the Earth at 3200 miles per second (5100 km/s). These are compression or "push–pull" waves, moving through their medium by alternately compressing and releasing the particles in their path, in the same manner that energy travels through a stretched and released spring. P waves are not damaging themselves, but are harbingers of disaster to come. Next in line are S – secondary or shear – waves, which eel their way side to side through the Earth like so many traveling snakes, at speeds of 1900 miles per second (3000 km/s). S waves are quite capable of shaking things apart, but the real troublemakers are the following surface waves. These, true to their name, travel on the surface of – rather than below – the ground, heaving rock and soil up and down in crests and troughs in the same manner that waves move across the surface of water, wreaking havoc on bridges, highways, gas and electric lines, and building foundations.

According to Douglas Adams, Earth's entry in the inter-planetary guidebook *The Hitchhiker's Guide to the Galaxy* reads "Mostly harmless." From the standpoint of those of us living on it, however, its primal elements are forces to be reckoned with. The Earth is Jekyll and Hyde: a solid and responsible citizen who periodically goes berserk.

Most of us, when we think elemental earth, think rocks. Our associations with Earth are of permanence and security; earth may be sand, soil, mud, and iron ore, but first and foremost it's rock, the sturdiest of elements, planted defiantly and immove-ably in the path of wind and waves. Rock, unless beset by seismic stresses, neither shakes, rattles, nor quivers like jelly, which is why institutions of notable dependability are invari-ably compared to it and anything of particular stability is said to be built upon it. The house I presently live in has a foun-dation of rock; the basement, much of it sledge-hammered out of granite ledge, is cold in winter, and in summer has a distressing tendency to harbor snakes, but it's reassuringly solid. Barring aerial bombardment or earthquake, this place isn't going anywhere.

Rock, to a scientist, is any naturally occurring solid mineral matter in the crust of the Earth – and by "solid," geologists mean just that, the sort of stuff that can only be broken by a vicious smack with a hammer. Rocks are made of minerals, which the

Oxford English Dictionary defines poetically as "products of the bowels of the earth." Most minerals are chemical compounds, though a few exist in nature as pure elements, among them gold, silver, copper, and sulfur. The first known attempt at a classification of minerals dates to the fourth century BCE, one of an immense body of works by the prolific Theophrastus of Eresos, a student of Plato and Aristotle, and the latter's successor as head of the Lyceum in Athens. Titled *De Lapidibus*, or *On Stones*, the book – which survives solely as a ten-page fragment – discusses the characteristics, sources, and uses of various rocks and minerals. Theophrastus is quoted admiringly by Pliny the Elder, whose 37-volume *Natural History*, written in the first century CE, includes five books on various aspects of geology, including metals, pigments, stones, and precious gems. Book XXXVI, "The Natural History of Stones," includes detailed lists of stones with medicinal and other uses (pumice, for example, was recommended as a palliative for eye ulcerations, an emollient for smoothing the skin, a toothpaste, and a polish for books), as well as discourses on stones as building material, and assorted digressions into "Marvellous Works in Egypt," "Buildings Erected without the Use of Nails," "Cisterns," "Pavements," and "The Origin of Glass."

In ancient and medieval literature, earth was medicine: stones appear frequently as pharmaceuticals. Somewhere between 50 and 70 CE, the Greek physician Dioscorides published *De Materia Medica*, the world's first comprehensive pharmacopeia, describing over 1000 medicinal substances of animal, plant, or mineral origin. Book V of this opus deals with the curative properties of wine, seawater, metal salts, and some 200 different kinds of stones, among them lodestone, marble, hematite, malachite, corundum, and asbestos. *De Materia Medica*, the equivalent of the modern *Physician's Desk Reference*, remained a medical authority for the next 1500 years, translated into Latin, Arabic, and Armenian, and disseminated across Europe and the Middle East. Between 1478 and 1600 it went into at least ninety-six separate printings in eight different languages, many of them elaborate folio editions with woodcut illustrations. Later herbals and lapidaries, building upon Theophrastus, Pliny, and Dioscorides, continued to promote the medical efficacy of minerals. Aetites – probably geodes, but said to be found in storks' and eagles' nests – were recommended to help women in childbirth; an example, set with gold and jewels and stored in a case of cypress wood, was listed in an inventory of the possessions of King Charles V of France in 1379. Glossopetra, or "tongue stones" – most likely fossilized shark teeth – were a treatment of choice for poisonous animal bites; hematite or bloodstone, for hemorrhage (particularly effective, claimed Pliny, if powdered and mixed with pomegranate juice). Some earths are still legitimate constituents of the modern-day pharmacopoeia. Such stomach-settling preparations as Maalox, Mylanta, and Milk of Magnesia all hark back to ancient Greek clays from the province of Magnesia, heavy in magnesium oxide and hydroxide. Rolaids and Tums, also stomach-settling antacids, are primarily calcium carbonate, the main component of chalk; and a major ingredient of kaopectate – a common prophylactic for gastrointestinal disease – is kaolin, a white aluminosilicate clay, also used for porcelain, plaster, bricks, and Portland cement.

Given all this, perhaps it's not surprising that the father of mineralogy was a physician. In 1546, Georgius Agricola, sometimes nicknamed the "Saxon Pliny," then practising medicine in the rich German mining town of Chemnitz, published *De Natura Fossilium* (*On the Nature of Fossils*). In a portrait, Agricola wears a high-collared coat, a pancake-shaped hat tipped at a rakish angle, and a narrow-eyed skeptical expression, as befits a man on the brink of modern science, noted for his emphasis on observation, analysis, and critical thinking. In his last work, published posthumously in 1556, he wrote what is still good advice for the modern investigator: "I have omitted all those things which I have not myself seen, or have not read or heard of from persons upon whom I can rely. That which I have neither seen, nor carefully considered after reading or hearing of it, I have not written about." Agricola, though admirable, is a lot less fun to read than the engaging

Pliny; he's clearly an earth personality: practical and methodical, humorless, but a monument of commonsense. The Latin
Fossilium of his title refers not only to fossils but to anything
dug out of the ground, and thus includes rocks, minerals,
and gemstones, all carefully classified according to such basic
properties as color, taste, odor, place of origin, hardness, shape,
geometrical form, and size. All were shuffled into five broad
mineral categories: "earths," such as chalk and clay; "solidified juices," such as rock salt and sulfur; "compounds," such
as iron pyrite and galena; metals; and gemstones.

Today scientists have identified some 3500 minerals, though
a mere twenty or so of these – mostly silicates – make up the
bulk of the world's rocks. Silica (silicon dioxide, SiO_2) is, in
fact, the most common compound in the Earth's crust. Beach
sand and quartz are almost entirely composed of silica; and
feldspars, which comprise 54 per cent of crustal minerals – the
name feldspar comes from the German for field, practically
any one of which feldspars can be found lying about in – are
primarily aluminum silicate.

Rocks today are commonly categorized not only by chemical
composition, but by mode of formation. The rock that bubbles
out of the rifts at the mid-ocean ridges is igneous rock, from
the Latin *igneus*, or "fiery." Igneous rocks, born of fire, are the
solidified offspring of molten magma from the interior of the
Earth. Those formed from magma spat out in volcanic eruptions
are called extrusive rocks, which are what make up the bulk
of the ocean floor, the dark splotches – the maria or "seas"
– on the face of the Moon, and most of the surface of Iceland.
Intrusive rocks, in contrast, form from magma that never quite
makes it to the top, but instead solidifies underground to form
massive subterranean plugs. A common example of such rock
is granite, a pink or charcoal-gray mix of quartz, feldspar, and
mica that forms the bedrock of the continents, the core of the
Rocky Mountains, and the towering Half Dome of Yosemite
National Park.

Sedimentary rock – from the Latin *sedimentum*, to settle
down or sink – is a more patient construct. Rather than solidi-

fying from superheated sludge, sedimentary rock is built from pebbles, particles, dust, grit, mineral grains, and microscopic scraps of shell that are dropped, washed, or blown into lakes, ponds, seas, deserts, deltas, and rivers. Over millennia, these crumbs, flecks, and fragments accumulate in ever-thickening layers, eventually weighty enough to crush and compress themselves into sandwich-like slabs of rock. Often such rock contains fossils, as ferns, fish, or dinosaurs expire on sand, clay, or bottom muck, eventually to be incorporated into the solidifying whole.

The third major class of rocks – metamorphic rock – is a case of Mother Nature refusing to let well enough alone. The word metamorphosis comes from the ancient Greek and means "to change shape;" in biology, it is the process by which tadpoles develop into frogs and caterpillars into butterflies. In geology, it refers to the processes by which igneous or sedimentary rocks are heated, compressed, contorted, and converted into something else. The four most common kinds of metamorphic rock are the compacted descendants of sedimentary mudstone or shale: slate, phyllite, schist, and gneiss (pronounced "nice"). If rocks were hamburgers, these four would correspond to rare, medium, well-done, and burned to a crisp: slate, in this metamorphic hierarchy, has been exposed to the least amount of heat and pressure, gneiss to the most. Limestone, heat- and pressure-treated, becomes metamorphic marble, the building stone of the Parthenon, the Taj Mahal, Michelangelo's David, and the Washington Monument.

Rocks – whether igneous, sedimentary, or metamorphic – remain rock as long as they're fused to the Earth's crust. Once a chunk ceases to be part of the main, however, the freed fragment – even if it's the size of a Volkswagen bus – becomes a stone. Thus cliffs, mountains, and canyons, the Rock of Gibraltar, and Australia's Ayers Rock – a sandstone behemoth rising out of the Great Australian Desert south of Alice Springs – are all rocks, while the Blarney Stone, the Rosetta Stone, and the sadly misnamed Plymouth Rock are stones. Our word "stone" comes from the Old English *stan* (in

Old Teutonic, *stein*); and stone and its permutations take up a good five and a half pages in the *Oxford English Dictionary*, beginning with definition number one – "a piece of rock" – and proceeding through such stone combos as cornerstone, hearthstone, milestone, tombstone, grindstone, hailstone, lodestone, and gallstone. The vulgar among us, according to definition number 11, use stone as a synonym for testicle; and the quantitative use stone as a measure of weight, equal to 14 pounds. Stone is the comparison of choice for anything unfeeling, unresponsive, or unemotional: the stone-hearted are Dickensian villains who send widows and orphans to the poorhouse; and the stone-faced are the last word in intimidating expressionlessness.

Stones figure in a long list of traditional proverbs. We're all aware, for example, that you can't get blood from a stone and that a rolling stone gathers no moss, and we also all know what people who live in glass houses shouldn't go around throwing. If you've left no stone unturned, you've exhausted every possibility; if you've killed two birds with one stone, you've accomplished two goals with a single action; and if you *are* stoned – at least in this day and age – you're drunk or drugged. To be stoned in ancient times, on the other hand, was a particularly brutal form of corporal punishment. The Old Testament prescribes it for a number of offenses – notably adultery; and the practice continues today in fundamentalist Islamic nations, among them Nigeria, Sudan, and Iran. In the fourteenth century, the legally recalcitrant could also be pressed to death by stones, a treatment known as *peine forte et dure*, used both as a punishment and a means of extracting confession. Victims were laid on the ground and slowly piled with heavy stones until their rib cages were crushed.

Earth is the element of science and technology, art and architecture, and the military–industrial complex. Our first tools were stones; and the long period during which we were fashioning these is commonly known as the Stone Age, a prolonged era usually subdivided by anthropologists and archaeologists into Old, Middle, and New Stone Ages, or – elegantly Latinized – Paleolithic, Mesolithic, and Neolithic periods. The Paleolithic commenced some two and a half million years ago, with the appearance of the first deliberately shaped – as opposed to serendipitously picked up off the ground – stone flakes and choppers. The Mesolithic clicked in sometime around 12,000 BCE with the development of a battery of more sophisticated tools, suitable to the more complex environment of the late Ice Age; and the Neolithic began around 8000 BCE, with the birth of agriculture. Of the species of modern man, the earliest, *Homo habilis*, is named for his/her inventive cleverness with stones: *habilis* means toolmaker.

The most versatile of Stone Age tools was the handaxe, a roughly fist-shaped chunk of shaped stone with a chipped edge, suitable for slicing meat, scraping skins, smashing bones, digging holes, and even chopping down small trees. These were made in such quantities that anthropologists suggest every Stone Age citizen had one of his/her own, in much the same manner that modern business people tote Palm Pilots. The ancient Greeks, finding abandoned handaxes still lying about millennia later, identified them as thunderbolts hurled by Zeus; and medieval Christians hypothesized that the Stone Age axes were missiles left over from the war in Heaven that resulted in the fall of Lucifer.

By the Upper Paleolithic – 40,000 to 12,000 years ago – stone tool technology had expanded to encompass a raft of specialized scrapers, blades, spear- and arrowheads, and serrated harpoon points. It was during this technological boom period that the Cro-Magnon culture (named for the French site where their fossils were first discovered in 1868) flourished in Europe, producing some of the world's earliest art, much of which was painted on stone: charcoal and ochre depictions of bison, mammoths, rhinoceroses, panthers, and bears.

A stone's throw as a measure of length dates at least to the 1300s (when it was commonly known as a stone's cast).

Then, as now, it referred to an unspecified shortish distance, about as far as an average person could hurl an average stone; and the pointy stone, hurled with malevolent intent, is almost certainly the progenitor of modern war. War, most anthropologists argue, is both a near-universal activity, and a very ancient one. The earliest known evidence of organized warfare comes from a 12,000-year-old burial side at Jebel Sahaba in Sudan, where archaeologists unearthed the bodies of fifty-nine men, women, and children, over half riddled with stone spearpoints. Neolithic cave paintings – prehistoric versions of *Guernica* – show armed warriors in battle lines; and the world's oldest cities, Israel's Jericho and Catal Huyuk in Turkey, respectively established around 8500 and 7000 BCE, were both fortified. The famous walls of Jericho, 10 feet (3 m) thick and 13 feet (4 m) tall, are thought to have been built before the settlement's harried inhabitants learned to farm.

By the Neolithic period, the lethal potential of stone-throwing had been greatly increased by the invention of the bow and the even deadlier sling, the weapon with which David reputedly downed Goliath. The properly slung stone could travel twice as far as an arrow and inflict even more damage; slings, loaded with tennisball-sized stones, were standard military equipment in both the Greek and Roman armies, and their use was an essential aspect of training for all would-be legionnaires.

The ancient Greeks are generally credited with elevating stone-throwing to its military peak with the invention of the catapult, a nefarious device capable, depending on its design, of firing Brobdingnagian arrows or hurling stones the size of television sets over distances of 200 yards (180 m). Versions of the catapult, soon a staple of siege warfare, spread quickly from culture to culture. Archimedes used catapults to beat off the Roman invasion of Syracuse in 214 BCE; Alexander the Great used them to conquer the world. England's King Edward I – known from his lanky physique as Edward Longshanks and from his aggressive behavior as the Hammer of the Scots – used them with impressive, though temporary, effect in his interminable battles in Scotland. Legendary among these was a monumental catapult called the Warwolf used at the siege of Stirling Castle in 1304, a weapon so intimidating that the resident Scots – who spent several beseiged weeks contemplating it in the course of its construction – surrendered before it could fire a shot. Edward, furious, refused to accept the surrender, ordered the Scots back into their stronghold, and launched his missile. The first blow reduced the castle walls to rubble.

The last recorded use of a catapult in battle dates to 1521, when Hernando Cortez, having run out of cannonballs, built one to attack Montezuma's Mexico City. He miscalculated the angle of fire; the first rock plummeted back down on the machine and smashed it to bits.

There's something impressive about big stones. Any stone of notable size in a landscape is almost certain to have some human association attached to it: chances are, at one time or another, people worshipped it, camped beside it, met upon it, used it as a landmark, or chiseled their names upon it. The much-disputed Temple Mount in Jerusalem is built upon just such a rock, said to be the very rock upon which Abraham prepared to sacrifice his son Isaac. Australia's cave-pitted Ayers Rock or Uluru, a 1141-foot (348-meter) red sandstone monolith rising out of the central desert, has been sacred for millennia to the aborigines. A massive rock on the coast of Egypt near the city of Matrouh is known as Cleopatra's Rock (the queen, local legend has it, used to swim there); and the bizarre Brimham Rocks of Yorkshire, England – monstrous mushroom-like formations of eroded sandstone scattered across the moors – are mentioned in the Domesday Book. Nebraska's unmissable Chimney Rock, a 325-foot-tall (99-meter-tall) chunk of naked rock with a 120-foot (37-meter) spire (the chimney) poking out of the top, was the most famous landmark on the California–Oregon Trail, mentioned in diaries, journals, and letters by thousands of passing pioneers. Centuries before their time, it was a landmark for the American Indians, who knew it as the Elk's Penis.

Natural stones, towering monumentally over prehistoric terrain, provided inspiration for early humans struggling to erect stone structures of their own. From the fifth millennium BCE, Stone Age humans, equipped with flint axes, deer-antler picks, and shovels made from the shoulder bones of oxen, erected thousands of megaliths throughout Western Europe, many of awesome scope. The most famous of these structures is almost certainly Stonehenge, erected between 1900 and 1600 BCE on Salisbury Plain in Wiltshire, an awesome circle of 30-ton sandstone uprights topped by 20-ton lintel stones. Written legends about it date at least to the twelfth century; and one Henry of Huntingdon, who deemed it the second-best marvel in England, described it as a place "where stones of a wonderful size have been erected after the manner of doorways," though no one "can conceive by which art such great stones have been raised aloft, or why were there constructed." Many, however, were willing to venture a guess, variously attributing it to Merlin, the Druids, the Egyptians, the Romans, the British queen Boudicca, and the Danes. Uther Pendragon and the Emperor Constantine were both said to be buried under it. Diarist John Evelyn, visiting in 1654, called it a "stupendous monument" and tried to break a piece off it with a hammer.

Cathedrals, castles, temples, and tombs are traditionally built of stone. Of the Greeks' designated Seven Wonders of the Ancient World, six were made of stone (the exception, the Colossus of Rhodes, was cast in bronze). Foremost among the six was the Great Pyramid of Giza, built in the twenty-fifth century BCE; when finished, it contained 2,300,000 two-ton limestone blocks and was the tallest manmade structure on Earth, a record not to be broken for 4300 years. Stonehenge, unknown to the Greeks, was dubbed a wonder by later historians, along with such other overlooked stone edifices as the Parthenon, the Colosseum, the Taj Mahal, Cambodia's Angkor Wat, Peru's Machu Picchu, the Moai heads of Easter Island,

and the Great Wall of China. The 788-item UNESCO World Heritage list is rich in stone monuments, though number two on the list, Afghanistan's stone-carved Buddhas of Bamiyan, are now gone, victims of tank fire and explosives, demolished by the Taliban under Mullah Mohammed Omar, who ordered their destruction in 2001.

Barring artillery fire, dynamite, or acid rain, however, stone lasts; our monuments generally survive a good deal longer than we do. Earth is the element of memory; stone is our best souvenir. If something is ephemeral, we say that it is written in wind or water. If we intend something to be permanent, we say confidently that it is written in stone.

My first vision was of metals, dozens of them in every possible form: rods, lumps, cubes, wire, foil, discs, crystals. Most were gray or silver, some had hints of blue or rose. A few had burnished surfaces that shone a faint yellow, and then there were the rich colors of copper and gold.

Oliver Sacks
Uncle Tungsten

Earth is the element of chemistry; researchers these days should wear its symbol on their lab coat pockets. Speculation and imagination were equally spurred by water, air, and fire, but experiment, investigation, and analysis – all the panoply of the modern laboratory – were born of earth, in the form of metals: gold and silver, copper, tin, and iron ore. Alchemy began as a quest for gold. The word alchemy is of uncertain origin – one theory holds that it comes from the ancient Greek *chyma*, referring to the casting of metals; another that it derives from the ancient Egyptian *kemi*, "the black land," where the practice may have originated. All agree that it was the Arabs who added the *al*, which in Arabic means "the."

The flashier and more obvious metals have been known since prehistoric times. Gold, silver, and copper doubtless caught the prehistoric eye because all occur in nature in elemental form and can be found in the shape of gleaming nuggets scattered temptingly and obviously about on the surface of the ground. Copper beads 10,000 years old have been excavated from archaeological digs in northern Iraq; gold necklaces, copper scimitars, and silver vases have been found in 5000-year-old Sumerian grave sites; and the wealth of metal objects found by Howard Carter in Tutankhamen's tomb – including a gold lion, a gold cobra, a gold throne, gold statuettes, gold jewelry, and a spectacular gold and carnelian funeral mask – show that the Egyptians were master metal-workers by the fourteenth century BCE.

Unlike the forthright gold/silver/copper trio, however, most metals are camouflaged and consequentially more difficult to find. Most, left to themselves, react with oxygen in the air to yield unremarkable-looking oxides, deceptive lumps with no apparent redeeming features, not worth the effort – from a Neolithic point of view – to haul home. The step from the discovery of raw metal to the extraction of metals from concealing ores accordingly took several thousand years, and most likely occurred serendipitously when someone dropped a metal-containing rock into a cooking fire. There's no way of knowing what this crucial sample may have been, but some historians suggest it was malachite, a copper ore, which has the curiosity-provoking advantage of being green.

The release of gleaming metals from undistinguished lumps of ore must have looked, to the startled observer, like an act of sheer magic; and early metallurgy was therefore closely linked to the supernatural, with attendant metal-wielding gods. The Greek god of the forge was the crippled Hephaestus (born with his feet facing backward), long-suffering husband of Aphrodite, noted for making Athena's spear and Zeus's thunderbolts. His Roman alter ego was Vulcan, who reputedly plied his trade inside the smoking volcano of Mount Etna. In Norse mythology, the dwarf-like metalsmiths of the gods lived in underground caves, supervised by Odin's son Tyr the One-armed; their specialty was the making of swords. Even the early Christians, though sternly rejecting the magical fantasies of paganism, produced a patron saint of metalworkers: Saint Eligius of France, a skilled smith, elevated to sainthood for his cheerful demeanor, zealous preaching, enthusiastic building of churches, and generosity to the poor.

Gold, according to Plato, who mentions it somewhat confusingly in the *Timaeus*, was an ultra-dense congealed form of water, a view which harks back to Thales and which some geologists have generously suggested may have been based on observations of deposits of mineral precipitates at the mouths of hot springs. Aristotle, who had expanded the four-element theory to encompass qualities of hot, cold, moist, and dry, argued that hot and dry permutations of earth generated rock and stones, while moist aspects made metals; and Theophrastus, in *De Lapidibus*, stated flatly that metals ("such as silver, gold, and so on") came from water, whereas stones ("including the more precious kinds") came from earth. Pliny somewhat begged the question of origin, merely pointing out – after a brief aside to discuss gold-digging ants in India and gold-mining griffins in Scythia – that gold can be either dug out of the ground or washed out of the detritus of rivers.

The prevalence of gold in certain stream- and riverbeds – what modern prospectors call placers – may have reinforced the theory of the watery origin of metals. The ancient Greeks submerged sheepskins in running water to capture the tumbling flecks and particles of gold; these mats of gold-impregnated wool doubtless inspired some prehistoric storyteller to invent the tale of Jason and the Golden Fleece (though the fleece in the story, strictly speaking, was solid gold and came from the back of a flying ram once belonging to the god Hermes). During the two-year period that we lived in Colorado, my husband and I took our children panning for gold. These days on the commodities market gold is worth roughly $350 an ounce; and at that price, we optimistically figured, a mere 62 pounds (28 kg) – an amount that could fit in three pint jars – should pay for three college educations, possibly with a bit left over. The only problem with this scheme is that we failed to follow step one in our pamphlet of gold-panning instructions. "There is no use," the pamphlet stated coldly, "in panning for gold in a stream where there is no gold."

Gold is marvelous stuff, heavy, soft, and lustrous, with a rich dull yellow sheen that gives it its name: *gold* comes from the Old English *gelo*, which means yellow. It's also chemically inert, which means that anything made of it lasts essentially forever: gold bracelets and headdresses emerge from Egyptian pyramids and Greek tumuli as bright and perfect as the day they were buried. Perfect, that is, unless something weighty has been placed upon them, in which case they will squash. Though gold is heavy – a block the size of a footstool weighs half a ton – it is also amazingly malleable: a pellet the size of a grain of rice can be hammered into a tissue-thin sheet over 10 feet square (a meter square) or spun into a fine wire over half a mile (0.8 km) long.

Gold, the seventy-third most abundant element in the Earth's crust, is by no means the rarest of elements, but it's hard to get at, which means that, comparatively speaking, there's not all that much of it around. Scarcity, durability, and general good looks combine to ensure gold a prime position as a status symbol, treasure, and measure of wealth. Ancient history is laden with accounts of conspicuous gold consumption, intended to elevate the mighty and awe the populace. Queen Hatshepsut of Egypt powdered her face with gold dust; King Darius of Persia bathed in a golden tub; and royal crowns, ostentatious representations of state and personal power, invariably were made of gold. Gold has always been much in demand – Pliny, who voiced sanctimonious strictures against people who used gold cups engraved with "libidinous subjects," bemoaned the avarice with which people so indefatigably chased after it.

People also tried to make it. It can be argued that Plato first broached the idea by equating the four elements with polyhedra: earth with the cube; water, the icosahedron; air, the octahedron; and fire, the tetrahedron. By dicing these shapes into triangles and then reassembling them in varying ratios,

Plato argued, one geometrical element could be turned into another, a process known as transmutation. Aristotle also believed that the fundamental elements, by undergoing various physical processes, could segue from one to the other; further-more, he suggested that metals grew in the ground like vegeta-bles, the baser and less valuable among them slowly ripening over time into gold. The idea that base metals had the potential to turn into something far better, and the concomitant realiza-tion that human beings, properly equipped, might be able to expedite the process was the root of alchemy. (Not everybody fell for this: German chemist Georg Stahl (1660–1732) pointed out acidly that modern British tin was still exactly the same stuff that had been mined by the Phoenecians, which meant that either Aristotle was wrong or there was something wrong with tin – "a peculiar kind of addled egg which will not be hard-boiled.")

Alchemy, perhaps the world's most long-lived get-rich-quick scheme, was from its inception a muddle of secrecy and fraud, since the stakes were high. Repeated claims of success and subsequent dismal proof of failure inevitably gave alchemy and its practitioners a bad name. The Roman Emperor Diocletian, suspecting that alchemists were counterfeiting coins, outlawed alchemy and ordered all books on the subject burned. Chaucer portrayed alchemists as shady tricksters in the *Canon's Yeoman's Tale*: the yeoman, the teller of the tale, has spent seven years as servant to the canon, an alchemist. The narrative describes the canon's fraudulent claims of turning mercury and coal into silver, and adds, in a cautionary caveat – clearly indicating that no man is a hero to his laboratory assistant – that "al thyng which that shineth as the gold/Nis nat gold." In *The Inferno*, Dante consigned alchemists to hell; and in 1323, Pope John XXII excommunicated them.

Belief in the alchemical transmutation of elements persisted for centuries. Isaac Newton dabbled in alchemy; and France's Cardinal Richelieu investigated it as a means of producing enough gold to pay the heavy costs of the Thirty Years' War. The foolishly hopeful and financially desperate continued to support alchemical transmutation schemes well into the nineteenth century. In the 1860s, gullible investors lost their shirts on a Hungarian process said to turn bismuth and aluminum into silver; and numerous gold-making schemes were proposed through the end of the century, including one, in 1894, devised by playwright August Strindberg. (Analysis showed his samples to be nothing more than suggestively colored iron compounds.) Even into the 1920s a pair of Dutch professors were still soldiering on, inspired by a report claiming that mercury could be transformed into gold using a modified ultraviolet lamp.

Transmutation was ultimately vindicated, though not in the lucrative fashion that early alchemists had so hopefully envisioned. The secret to the process lay in the atomic nucleus, notably in those untrustworthy nuclei that chemists and physicists deem "unstable." Such unstable nuclei form the centers of radioactive elements, or radionuclides, of which sixty or so are found in nature and well over a thousand have been produced in the laboratory with the inquisitive help of man. The first of these to be characterized was element 88, radium, an intensely radioactive metal that glows a lovely electric blue in the dark. This lethal blue – discoverer Marie Curie innocently compared it to "fairy lights" – was transmutation in action. Radium is radioactive because it possesses an oversized nucleus. Its subatomic fat renders it unstable, compelling it to jettison a two-proton/two-neutron chunk known as an alpha particle. The result is a new element, radon, a colorless and essentially inert gas, still intensely radioactive and the bane of householders and miners, since in nature it seeps from the Earth's bedrock to accumulate in basements and tunnels. Radon, in turn, ejects an alpha particle, transmuting itself into polonium, and this process, known as a radioactive decay or a disintegration series, continues until a final alpha emission produces, ironically enough, the traditional alchemical starting point: element 82, lead.

Transmutation as effected by natural radioactivity runs downhill: that is, larger elements lose particles and turn into smaller elements. The alchemists at least were unwittingly headed in the right direction: lead, element 82, which commonly contains 82 protons and 124 neutrons, is larger than gold, element 79 (79 protons and 118 neutrons). In 1980, Berkeley chemist Glenn Seaborg and colleagues finally successfully transmuted a sample of lead into gold in the particle accelerator at Berkeley's Lawrence Livermore Laboratory. The process cost $10,000 and the resultant gold was worth one-billionth of a cent.

Earth, with its complement of gold, silver, platinum, and precious stones, is the element of wealth; concomitantly, it's the element of trade, profits, taxes, deficits, and debtor's prison. Pirates, if they had had the opportunity to choose a signature element, would have thrown in their lot with earth, as would prospectors, bankers, and Alan Greenspan. Earth is the element of money: of the gold doubloon Ahab nailed to the *Pequod*'s mast, of Judas's thirty pieces of silver, of Billy Bones's moidores and pieces of eight, of "Sing a song of sixpence" and "A penny saved is a penny earned."

Gold's transition from a gaudy medium of ornamentation – gold bathtubs – to a medium of exchange – money – is attributable in part to its sheer heft. Gold is impressively heavy; at 19.3 g/cc, it's nearly twice as dense as lead and nineteen times as dense as water. This means that small amounts of gold, physically, are equivalent in value to a substantial number of goods and services. You can carry a tiny lump of it in your pocket and use it to buy a cow.

Money, according to the *Oxford English Dictionary*, is "current coin; metal stamped in pieces of portable form as a medium of exchange and measure of value." That metal, ideally, is gold or silver, although throughout history any number of baser metals have been pressed into monetary service, among them bronze, tin, lead, copper, nickel, zinc, and iron. The earliest known money on Earth seems to have been gold bars, cast in Egypt around 4000 BCE and stamped with the name of the pharaoh Menes. In Mesopotamia, by 2500 BCE, the going currency was silver, packaged in the form of coils, some delicate spirals as thin as bell wire, and some – fortunes, hoarded by Sumerian Bill Gateses – the size of bedsprings, weighing over a pound (450 g) apiece. Using standard measures of silver – the Mesopotamian shekel, for example, a month's wage for a laborer, weighed one-third of an ounce (8 g), about as much as three thin dimes – the citizens of Ur paid their taxes and their bills, and went on shopping sprees for honey, perfume, wool, linen, harps, trussed ducks, and jewelry inlaid with carnelian or lapis lazuli. Those at socioeconomic rock bottom, too poor for silver, paid their debts with barley.

The *OED* goes on to explain that the term money, by extension, can be applied to "any objects, or any material, serving the same purposes as coin" – barley, for example – as well as, over the course of history, a wildly mixed bag of designated valuables. The Africans and ancient Chinese dealt in cowrie shells; the Fiji islanders in whales' teeth; and the North American Indians in wampum – white and purple beads carved from the shells of quahog clams. Slabs of salt were used as a medium of exchange in Ethiopia and Tibet; the ancient Mexicans traded in cacao beans and copper axe heads; and other cultures have variously fueled their economies with feathers, nails, kettles, hunks of amber, cattle, pigs, vodka (Russia), and tobacco (colonial Virginia). Though workable, all such schemes were ultimately awkward, beset with debates over equivalent weight and quality. The solution was coins.

The world's first coins are generally agreed to have been invented in Lydia, now central Turkey, in the seventh century BCE. They were made of electrum, a lemon-colored alloy of gold and silver, found in nuggets in the silt of the River Pactolus which ran through the Lydian capital city; according to legend, these deposits originated when King Midas bathed in the river to rid himself of the burdensome golden touch. The nuggets, under the stern direction of the state, were converted into round coins, of uniform shape and weight, each bearing an official lion's-head stamp. The Lydian King Croesus, whose name to this day is synonymous with immense wealth, subsequently expanded the idea to a bimetallic currency: a dual issue of both gold and silver coins, the silver representing denominations too small for gold. This clever innovation rendered money much more versatile on an everyday basis. If you wanted to buy the Lydian equivalent of a sandwich, for example, you didn't want to have to pay for it in gold, the Lydian equivalent of a thousand-dollar bill.

By the first century BCE, Publilius Syrus of Rome was already opining that "Money alone sets all the world in motion." Generally such phrases have a condemnatory Scrooge-ish tone. While an inordinate love of money is generally viewed as bad – the Bible refers to it tellingly as "the root of all evil" – its overall performance has been impressive. The invention of cold cash supported the growth of industry, the expansion of trade, the proliferation of specialized professions, and enough leisure time to generate the sciences and the arts. Earth is the element of aspiration and ambition. Most of us are able to do what we do today because somebody, somewhere, has funds.

Supernatural beings are fond of gold. Depending on who or what they are, they bestow gifts of it, spin straw into it, live in palaces made of it, or bury pots of it at the end of the rainbow. Iron, on the other hand, is supernatural anathema. According to Celtic folklore, an iron horseshoe nailed over the door will keep a home safe from mischievous little folk; an iron nail in the pocket is sure-fire insurance against kidnapping by elves; and an iron knife buried beneath the doorstep will ward off witches. In India and Morocco, iron is considered a protection against demons; and in Burma, iron amulets were once worn by rivermen to prevent attacks by crocodiles. Iron was banned from Greek temples – swords, spears, and daggers had to be discarded at the door; and the

Hebrews decreed that no iron tools could be used for the building of altars.

The Greek poet Hesiod in his *Works and Days*, written in the eighth century BCE, described five ages of man, beginning with the ancient age of gold, when men lived like gods, free from "toil and grief" and exempt from the depredations of "miserable age." This was followed by the less god-like age of silver, and the even worse age of bronze, during which men were devoted to "the lamentable works of Ares and deeds of violence." They committed these violent deeds with the help of their signature metal, bronze, an alloy of copper and tin: "Their armor was of bronze, and their houses of bronze, and of bronze were their implements: there was no black iron." There was no black iron in the fourth age either, an age of noble heroes, many of whom, still armed with bronze, fought in the interminable battles of the Trojan War. In the fifth age, however, "would that I were not among the men of the fifth generation," moaned Hesiod, for "now truly is a race of iron, and men never rest from labor and sorrow by day, and from perishing by night; and the gods shall lay sore trouble upon them."

Iron is the fourth most common element in the Earth's crust, accounting for some 5 per cent of the planet's total mass, outstripped only by oxygen (46.6 per cent), silicon (27.7 per cent), and aluminum (8.1 per cent). Today world output amounts to some 500 million tons per year. Over 90 per cent of all metal refined is iron; and iron and its various offspring have more uses than any other metal on the planet.

Prior to about 1000 BCE, however, iron was rare. Its scarcity was a matter of heat, rather than supply: there was plenty of iron ore around, but nobody could manage to get any iron out of it. While copper and its tougher alloy, bronze, melt at approximately 1000°C – a temperature just attainable in a really roaring wood fire – the melting point of iron is 1535°C, well beyond the reach of primitive technologies. The first people to successfully extract iron from its confining ore seem to have been the Hittites, whose substantial empire stretched through what is now Turkey, Iraq, and Syria between 1600 and 700 BCE. Their secret was a mix of enhanced fuel and improved furnace: charcoal was packed around iron ore in an enclosed crucible, and the fire fanned to heroic heat by a bellows frantically pumping air through a narrow inlet port. The result was a black puddle of iron on the furnace floor, a spongy Rorschach blot of impure metal known as a bloom.

Iron, writes Pliny, is "at the same time the most useful and the most fatal instrument in the hand of mankind." Iron, he continues, is used for plowing, planting trees, pruning vineyards, building houses, and cleaving rocks; some nice statues have been made out of it, and iron nails, taken from a tomb and driven into the threshold of a door, prevent nightmares. However, "it is with iron also that wars, murders, and robberies are effected, and this, not only hand to hand, but from a distance even, by the aid of missiles and winged weapons, now launched from engines, now hurled by the human arm, and now furnished with feathery wings. This last I regard as the most criminal artifice that has been devised by the human mind; for, as to bring death upon man with still greater rapidity, we have given wings to iron and taught it to fly." None of this is iron's fault, he adds considerately. It's all the fault of man.

It's certainly not the fault of bloom iron, which is brittle to the point of uselessness. A repetitious and labor-intensive process of heating, hammering, and folding is necessary to remove enough contaminating slag to convert bloom iron to the tougher and more malleable wrought iron; and if the hammering and folding process is repeated long enough, iron becomes even harder and turns into steel.

Iron was earth's signature metal for over three millennia of human history. Iron is swords and spears, battleships and bridge cables, skyscrapers and automobiles, soup cans, screwdrivers, scissors, staples, and sledgehammers, of the sort wielded by John Henry, the steel-driving man. English abounds with iron imagery. We are urged to strike when the iron is hot – a proverb in use at least since the thirteenth

century, meaning then as now to act quickly at an opportune
moment. The Holy Roman Emperor Charles V is credited with
coining the ominous phrase "iron hand in a velvet glove;" and
any prominent person noted for firm convictions or implac-
able behavior is likely to be described in terms of iron. Thus
Otto von Bismarck, the iron-fisted founder of the German
Empire, was called the Iron Chancellor; Margaret Thatcher,
Great Britain's formidable first female prime minister, was
known as the Iron Lady; and Arthur Wellesley, redoubtable
first Duke of Wellington, who defeated Napoleon at the battle
of Waterloo, was called the Iron Duke. Traditionally iron represents power and strength: bodybuilders style themselves "men
of iron," and the quintessential strong man – Superman, who
came to Earth from the distant planet Krypton – was known
as the "Man of Steel," which he was able to bend with his bare
hands. In ancient and medieval alchemical texts, iron was associated with the warlike planet Mars. On the other hand, iron
is also the element of plowshares, pruning hooks, and garden
spades.

Though Earth yields diamonds and rubies, oil and iron, gold
and silver, and immense amounts of building stone, its most
valuable production may well be its humblest. Dirt, for most
of us, is a stunningly underappreciated commodity. The
language of dirt drips with scorn: we all know what kind of
people think dirty thoughts, play dirty tricks, or shamelessly
attempt to wash dirty linen in public; and when journal-
ists scurry about trying to dig up dirt on prominent public
figures, they're not talking about gardening. Dirt, if anything,
is thought to be overly ubiquitous. If something is viewed as
lowly and worthless, we castigate it as "common as dirt."

Of all the Greek elements, claims Peter Warshall in *Whole
Earth* magazine, earth is the "most distant from the intellect
and ethics of contemporary humans." Air and water remain
intensely personal. We continually breathe one and drink the

other, and erupt in vociferous protest when either is compromised. We're all aware of the evils of second-hand cigarette smoke, automobile exhaust, factory smoke, and groundwater contaminants; and certainly there's no missing it when the air smells acrid or the water is full of sludge. On the other hand, in most of the developed world, most of us are divorced from dirt. "There are two spiritual dangers in not owning a farm," writes environmentalist Aldo Leopold. "One is the danger of supposing that breakfast comes from the grocery store and the other that heat comes from the furnace." The great majority of Americans, claims Warshall, have never laid eyes on the soil that grows their food.

To appreciate the complexity of dirt, imagine assembling it from scratch. The first step is to decide just what sort of soil to make. This is no easy trick, since soil scientists, officially known as pedologists, recognize twelve distinct orders of soils. Gelisols are the cold soils, the dirt of Alaska, the Yukon, and the Siberian tundra. These tend to be high in organic carbon since decomposition at low temperatures proceeds very slowly – in other words, anything that dies in a gelisol tends to hang around for a long time. Histosols – the mucky soils of swamps, bogs, and marshes – are also high-carbon soils, typically forming in soggy areas where poor drainage prevents rapid decomposition of dead plants and animals. Peat, a favored fuel of Ireland, is a histosol. Spodosols are acidic soils typically forming on the floors of evergreen forests – Maine, the Pine Tree State, for example, is blanketed with spodosols.

Andisols form from volcanic ash, and are found throughout Iceland, the Aleutian Islands, the American Northwest, Hawaii, and Japan – in fact, all the territory corresponding to the volcanic Pacific Ring of Fire. Andisols tend to be high in iron and are often red in color, as are oxisols, the weathered soils of the tropics. The gaudy red soils in Gauguin's Tahitian paintings are probably oxisols.

Vertisols are rich in clay and will shrink or swell in, respectively, dry or wet periods. The geometric crack patterns that develop in dry ground during hot rainless summers in Texas and eastern Australia are characteristic of vertisols. Even drier are aridosols, the parched soils of Chile and Argentina, western Australia, north Africa, and the American Southwest. Sagebrush, saguaro cactus, and tumbleweed thrive in aridosols.

Ultisols are temperate and tropical forest soils, generally of relatively low fertility, often colored yellow or red from iron oxides. These soils are found throughout Southeast Asia and the southeastern United States; the red earth of Tara, Scarlett O'Hara's plantation in *Gone with the Wind*, was almost certainly an ultisol. Mollisols are the soils of grasslands, the dark fertile soils of the North American Great Plains and the Eurasian steppes. Attila the Hun and his followers galloped across mollisols; and these are the soils that made the Ukraine the "Breadbasket of Europe" and the American Midwest the "Breadbasket of the United States."

Alfisols, like ultisols, are forest soils, but of higher fertility, common throughout Europe, the Horn of Africa, and parts of the American Midwest. Inceptisols are young soils, found worldwide and in a wide range of settings, though notably in mountainous areas – alfisols, for example, are found in the Appalachian Mountains of the American Middle Atlantic states. Entisols are younger yet, recently formed soils as yet relatively undifferentiated, found in mountains and river valleys.

Each of these twelve soil orders has a large and complex family, a populous taxonomy of dirt, including suborders, groups, subgroups, families, and series, each with characteristic allotments of minerals and organic matter. There are, all told, an estimated 20,000 types of soils worldwide. Here in Vermont, the stuff that we mold into mudpies or struggle to scrub off the knees of our overalls isn't just dirt; it's a spodosol, of the Suborder Orthods, Great Group Haplorthods, Typic Subgroup, a member of the Tunbridge Soil Series, with a complex structure and composition all of its own.

Good soil is about 45 per cent minerals, in the form of particles that range in size from coarse sand grains to microscopic specks of clay. The tiniest particles are the most important. In dirt, these exist in a state known to chemists as a colloid – that is, a two-phase system, in which one form of matter is dispersed within another. The word comes from the ancient Greek for "glue." It's simpler than it initially sounds. Ink, blood, toothpaste, and paint, for example, which all consist of small solid particles dispersed in a liquid, are colloids. So is the dollop of whipped cream on a mug of Irish coffee or the foamy head on a glass of beer, in both of which a gaseous phase – air bubbles – is dispersed in a liquid. In dirt, a solid, in the form of small clay particles, is dispersed in water or air. Collectively these tiny flecks possess an enormous surface area – the mineral particles in a single ounce (28 g) of soil, according to one study, have a surface area of 6 acres (2.5 hectares) – and are capable of attaching immense numbers of water molecules as well as ions such as calcium, magnesium, and potassium. Clay, in effect, acts as a slow-release food dispenser, binding water and nutrients, then judiciously doling them out as needed to the roots of growing plants.

About 5–9 per cent of fertile soil is organic matter – flecks of dead leaves, decayed plant and animal remains, and dung – that binds with mineral particles to form aggregates. It's the organic material that gives good dirt the crumbly chocolate-cake texture so prized by gardeners; and it's these organic aggregates that most effectively allow soil to hold water after a rain. If soil contains too much clay, it packs when wet into an airless brick. This makes an excellent building material – 2000-year-old pueblos in the American Southwest are built of mud bricks, and the prime building material of Mesopotamia was mud from the Tigris and Euphrates riverbanks – but is disastrous for budding crops. Fertile soil isn't solid; it's about 50 per cent empty space. The good stuff feels spongy when you step in it.

Soil owes its open structure to traffic. Dirt is crammed with inhabitants, in numbers that put human urban centers to shame. Compared to a mere teaspoonful of dirt, Tokyo – generally considered the largest city in the world, with a population of 34 million people – is as empty as the Moon. A gram of good soil contains up to 10 billion bacteria and 100 million fungi; a square meter is packed with 1000 each of ants, spiders, woodlice, and beetles, 2000 each of earthworms, millipedes, and centipedes; 8000 slugs and snails, 40,000 springtails, 120,000 mites, and 12 million nematodes. All of these, roiling about beneath the ground, eating, mating, and excreting, continually churn and stir the soil, a process known as bioturbation. And some do far more. Take, for example, nitrogen-fixing bacteria.

Nitrogen – element 7 on the Periodic Table – is essential for life, a fundamental building block of proteins and nucleic acids such as DNA. Luckily we're well provided with it – it comprises 78 per cent of the air we breathe – but perversely most living things cannot absorb nitrogen as it appears in gaseous form in the atmosphere. We breathe it in and breathe it right back out again, pristine and untouched. Nitrogen is chemically inert, existing – in free form, in nature – as a diatomic molecule, two atoms connected by a nearly unbreakable triple bond. If nitrogen were a food, it would be a coconut – richly nutritious, but practically impossible to crack.

Before we can use it, nitrogen must be fixed – that is, converted into some usable, absorbable form. Each year small amounts are fixed spectacularly by lightning, which blasts atmospheric nitrogen into pieces and reforms it as ammonia. The bulk, however, is fixed by bacteria. "If all the elephants in Africa were shot, we should barely notice it," states Hans Jenny, the Swiss-born soil conservationist, "but if the nitrogen-fixing bacteria in the soil … were eliminated, most of us would not survive for long because the soil could no longer support us."

"Don't sweat the small stuff" has been a cautionary motto for modern over-stressed society – it means don't drive yourself bonkers over such unimportant annoyances as spilled milk, traffic jams, and misplaced umbrellas. The small stuff, on the other hand, is often where we make our greatest mistakes. Sometimes the small – even the flatly invisible – is

globally important; and failing to respect it can have catastrophic consequences. An impressive example of this, in
experimental microcosm, was the unexpectedly tragic history
of Biosphere 2.

Biosphere 2 is the world's biggest greenhouse. The immense
glass-and-steel-frame edifice, designed by Peter Pearce, a
student of Buckminster Fuller, covers three acres (a bit over
a hectare) in the Arizona desert at the foothills of the Santa
Catalina Mountains, 20 miles (32 km) north of Tucson. The
$200-million project, funded by Texas oil billionaire Edward
P. Bass, was intended to give scientists new insights into
the operations of the Earth's biosphere – that is, the big one,
Biosphere 1 – and to establish guidelines for the construction
of biosphere-like habitats for space colonists on the Moon or
Mars.

For two years, between 1991 and 1993, its glittering glass
panels enclosed a population of eight biospherians, four
men and four women, committed to living off their elaborately designed faux land. Described as part Noah's Ark, part
terrarium, Biosphere 2 featured a landscape of seven varied
biomes, among them a tropical rainforest, a savannah, a
marsh, a desert, and an ocean, simulated by a million-gallon
(14-million-liter) saltwater aquarium, variously inhabited by
over 3000 species of plants, animals, fish, and insects. Soils
were mixed for each mini-environment, using thirty different
recipes: a total of 30,000 tons of carefully crafted dirt, covering
the concrete foundation anywhere from 6 inches to 20 feet
(15 cm to 6 m) deep. Misting machines provided humidity; a
system of pipes, valves, and storage tanks recycled water from
the wilderness into a stream, pool, and artificial waterfall. The
hope of the project's designers was that in time the miniature
ecosystems would interact and self-regulate, reaching an
Eden-like homeostasis comparable to, or perhaps even better
than, that of the outer planet.

Instead, all went wildly awry. First the Biosphere's carbon
dioxide level skyrocketed to twice the normal atmospheric
concentration, the result, investigators hypothesized, of

rapidly multiplying and metabolizing soil bacteria. Oxygen dropped to dangerously low levels; nitrous oxide levels rose. Biosphere 2 grew unpleasantly stuffy. Most of the animals and insects rapidly became extinct, with the exception of cockroaches, katydids, and a hyperactive species of Arizona greenhouse ants known as crazy ants. Crops failed; and the biospherians got skinny. Supplementary oxygen, pumped in from the outside, failed to reverse the trend. Biosphere 2, rather than attaining harmonious environmental balance, had reached a state of environmental meltdown.

Nature is complicated. The soil that contributed so dramatically to Biosphere's downfall is unfathomably complex; and things haven't changed all that much since Leonardo da Vinci wrote in the sixteenth century, "We know more about the movement of celestial bodies than about the soil underfoot." The subject interested no less a personage than Charles Darwin, who, beginning in 1837, spent decades studying the formation of soil. It was, in fact, the subject of his last book, published in 1881, titled *The Formation of Vegetable Mould, through the Action of Worms, with Observations of Their Habits*. It outsold *Origin of Species*.

Depending on place and circumstances, it can take a good 500 years or more to make an inch of soil, beginning with the slow erosion of bedrock by wind and water. Beset by weather, rock is gradually ground down to pebbles, sand, and clay. Plants take hold, slowly splintering the rocky substrate with their roots; dying and adding organic matter to the mix. Microorganisms flourish; animals arrive. Worms alone, munching their way daily through their own weight in organic junk and dirt, can, according to Darwin, add a fifth of an inch (0.5 cm) to topsoil in a year. That increment is primarily in the form of castings, worm end-products, squiggly little piles like the squeezings from miniature toothpaste tubes, delicately deposited on the soil's surface. Gradually mixed into the earth, castings improve soil structure; impressively enriched in nitrogen, phosphorus, and potassium, they are also a lush source of food for plants. Soon the system flourishes with life, burning carbon, breathing out carbon dioxide, and generating energy. An acre of soil, writes William Bryant Logan in his master work, *Dirt*, puts out one horsepower of energy per acre every day.

How much of this is there on the planet? A children's exercise demonstrates with an apple. Cut the apple in quarters and set three aside. These represent the three-quarters of the globe occupied by the world's oceans. Slice the remaining quarter in half. One half represents land too inhospitable to be inhabited by humans: the frigid wastes of Antarctica, waterless deserts, the steep slopes and peaks of mountains. Divide the other half – now representing one-eighth of the globe, the portion lived upon by people – into four pieces. Set three of these aside: these are all unsuitable for growing food, being too cold, too rocky, too wet, or too inaccessible, having been utilized for houses, roads, and other space-consuming aspects of urban development. The final slice – just one-thirty-second of the Earth's total area – represents arable land, the portion of the globe that all six billion of us depend upon for food. Peel it. Fertile soil is a thin skin, in most places no more than 5 feet (1.5 m) deep. "A cloak of loose, soft material, held to the Earth's hard surface by gravity, is all that lies between life and lifelessness," writes Wallace Fuller in *Soils of the Desert Southwest*.

Fertile soil – that one thin fraction of the global apple – is the true earth mother. And, should there be any doubt, history shows us exactly what happens if we lose her. Malignant combinations of climatic change and ill-considered farming practices are death to fertile soil; and the archaeological record is a dreary progression of what happens to one's civilization when the two intersect. Soil erosion and exhaustion contributed to the fall of Greece, the collapse of the Mayans, the disappearance of the ancient Pueblo culture of the American Southwest, and the devastation of the Great Plains Dust Bowl of the 1930s. Today desertification, urged along by overcultivation, overgrazing, and deforestation, is rampant worldwide, expanding at the rate of 30 million acres (12 million hectares) per year.

In Frank Herbert's science-fiction saga *Dune*, the action takes place on Arrakis, a desert planet so fearsomely arid that even a single tear is precious. Except for the presence of locomotive-sized giant sandworms, Arrakis, a vast expanse of rock, sand, and wandering dunes, could be the model for any of the Earth's great deserts. Earth's largest, the Sahara of north Africa, a swathe of sand the size of the continental United States, covers over 3.5 million square miles (9 million sq km). Runner-up, the Arabian Desert, occupies 690,000 square miles (1.8 million sq km); and number three, the Gobi of Mongolia, 500,000 square miles (1.3 million sq km), about twice the area of the state of Texas.

The name Gobi means "waterless place," which is appropriate since lack of water is the defining feature of deserts. Deserts may be hot or cold, located on coasts or in the centers of continents, spotted with specialized vegetation or barren seas of sand, but to be a desert you have to be dry. Less than 10 inches (25 cm) of rain a year qualifies a region for desertdom; and some deserts receive only a fraction of that amount. Antarctica, for example, the world's driest continent, barely gets two inches (5 cm) of annual precipitation.

According to the United States Geological Survey, about one-third of Earth's land surface is desert. Some deserts are the result of land being in the wrong place at the wrong time. Anything in the trade wind belt, for example, has a good chance of being a desert. This crucial territory is located approximately between 15° and 35° latitude, north and south, centered respectively over the Tropics of Cancer and Capricorn. Hot air rising over the equator, like a massive plume of steam, cools as it rises to higher altitudes, condenses, and drops its moisture as rain, feeding the lush growth of the equatorial rainforests. The cool air, stripped now of its moisture, then begins to fall, typically bottoming out at about 30° north or south. The result is a determined dry wind, beating cloud cover away, exposing the land beneath it to the pitiless rays of the sun. The same steady winds that propelled sixteenth-century merchant ships to the rich New World also created the blazing sands of the Sahara.

Deserts, furthermore, don't always stay put. Each year several hundred million tons of African desert dust are hoisted into the atmosphere by summer storms and blown west across the Atlantic to coat Central and South America and the islands of the Caribbean. The dust, with its attendant load of bacteria, fungi, and viruses – researchers estimate some 10 billion microbes per ton of dust – is thought to be the culprit behind the recent demise of Caribbean coral reefs, the decimation of certain marine organisms – among them the delicate sea fan and *Diadema antillarium*, the long-spined sea urchin – and increased incidences of respiratory diseases and allergies in humans. Spring winds also whip up dust from Asia's Gobi and Takla Makan deserts, swirling it into monstrous yellow-brown clouds over 1200 miles (2000 km) long. Such clouds have staying power – capable of crossing the Pacific Ocean, North America, and a good stretch of the Atlantic, raining debris and pollutants as they go, before finally dissipating.

An ancient Chinese saying tells us that "Soil is the mother of all things." The converse may be the Bible's ominous "Dust thou art, and unto dust thou shalt return" – which makes it abundantly clear where, if we're not careful, we're going.

Many religious traditions offer a hope of something other than dust beyond the grave; and the belief in an appealing afterlife was a prominent factor in earth burial – that is, interment of the dead in the ground. Burial of the dead may pre-date our species: the first to practise it seem to have been the Neanderthals, who inhabited Europe and Eurasia for over 150,000 years, vanishing around 30,000 BCE. Other than the suggestive act of burial, little evidence survives to reveal the Neanderthal concept of a life hereafter, if there was one; the burials of the Cro-Magnon people, however – our own ancestors, who wandered into Europe between 30,000 and 40,000 years ago – clearly indicate an expectation that the departed were going somewhere where they'd need supplies. Paleolithic graves contain food, stone weapons, mammoth-ivory beads, bracelets, and animal figurines. Bronze Age Greeks were interred with swords, spears, armor, and painted pots; Sumerian royalty with gold vessels, jewelry, and dozens of slaughtered courtiers and soldiers, destined to continue their service in the world beyond the grave. The North American Hopewell people, who dominated the Mississippi valley from 200 BCE to 400 CE, furnished their elaborate burial mounds with pearl beads, copper bracelets, stone atalatl weights, sea-snail cups, pipes, and grizzly bear teeth. The ancient Egyptians, obsessively concerned with provisioning themselves for life after death, spent much of their time on Earth industriously preparing, as evidenced by the fabulous and comprehensive contents of the pyramids and the hidden tombs in the rocky cliffs of the Valley of the Kings.

Burial, as well as thoughtfully providing the deceased with the wherewithal for continued existence, also protected the living from potentially angry ghosts. Fear of the dead was a governing factor in many early burial practices. In Rome, for example, burials, by governmental decree, took place outside the city walls, which is why all ancient roads leading to and from Rome are lined with tombs. The angry souls of the dead – referred to by the Romans as *Lemures* – were terrifying beings, so frightening that propitiatory feasts were held in their honor each May. During these feasts, temples were closed, marriages were forbidden, and the particularly nervous threw black beans on graves in the belief that restless spirits, who reportedly found beans unbearable, could thus be banished. According to legend, these May festivities were initiated by Romulus, co-founder of Rome, who, having murdered his twin brother Remus, was understandably apprehensive about the return of his furious ghost.

Spirits also walked if bodies were improperly buried; thus

funeral rites, to the ancients, were not only a mark of respect, but a security measure. In Sophocles's *Ajax*, written in the mid-fifth century CE, Ajax, a hero of the Trojan War, goes berserk and attempts to slaughter a number of his erstwhile companions, then, in a fit of shame and remorse, commits suicide. Menelaus, appalled by such treachery, wants to leave the corpse where it lies to be eaten by birds, but Ajax's brother Teucer insists that Ajax have a proper burial. Not only is it a moral obligation, he points out, but it's a sensible precaution: those who "act wickedly towards the dead" will live to rue the day. In another version of the same tale, it's the canny Odysseus who urges burial, intimating that to do otherwise would be to risk the safety of the entire Greek army.

With the rise of Christianity, proper burial came to be seen as a prerequisite for resurrection: many believed that those buried in the wrong place or, worse, not buried at all, would be left behind on the Day of Judgement when the graves gave up their dead. This was such a hideous prospect that the relatives of those denied burial in consecrated ground went to great lengths to reverse the Church's decision: Philippe Aries in *The Hour of Our Death* cites the case of an excommunicated cleric whose body spent eighty years in storage in a lead coffin while his family sued for clemency.

"This be the verse you grave for me," wrote Robert Louis Stevenson in his poem "Requiem," which indeed became his epitaph. "Here he lies where he longed to be/Home is the sailor, home from the sea/And the hunter home from the hill." We come full circle. Earth, at last, is again a mother; at the end, we speak of coming home.

Part VI

We Are the Elements

I yam what I yam.

Popeye

Socrates's injunction "Know thyself" is tougher than it sounds. Knowing ourselves, much less our families, friends, and neighbors, is a colossal endeavor, the magnitude of which explains why psychology is such an acrimonious discipline today. What makes us all the intelligent, insightful and delightful people we hope we are is a dauntingly complex mix of genetic, physiological, psychological, evolutionary, and environmental factors, melded in some unspecified fashion to make each of us unique. Unique, of course, is relative: even the quirkiest among us isn't entirely odd. For all our differences, humans share many behavioral, emotional, and cognitive traits which predict, more or less, how we're likely to learn, adapt to changing environments, and interact in social situations. Such traits allow psychologists to categorize our personalities into related groups, based on our personal preferences – for example, our avoidance of solitude or crowds, our fondness for eating or exercise, and our predilections for problem-solving by emotional intuition or intellectual analysis.

Personality study, though new in its present incarnation, dates back at least to the ancient Greeks. The Greek four elements, historically, were not only the fundamental substances of matter, but also the raw material of human nature. The idea of the elements as determinants of behavior was an outgrowth of the theory of the four essential body fluids, or humors, proposed by Hippocrates. These were phlegm, associated with the element water; blood, associated with air; yellow bile, with fire; and black bile, with earth. One's dominant humor was believed to determine one's personality type, a theory first mentioned in *On the Nature of Man*, a work by Hippocrates's son-in-law, Polybus, and expanded in the second century CE by Galen in *De Temperamentis*, or *On Temperament*. In the Polybus/Galen scheme, the element earth was a psychological downer: an individual high in black bile was of melancholic disposition, droopy and dismal, the human equivalent of Eeyore, the mournful donkey in *Winnie-the-Pooh*. Elevated yellow bile (fire) rendered its possessor choleric, violent, vengeful, and prone to fits of anger – the sort of disposition that led John McEnroe to throw his tennis racket and Mike Tyson to bite off a chunk of Evander Holyfield's ear. (According to Polybus, the tantrum-throwing Achilles had too much yellow bile.) High blood (air) engendered a sanguine temperament, and produced amorous, cheerful, and generous people – Falstaff, for example, with a touch of Santa Claus; and phlegm (water) made for phlegmatic dispositions – dull, pale, timorous types, on the order of Uriah Heep.

Galen's theory of personality was accepted as dogma in Western medicine until well into the seventeenth century. Robert Burton's *Anatomy of Melancholy*, published in 1621, still ascribes a melancholic disposition to an imbalance among the humors – though Burton's view of mental illness is more complex, and melancholies are also blamed on evil angels, unfortunate heredity, old age, the loss of a loved one, and spicy food. By the twentieth century, the humors had vanished from medicine, but psychological theory continued to reflect ancient Greek constructs. In Carl Jung's *Psychological Types*,

for example, published in 1921, his famous four basic personality functions – feeling, thinking, intuition, and sensation
(each with introverted or extroverted aspects) – echo water,
air, fire, and earth. William Sheldon's *Varieties of Temperament*
(1942) cuts the personality categories to three: viscerotonics,
who are fat, friendly people, fond of kids and parties; soma-
totonics, competitive athletic types, the sort of people who
drive too fast and shout everybody down in meetings; and
cerebrotonics, shy, skinny, nervous persons who spend a lot
of time in their rooms. New Agers today claim that the four
elements are still a natural part of our mental make-up, though
in each of us only one predominates. Water people, in New
Age terms, are compassionate, sensitive, and empathetic. Air
people are rational and bookish, but deficient in commonsense – Sheldon's cerebrotonics, in other words: absentminded
professors and computer geeks. Fire people are dynamic and
flamboyant, as in rock stars, televangelists, and charismatic
politicians; and earth personalities – viscerotonics all – are
homebodies, conservative couch potatoes, and good cooks.

None of this quite agrees with Polybus and Galen; and
much of it is stereotyped and simplistic. There are fat people
who are not in the least jolly or fond of children – W. C. Fields
leaps to mind; football jocks who are shy and read Proust; and
intellectuals who make wicked chocolate cheesecake. Little of
this applies to professional personality assessment today, now
a much more sophisticated and mathematically demanding
field. Still, there's a lurking appeal to the Greek view, to the
idea that a single one-word answer can tell us something
about what we are. Lured, I took an online quiz to identify my
defining element.

It was air.

While the nature of our psychological elements is open to
debate, the nature of our material elements isn't. Chemically
speaking, we know just what we're made of. "We stand on the

elements, we eat the elements, we *are* the elements," writes chemist P. W. Atkins in *The Periodic Kingdom*. We are indeed, though, as elements go, ours are stunningly atypical. Of all the known elements, the universe consists almost entirely of just two – hydrogen and helium – which respectively constitute about 90 per cent and 9 per cent of the whole. All the rest, from the carbon so integral to organic molecules to the copper that edged us out of the Stone Age, the iron that fueled the Industrial Revolution, and the silicon that sent us into cyberspace, are chemical rarities. Our lives depend on the residual 1 per cent, the elemental needles in the cosmic haystack.

A diamond consists of just one element; table salt, two; sugar, three. A cell phone contains forty-two. In shaming contrast, human beings are cobbled together from a mere thirty – predominately oxygen (65 per cent), carbon (18 per cent), hydrogen (10 per cent), and nitrogen (3 per cent), plus assorted bit players such as calcium, phosphorus, potassium, and sulfur. Boil us down to our elemental constituents and we're worth about $4.50 max, according to one estimate, and some cynics price us out at less than a dollar. This seems economically shortsighted, since our personal elements – except for the plebeian hydrogen – are so exceptionally rare. They're also of impressive pedigree. We are the stuff of stars.

Our Sun, though it consists primarily of the inevitable hydrogen and helium, also contains a smattering of heavier elements: oxygen, carbon, nitrogen, silicon, magnesium, neon, iron, and sulfur. Their telltale presence indicates that the Sun is at least a second-generation star. Our entire Solar System is in effect a junkyard project, recycled and reconstituted from the remains of an ancient supernova. While the matter of the Universe is generally believed to have been dispersed some 14 billion years ago by a spectacular event known prosaically as the Big Bang, the elements in our particular segment of it are the result of a smaller secondary boom.

The demise of stars is ultimately the result of depleted fuel. Stars ordinarily run on hydrogen, converting it, via nuclear fusion, into helium, with concomitant massive release of energy. As supplies of hydrogen are exhausted, however, the star begins to sputter to a halt. The drop in energy output initially causes the stellar core to contract; simultaneously the star's outer layers expand, ballooning outward to form a bloated monstrosity known as a red giant. Contraction increases temperatures at the core from 10 million to 100 million degrees C, which blazing boost allows the star to burn its helium, fusing it to form beryllium, then carbon, then oxygen. A Sun-sized star from here has hit its celestial glass ceiling and can go no further; its outer layers dissipate, leaving behind a solid core of slowly cooling crystallized carbon. In larger stars, however – those three times the size of the Sun or more – greater mass allows the core to contract still further, raising temperatures to the point where the now desperate star is able to fuse its reserves of carbon. The process continues – a runaway gallop into the grave – as the star, in increasingly rapid succession, creates new and heavier elements, and then exhausts each new fuel. At last, with one final brief Herculean effort, it fuses silicon, producing iron.

Iron, in stellar terms, is the cup of hemlock, the coup de grâce, and the end of the line. At this point, in cross-section, the star looks like a fiery onion, its iron middle surrounded by concentric layers in which other fusion reactions – making other elements – soldier on. Iron, however, is too stable for the star to fuse. Now deprived of energy and crushed inward by its own gravitational force, the star collapses. Internal temperature rockets to billions of degrees; the onion-like layers, instantaneously compressed, slam violently into the core and bounce back again, flinging their contents into space in a blinding explosion as bright as a billion suns. In the course of the star's spectacular death throes, violent atomic collisions generate even heavier elements – anything heavier than iron, in fact, is a product of this awesome stellar swan song.

The force of the blast propels the elements en masse into space, chemical shrapnel carried on a glowing wave of gas. Supernovas, astronomers estimate, occur at a rate of about two per century per galaxy, and each, like a vast conflagratory mushroom showering spores, seeds its surroundings with the raw material of life. We are, quite literally, stardust, the products of cosmic infernos, children of nebulas. It's a wondrous and briefly enobling thought. Briefly, that is, because a moment's reflection reveals that our glorious origin is shared by slugs, slime molds, and driveway gravel.

In the millennia since Thales of Miletus posited a world made wholly of water, the elements have been counted, named, weighed, numbered, analyzed, and, in particularly determined laboratories, created. The Periodic Table continues to expand; in early 2004, researchers from California's Lawrence Livermore National Laboratories and Russia's Joint Institute for Nuclear Research spent a month peppering americium with calcium atoms, finally generating four atoms of two brand-new elements: element 115, which persisted for 1/100,000 of a second, and then spat out an alpha particle to form element 113. Our elucidation of the nature of the elements allows us, with ever-increasing precision, to understand the world around us. So why do the clumsy and clearly non-elemental four continue to resonate in the human mind?

In a sense, it's a matter of time and tradition. The four, after all, dominated human thought for over 2000 years. Over millennia, they acquired a rich array of images and associations, and a symbolism all their own. "The four elements are not a conception of much use to modern chemistry – that is, they are not the elements of nature," writes critic Northrop Frye, but "earth, air, water and fire are still the four elements of imaginative experience, and always will be." Perhaps the closest analogy is our attitude toward heliocentrism. Since Copernicus (on his deathbed) in 1543 published *De revolutionibus orbium coelestium* (*Concerning the Revolution of the Heavenly Spheres*), we've known that the Earth is not the center of the universe, but merely one of several insignificant planets orbiting a rather mediocre central star. The Sun doesn't circle us; instead, we, sheeplike, circle it, whirling like a dervish as we go, though you wouldn't know it to hear us talk. We still speak of sunrise and sunset; and images of the Sun's passage across the sky are embedded in our collective vocabularies. Our symbols change slowly. After decades of keypads, we still claim to be dialing the telephone.

Or perhaps we don't so much alter our old symbols as incorporate and embellish the new. The Periodic Table, once the esoteric tool of a handful of progressive chemists, is an increasingly familiar image. It appears on T-shirts and coffee mugs; it has been touted as abstract art. It has become an icon of classification: we now have Periodic Tables of music, dance, pasta, candy, vegetables, condiments, desserts, animals, presidents, beer, and college basketball. There are illustrated and animated Tables, comic book Tables, and tongue-in-cheek Tables of Rejected Elements and Subversive Elements. Humorist Tom Lehrer has set the entire Table to music, to the tune of Gilbert and Sullivan's "I Am the Very Model of a Modern Major General." In 2002, Theodore Gray won an Ig-Nobel Prize for building a literal Periodic Table in walnut.

Still, the Periodic Table is new. The four elements are our past. Just as literature grew from ancient fireside tales of heroes, talking animals, and children lost in magical woods, so the intricate edifice of modern science grew from the fundamental four. Water, air, fire, and earth exert their influence at the mental border where the inner world of fantasy and imagination meets the outer world of fact. Our perspective still shifts between the two, wavering from subatomic particle to water sprite and back again – different ways of seeing, just as an optical illusion can show us first a goblet, next a pair of human profiles. Pair the Periodic Table and the classic Greek four and we see what Carl Sagan meant by science's "marriage of skepticism and wonder" – an analytical eye coupled with a sense of awe, and a respect for the long road we've traveled to get us where we are today.

The four elements are also the template through which

we view the world. In Thornton Wilder's play *Our Town*, set in Grover's Corners, New Hampshire, in which all the joy and tragedy of life is encapsulated in three acts, Emily Gibbs – newly dead – cries out, "Do human beings ever realize life while they live it? Every, every minute?" The answer, replies the omniscient Stage Manager bluntly, is no. Only one person in a million, wrote Thoreau, is truly awake to the wonders of daily living. "Few adults can see nature," said Ralph Waldo Emerson. Perhaps they're right. But I prefer to think that the sheer marvels of the planet – of water, fire, air, and earth – occasionally flip the blindfolds off our preoccupied eyes. We are the species, after all, that sometimes manages to see the world in a grain of sand and heaven in a wildflower, to see every common bush afire with God. Sometimes, for all the sorrow and pain the Earth dishes out, we see what a peach of a planet this is and get a glimpse of how lucky we are to be here.

We see, in fact, what the Greek natural philosophers saw.

Sky and ocean, sun and stars. Rain and rocks and mountains.

Major Sources

Ackerman, Diane. *A Natural History of the Senses*. Random House, 1990.

Agosta, William. *Bombardier Beetles and Fever Trees: A Close-up Look at Chemical Warfare and Signals in Plants and Animals*. Addison-Wesley, 1996.

Allegre, Claude. *From Stone to Star: A View of Modern Geology*. Harvard University Press, 1992.

Allegre, Claude J., and Stephen H. Schneider. "The Evolution of the Earth." *Scientific American*, October 1994, pp. 6–11.

Allerman, J. E., and B. T. Mossman. "Asbestos Revisited." *Scientific American*, July 1997, pp. 70–75.

Amato, Ivan. *Stuff: The Materials the World is Made of*. Basic Books, 1997.

Aries, Philippe. *The Hour of Our Death*. Alfred A. Knopf, 1981.

Armbruster, Peter, and Fritz Peter Hessberger. "Making New Elements." *Scientific American*, September 1998, pp. 72–7.

Asimov, Isaac. *On Chemistry*. Doubleday, 1974.

Asimov, Isaac. *The Search for the Elements*. Basic Books, 1962.

Atkins, P. W. *The Periodic Kingdom*. Basic Books, 1995.

Bachelard, Gaston. *The Psychoanalysis of Fire*. Beacon Press, 1964.

Ball, Philip. *The Ingredients: A Guided Tour of the Elements.*
Oxford University Press, 2002.

Ball, Philip. *Life's Matrix.* Farrar, Straus & Giroux, 1999.

Barash, David P. "Mountain Climbing" in *The Great Outdoors.*
Carol Communications, 1989.

Barber, Lynn. *The Heyday of Natural History.* Doubleday, 1980.

Bernstein, Peter L. *The Power of Gold: The History of an
Obsession.* John Wiley & Sons, 2000.

Broad, William J. *The Universe Below.* Simon & Schuster, 1997.

Brock, William H. *The Norton History of Chemistry.* W.W.
Norton, 1992.

Burr, Chandler. *The Emperor of Scent.* Random House, 2003.

Calvin, William H. *The Ascent of Mind: Ice Ages and the
Evolution of Intelligence.* Bantam, 1990.

Calvin, William H. *A Brain for All Seasons.* University of
Chicago Press, 2002.

Calvin, William H. "The Great Climate Flip-Flop." *Atlantic
Monthly,* January 1998, pp. 47–64.

Carson, Rachel. *The Sea Around Us.* The New American
Library, 1961.

"Celebrating Soil." Various authors; *Whole Earth,* Spring 1999.

Ceram, C.W. *Gods, Graves, and Scholars.* Alfred A. Knopf, 1954.

Chaisson, Eric, and Steve McMillan. *Astronomy Today.*
Prentice Hall, 1996.

Chown, Marcus. "The Last Supper." *New Scientist,* 13
November 1999, pp. 44-50.

Conway, W. Fred. *Firefighting Lore.* Fire Buff House
Publishers, 1994.

Cox, Tony. "Origin of the Chemical Elements." *New Scientist,*
Inside Science, 3 February 1990.

David, Elizabeth. *Harvest of the Cold Months: The Social History
of Ice and Ices.* Viking, 1995.

Davies, Glyn. *A History of Money: From Ancient Times to the
Present Day.* University of Wales Press, 2002.

DeBlieu, Jan. *Wind: How the Flow of Air has Shaped Life, Myth,
and Land.* Houghton Mifflin Company, 1998.

De Duve, Christian. "The Beginnings of Life on Earth." *American Scientist*, September/October 1995.

Dennis, Jerry. *The Bird in the Waterfall: A Natural History of Oceans, Rivers, and Lakes*. HarperCollins, 1996.

Denny, Mark W. *Air and Water: The Biology and Physics of Life's Media*. Princeton University Press, 1995.

Dietrich, R.V. *Stones: Their Collection, Identification, and Uses*. GeoScience Press, 1987.

Djerassi, Carl, and Roald Hoffman. *Oxygen*. Wiley-VCH, 2001.

Edinger, James G. *Watching for the Wind*. Anchor Books, 1967.

Ehrlich, Gretel. *A Match to the Heart: One Woman's Story of Being Struck by Lightning*. Penguin, 1994.

Emsley, John. *Molecules at an Exhibition*. Oxford University Press, 1998.

Emsley, John. *Nature's Building Blocks: An A–Z Guide to the Elements*. Oxford University Press, 2001.

Emsley, John. *The 13th Element: The Sordid Tale of Murder, Fire, and Phosphorus*. John Wiley & Sons, 2000.

Fagan, Brian. *Floods, Famines, and Emperors: El Niño and the Fate of Civilizations*. Basic Books, 1999.

Faraday, Michael. "The Chemical History of a Candle" (1860) in *The Modern History Sourcebook*, www.fordham.edu/halsall.

Farmer, Jack D. "Hydrothermal Systems: Doors to Early Biosphere Evolution." *GSA Today*, July 2000.

Ferrill, Arther. *The Origins of War: From the Stone Age to Alexander the Great*. Westview Press, 1997.

Fisher, P. J. *The Science of Gems*. Charles Scribner's Sons, 1966.

Fortey, Richard. *Life: A Natural History of the First Four Billion Years of Life on Earth*. Alfred A. Knopf, 1997.

Frank, Louis A. *The Big Splash*. Birch Lane Press, 1990.

Frazer, Sir James George. *The Golden Bough: A Study in Magic and Religion*. The Macmillan Company, 1958.

Freuchen, Peter. *Peter Freuchen's Book of the Seven Seas*. Julian Messner, 1958.

Gies, Frances & Joseph. *Cathedral, Forge, and Waterwheel: Technology and Invention in the Middle Ages.* HarperPerennial, 1995.

Glassman, James K. "Dihydrogen Monoxide: Unrecognized Killer." *Washington Post,* 21 October 1997.

Gohau, Gabriel. *A History of Geology.* Rutgers University Press, 1991.

Gould, Stephen Jay. *Hen's Teeth and Horse's Toes.* W.W. Norton & Company, 1983.

Gould, Stephen Jay. *The Lying Stones of Marrakech.* Three Rivers Press, 2000.

Gould, Stephen Jay. *Wonderful Life: The Burgess Shale and the Nature of History.* W.W. Norton & Company, 1989.

Gray, Harry B., John D. Simon, and William C. Trogler. *Braving the Elements.* University Science Books, 1995.

Green, Harvey. *Fit for America: Health, Fitness, Sport & American Society.* Pantheon Books, 1986.

Greenaway, Frank. *John Dalton and the Atom.* Cornell University Press, 1966.

Hardy, Alister C. "Was man more aquatic in the past?" *New Scientist,* 17 March 1960, pp. 642–5.

Harris, Marvin. *Cows, Pigs, Wars, and Witches: The Riddles of Culture.* Random House, 1974.

Hawkings, Gerald. *Stonehenge Decoded.* Fontana/Collins, 1972.

Hazen, Margaret Hindle, and Robert M. Hazen. *Keepers of the Flame: The Role of Fire in American Culture 1775—1825.* Princeton University Press, 1992.

Heiserman, David. *Exploring Chemical Elements and their Compounds.* TAD Books, 1992.

Hollenbach, D. F., and J. M. Herndon. "Deep-earth reactor: nuclear fission, helium, and the geomagnetic field." *Proceedings of the National Academy of Sciences* (USA): 98 (20) 2001, pp. 11085–90.

Holmes, Hannah. *The Secret Life of Dust.* John Wiley & Sons, 2001.

Huxley, Aldous. *Plant and Planet.* Viking Press, 1975.

Jay, Ricky. *Learned Pigs and Fireproof Women*. Villard Books, 1986.

Jellinek, Paul. *The Psychological Basis of Perfumery*. Thomson Publishing, 1997.

Judson, Sheldon, Marvin E. Kauffman, and L. Don Leet. *Physical Geology*. Prentice Hall, 1987.

Kerr, Richard. "Earth Seems to Hum along with the Wind." *Science* 283: 321, 1999.

Kious, W. Jacquelyne and Robert I. Tilling. "This Dynamic Earth: The Story of Plate Tectonics." U.S. Government Printing Office, 1996.

Kocsis, Richard N. "Arson: Exploring Motives and Possible Solutions." *Trends & Issues in Crime and Criminal Justice*, August 2002.

Korfman, Manfred. "The Sling as Weapon." *Scientific American*, October 1973, pp. 35–42.

Kunz, George Frederick. *The Magic of Jewels and Charms*. Dover Publications, 1997.

Kunzig, Robert. "The Physics of Fire: Infernal Combustion." *Discover*, January 2001; pp. 35–6.

Kurlansky, Marc. *Salt: A World History*. Walker and Company, 2002.

Lane, Nick. *Oxygen: The Molecule that Made the World*. Oxford University Press, 2002.

Laszlo, Pierre. *Salt: Grain of Life*. Columbia University Press, 1998.

Le Couteur, Penny, and Jay Burreson. *Napoleon's Buttons: How 17 Molecules Changed History*. Jeremy P. Tarcher/Putnam, 2003.

Leopold, Luna B., and W. B. Langbein. "River Meanders." *Scientific American*, June 1996.

Leslie, Jacques. "Running Dry." *Harper's Magazine*, July 2000, pp. 37–52.

Levy, Matthys, and Mario Salvadori. *Why the Earth Quakes*. W.W. Norton, 1995.

Lovelock, J. E. *Gaia: A New Look at Life on Earth*. Oxford University Press, 2001.

McCutcheon, Lynn. "What's That I Smell? The Claims of Aromatherapy." *Skeptical Inquirer*, May/June 1996.

McGee, Harold. *On Food and Cooking: The Science and Lore of the Kitchen*. Charles Scribner's Sons, 1984.

Manguel, Albert. *A History of Reading*. Viking, 1996.

McPhee, John. *Annals of the Former World*. Farrar, Straus & Giroux, 1998.

Mende, Stephen B., Davis D. Sentman, and Eugene M. Wescott. "Lightning Between Earth and Space." *Scientific American*, August 1997, pp. 56–9.

Morey-Holton, Emily R. "Gravity: A Weighty Topic" in *Evolution on Planet Earth*, Academic Press, 2002.

Morgan, Elaine. *The Scars of Evolution: What Our Bodies Tell Us about Human Origins*. Oxford University Press, 1994.

Morton, Ron L. *Music of the Earth: Volcanoes, Earthquakes, and Other Geological Wonders*. Perseus Press, 1996.

Nahm, Milton C. *Selections from Early Greek Philosophy*. Prentice Hall, 1964.

Nicholl, Charles. *The Chemical Theatre*. Akadine Press, 1997.

Novella, Robert. "The Physics and Fantasy of Firewalking." *The New England Journal of Skepticism*, Summer 1999.

Pappalardo, Robert T., James W. Head, and Ronald Greeley. "The Hidden Ocean of Europa." *Scientific American*, October 1999, pp. 54–63.

Parfit, Michael. "The Essential Element of Fire." *National Geographic*, September 1996, pp. 116–39.

Partington, J. R. *A History of Greek Fire and Gunpowder*. Johns Hopkins University Press, 1999.

Partington, J. R. *A Short History of Chemistry*. Dover Publications, 1989.

Pelham, David. *The Penguin Book of Kites*. Penguin Books, 1976.

Pellegrino, Charles. *Return to Sodom and Gomorrah*. Avon Books, 1994.

"The Periodic Table of Elements." Various authors, *Chemical and Engineering News*; 8 September 2003, pp. 27–190.

Pielou, E. C. *Fresh Water*. University of Chicago Press, 1998.

Pitman, Walter C., and William B. F. Ryan. *Noah's Flood: The New Scientific Discoveries about the Event that Changed History*. Simon & Schuster, 1999.

Plimpton, George. *Fireworks: A History and Celebration*. Doubleday, 1984.

Pliny the Elder. *Natural History: A Selection*. Penguin Books, 1991.

Pringle, Hester. "The Cradle of Cash." *Discover*, October 1998.

Pullman, Bernard. *The Atom in the History of Human Thought*. Oxford University Press, 1998.

Pyne, Stephen J. *World Fire: The Culture of Fire on Earth*. University of Washington Press, 1995.

Raymo, Chet. *The Crust of Our Earth*. Prentice Hall, 1983.

Revelle, William. "Personality Processes." *Annual Review of Psychology* 46, 1995, pp. 295–328.

Robertson, James C. *Introduction to Fire Prevention*. Prentice Hall, 2000.

Sacks, Oliver. "Mendeleev's Garden." *The American Scholar*, Autumn 2001, pp. 21–7.

Salzberg, Hugh W. *From Caveman to Chemist*. American Chemical Society, 1990.

Sandford, Scott. "Amorphous Ice." *Mercury Magazine*, January/February 1998.

Sarton, George. *A History of Science: Ancient Science through the Golden Age of Greece*. John Wiley & Sons, 1952.

Sarton, George. *A History of Science: Hellenistic Science and Culture in the Last Three Centuries B.C.* John Wiley & Sons, 1965.

Scerri, Eric R. "The Evolution of the Periodic System." *Scientific American*, September 1998, pp. 78–83.

Schama, Simon. *Landscape and Memory*. Alfred A. Knopf, 1995.

Spindler, Konrad. *The Man in the Ice*. Harmony Books, 1994.

Stanley, Steven M. *Children of the Ice Ages: How a Global Catastrophe Allowed Humans to Evolve*. Henry Holt, 1998.

Steingarten, Jeffrey. "Water" in *The Man Who Ate Everything*. Alfred A. Knopf, 1997, pp. 63–74.

Stevens, William K. *The Change in the Weather: People, Weather, and the Science of Climate*. Delacorte, 1999.

Strathern, Paul. *Mendeleyev's Dream: The Quest for the Elements*. Thomas Dunne Books, 2001.

Strauss, Stephen. "How Hot is Hell Anyway?" in *The Sizesaurus*. Kodansha International, 1995.

Stwertka, Albert. *A Guide to the Elements*. Oxford University Press, 1998.

Sullivan, Walter. "Looking for the Drift" in *Galileo's Commandment: An Anthology of Great Science Writing*. W. H. Freeman, 1997.

Szpir, Michael. "Bits of Ice XI and Ice XII." *American Scientist*, September–October 1996.

Taylor, F. Sherwood. *The Alchemists*. Collier Books, 1949.

Taylor, Robert. "The Sixth Sense." *New Scientist*, 25 January 1997, pp. 36–40.

Trefil, James. *Meditations at 10,000 Feet*. Charles Scribner's Sons, 1986.

Uhlman, John, and Peggy Heinrich. *The Soul of Fire: How Charcoal Changed the World*. American Fireworks News, 1987.

Van Dover, Cindy Lee. *The Octopus's Garden: Hydrothermal Vents and Other Mysteries of the Deep Sea*. Addison-Wesley, 1996.

Van Dyke, John C. *The Mountain*. University of Utah Press, 1992.

Vine, F. J., and D. H. Matthews. "Magnetic anomalies over oceanic ridges." *Nature* 199, 1963, pp. 947–9.

"Vital Fires." Various authors. *Whole Earth*, Winter 1999.

Volcanoes and the Earth's Interiors. Various authors. W. H. Freeman and Company, 1982.

Von Baeyer, Hans Christian. *Rainbows, Snowflakes, and Quarks*. Random House, 1984.

Walker, Jearl. "Boiling and the Liedenfrost Effect." *Scientific American*, August 1977, pp. 126–31.

Watson, Lyall. *Heaven's Breath: A Natural History of the Wind*. Hodder & Stoughton, 1984.

Watson, Lyall. *Jacobson's Organ and the Remarkable Nature of Smell*. W. W. Norton & Company, 1999.

Westroek, Peter. *Life as a Geological Force*. W. W. Norton & Company, 1991.

Whelan, Robert. *The Ecology of Fire*. Cambridge University Press, 1998.

Wolke, Robert L. *What Einstein Told His Cook: Kitchen Science Explained*. W. W. Norton & Company, 2002.

Wright, Karen. "Red Sky, Hot Nights, Red Sprites" in *Scientific American Presents the Weather*, May 2000, pp. 48–53.

Wright, Lawrence. *Clean and Decent: The Unruffled History of the Bathroom and the WC*. Viking Press, 1960.

Wysession, Michael. "The Inner Workings of the Earth." *American Scientist*, March–April 1995.

Young, Louise B. *Earth's Aura*. Alfred A. Knopf, 1977.

Zebrowski, Ernest, Jr. *Perils of a Restless Planet: Scientific Perspectives on Natural Disasters*. Cambridge University Press, 1997.

Zeilinga de Boer, Jelle, and Donald Theodore Sanders. *Volcanoes in Human History*. Princeton University Press, 2002.

Index